Lecture Notes in Physics

Volume 1010

The series Lecture Notes in Physics (LNP), founded in 1969, reports new developments in physics research and teaching - quickly and informally, but with a high quality and the explicit aim to summarize and communicate current knowledge in an accessible way. Books published in this series are conceived as bridging material between advanced graduate textbooks and the forefront of research and to serve three purposes:

- to be a compact and modern up-to-date source of reference on a well-defined topic;
- to serve as an accessible introduction to the field to postgraduate students and non-specialist researchers from related areas;
- to be a source of advanced teaching material for specialized seminars, courses and schools.

Both monographs and multi-author volumes will be considered for publication. Edited volumes should however consist of a very limited number of contributions only. Proceedings will not be considered for LNP.

Volumes published in LNP are disseminated both in print and in electronic formats, the electronic archive being available at springerlink.com. The series content is indexed, abstracted and referenced by many abstracting and information services, bibliographic networks, subscription agencies, library networks, and consortia.

Proposals should be sent to a member of the Editorial Board, or directly to the responsible editor at Springer:

Dr Lisa Scalone
lisa.scalone@springernature.com

Luca Lista

Statistical Methods for Data Analysis

With Applications in Particle Physics

Third Edition

 Springer

Luca Lista (iD)
Physics Department "Ettore Pancini"
University of Naples Federico II
Naples, Italy

INFN Naples Unit
Naples, Italy

ISSN 0075-8450   ISSN 1616-6361   (electronic)
Lecture Notes in Physics
ISBN 978-3-031-19933-2   ISBN 978-3-031-19934-9   (eBook)
https://doi.org/10.1007/978-3-031-19934-9

This Springer imprint is published by the registered company Springer Nature Switzerland AG
The registered company address is: Gewerbestrasse 11, 6330 Cham, Switzerland

Preface

This book started as a collection of material from a lecture course on statistical methods for data analysis I gave to PhD students in physics at the University of Naples Federico II from 2009 to 2017 and was subsequently enriched with material from seminars, lectures, and courses that I have been invited to give in the last years.

The aim of the book is to present and elaborate the main concepts and tools that physicists use to analyze experimental data.

An introduction to probability theory and basic statistics is provided mainly as refresher lectures to students who did not take a formal course on statistics before starting their PhD. This also gives the opportunity to introduce Bayesian approach to probability, which could be a new topic to many students. More advanced topics follow, up to recent developments in statistical method used for particle physics, in particular for data analyses at the Large Hadron Collider.

Many of the covered tools and methods have applications in particle physics, but their scope could well be extended to other fields.

A shorter version of the course was presented at CERN in November 2009 as lectures on statistical methods in LHC data analysis for the ATLAS and CMS experiments. The chapter that discusses discoveries and upper limits was improved after the lectures on the subject I gave in Autrans, France, at the IN2P3 School of Statistics in May 2012. I was also invited to hold a seminar about statistical methods at Gent University, Belgium, in October 2014, which gave me the opportunity to review some of my material and add new examples.

Notes on Third Edition

There are two main updates in the third edition of this book. I reviewed the introductory part to probability theory and statistics after a course I gave on statistical data analysis of experimental data at the University of Naples Federico II in 2021 and 2022. Furthermore, I also extended the machine learning part, which now is a chapter on its own. This edition gave me the opportunity to simplify and clarify the text in several parts, to add more examples, and to try to adopt a more uniform notation throughout the text.

Notes on Second Edition

The second edition of this book reflects the work I did in preparation of the lectures that I was invited to give during the CERN-JINR European School of High-Energy Physics (15–28 June 2016, Skeikampen, Norway). On that occasion, I reviewed, expanded, and reordered my material.

Further, with respect to the first edition, I added a chapter about unfolding, an extended discussion about the best linear unbiased estimator (BLUE), and an introduction to machine learning algorithm, in particular artificial neural networks, with hints about deep learning, and boosted decision trees.

Acknowledgments

I am grateful to Louis Lyons who carefully and patiently read the first edition of my book and provided useful comments and suggestions. I would like to thank Eilam Gross for providing useful examples and for reviewing the sections about the look elsewhere effect. I'd like to thank Robert Cousins for the permission to use his illustrations of Bayes' theorem. I also received useful comments from Vitaliano Ciulli, Riccardo Paramatti, Luis Isaac Ramos Garcia, and from Denis Timoshin (Денис Тимошин). I wish to thank Lorenzo Unich for bringing to my attention an interesting property of credible intervals during one of my lectures that is reported in the third edition. Finally, I appreciate and considered all feedback I received in the preparation of the second and third editions.

Napoli, Italy   Luca Lista

Contents

List of Figures

List of Tables

List of Examples

Introduction to Probability and Inference

1.1 Why Probability Matters to a Physicist

The main goal of an experimental physicist is to investigate Nature and its behavior. To do this, physicists perform experiments that record quantitative *measurements* of the physical phenomena under observation. Experiments measure several quantities, possibly with maximum precision, and compare the measurements with the prediction of *theory* that provides a mathematical model of Nature's behavior. In the most favorable cases, experiments can lead to the discovery of new physical phenomena that may represent a breakthrough in human knowledge.

Experimental data are often affected by *randomness* due to various effects, such as fluctuations in detector response due to imperfect resolution or efficiency, background, and so on. Moreover, natural phenomena under observation may be affected by intrinsic randomness. Though the outcome of a random measurement cannot be predicted, quantitative predictions can be done in terms of *probability*. Quantum mechanics predicts different possible outcomes of measurements of observable quantities. For each of the possible outcomes, the probability can be computed as the square of the wave amplitude of the observed process.

Experiments at particle accelerators produce large numbers of particles in collision processes. For each collision, different quantities are measured and recorded, such as the position or time at which a particle crosses detector elements, or the amount of energy it deposits in the detector, etc. Such measured quantities are affected by both the intrinsic randomness of quantum processes and fluctuations of detector response.

By comparing the data with theoretical predictions, questions about Nature can be addressed. For instance, experiment at the Large Hadron Collider confirmed the presence of the Higgs boson in their data samples in 2012. To date, the experiments provide no evidence yet of many supposed particles, such as Dark Matter. For many newly proposed physics models, experiments allowed to exclude ranges of theory parameters incompatible with data.

© The Author(s), under exclusive license to Springer Nature Switzerland AG 2023
L. Lista, *Statistical Methods for Data Analysis*, Lecture Notes in Physics 1010,
https://doi.org/10.1007/978-3-031-19934-9_1

In the measurement process, *probability theory* provides a *model* of the expected distribution of data. The interpretation of the observed data, applying the probabilistic model, provides answers to the relevant questions about the phenomenon under investigation.

1.2   Random Processes and Probability

Many processes in Nature have uncertain outcomes, in the sense that their result cannot be predicted in advance, even if they are repeated, to some extent, within the same boundary and initial conditions. These are called *random processes*. A possible outcome of such a process, in statistics, is also called *event*.[1] *Probability* is a measure of how *favored* one of the events is, compared with the other possible outcomes.

Most experiments in physics can be repeated in a controlled environment in order to maintain the same, or at least very similar, conditions. Nonetheless, it is frequent to observe different outcomes at every repetition of the experiment. Such experiments are examples of *random processes*. Randomness in repeatable processes may arise because of insufficient information about their intrinsic dynamics, which prevents to predict the outcome, or because of lack of sufficient accuracy in reproducing the initial or boundary conditions. Some physical processes, like the ones ruled by quantum mechanics, have intrinsic randomness that leads to different possible outcomes, even if an experiment is repeated precisely within the same conditions.

The result of an experiment may be used to address questions about natural phenomena. For instance, we may investigate about the value of an unknown physical quantity, or about the existence or not of some new phenomena. We may find answers to those questions by defining a model of the random process under investigation in terms of probability of the possible outcomes.

1.3   Different Approaches to Probability

Different definitions of probability may apply to statements which refers to repeatable processes, as the experiments described in the previous section, or not. Examples of probability related to repeatable processes are:

- What is the probability to extract one ace in a deck of cards?
- What is the probability to win a lottery game, bingo, or any other game based on random extractions?

---

[1] Note that in physics often *event* is intended as what in statistics is called *elementary event* (see Sect. 1.4). Therefore, the use of the word event in a text about both physics and statistics may sometimes lead to some confusion.

- What is the probability that a pion is incorrectly identified as a muon in a particle identification detector?
- What is the probability that a fluctuation of the background in an observed spectrum can produce by chance a signal with a magnitude at least equal to what has been observed by my experiment?

Note that the last question is different from asking: *"what is the probability that no real signal was present, and my observation is due to a background fluctuation?"* This latter question refers to a non-repeatable situation, because we cannot have more Universes, each with or without a new physical phenomenon, where we can repeat our measurement. This example introduces probabilities that may refer to statements about unknown facts, rather than outcomes of repeatable processes. Examples are the followings:

- Future events:
  - What is the probability that tomorrow it will rain in Geneva?
  - What is the probability that your favorite team will win the next championship?
- Past events:
  - What is the probability that dinosaurs went extinct because of the collision of an asteroid with Earth?
- Unknown events, in general:
  - What is the probability that the present climate changes are mainly due to human intervention?
  - What is the probability that Dark Matter is made of weakly interacting massive particles heavier than 1 TeV?

Different definitions of probability that address the above cases, either for repeatable processes or not, can be introduced. The most useful are *frequentist* probability, applicable only for repeatable processes, and *Bayesian* probability, named after Thomas Bayes, applicable on a wider range of cases:

- **Bayesian probability** measures *one's degree of belief* that a statement is true. It also provides a quantitative prescription to *update* one's degree of belief based on the observation of data, using an extension of Bayes' theorem, as discussed in details Sect. 5.2. Bayesian probability applies to a wide variety of cases, such as unknown events, including past and future events. Bayesian probability also applies to values of unknown quantities. For instance, as consequence of an experimental measurement, the probability that an unknown particle's mass is greater than, say, 200 GeV may have meaning in the Bayesian sense. Other examples where Bayesian probability applies, but frequentist probability is meaningless, are the outcome of a future election, uncertain features of prehistoric extinct species, and so on.
- **Frequentist probability** is defined as the fraction of the number of occurrences of an event of interest on the total number of trials of a repeatable experiment,

in the limit of very large number of trials. A more precise definition is given in Sect. 6.1. Frequentist probability only applies to processes that can be repeated over a reasonably long time but does not apply to unknown statements. For instance, it is meaningful to define the frequentist probability that a particle is detected by a device, since, if a large number of particles, ideally infinite, cross the device, we can count how many are detected, but there is no frequentist meaning of the probability of a particular result in a football match, or the probability that the mass of an unknown particle is greater than, say, 200 GeV.

In Sect. 1.4 below, the classical probability theory is discussed first, as formulated since the eighteenth century; then some fundamental aspect of probability theory based on a general axiomatic definition due to Andrey Kolmogorov are presented. Chapters 2 and 3 present properties of discrete and continuous probability distributions, respectively, valid under all approaches to probability. After an introduction in Chap. 4 to random number generators, which are a useful tool in many applications, Bayesian probability is discussed in Chap. 5, and frequentist probability is introduced in Chap. 6, before moving to more advanced topics, discussed in the following chapters.

1.4   Classical Probability

A *random variable* represents the outcome of an experiment whose result is uncertain. An *event* consists of the occurrence of a certain condition about the value of the random variable resulting from the experiment. For instance, events could be:

- *a coin is tossed and gives a head*;
- *a die is rolled and gives an even value*.

In 1814, Pierre-Simon Laplace gave the following definition of probability that we call today *classical probability* [1]:

> The theory of chance consists in reducing all the events of the same kind to a certain number of cases equally possible, that is to say, to such as we may be equally undecided about in regard to their existence, and in determining the number of cases favorable to the event whose probability is sought. The ratio of this number to that of all the cases possible is the measure of this probability, which is thus simply a fraction whose numerator is the number of favorable cases and whose denominator is the number of all the cases possible.

The probability P of an event E, which corresponds to a certain number of elementary favorable cases, can be written, according to Laplace, as:

$$P(E) = \frac{\text{Number of favorable cases}}{\text{Number of possible cases}} . \tag{1.1}$$

This approach can be used in practice only for relatively simple problems, since it assumes that all possible cases under consideration are equally probable. This may not always be the case in complex situations. Examples of cases where the classical probability can be applied are coin toss, where the two faces of a coin are assumed to have an equal probability of $1/2$ each, or dice roll, where each of the six faces of a die[2] has an equal probability of $1/6$, and so on.

Starting from simple cases, like a coin toss or a die roll, more complex models can be built using combinatorial analysis. In the simplest cases, one may proceed by enumeration of all the possible cases, which are in finite number, and again the probability of an event can be evaluated as the number of favorable cases divided by the total number of possible cases of the combinatorial problem. Example 1.1 applies this direct approach to case of rolling a pair of dice.

Example 1.1 - Probabilities for Rolling Two Dice

An easy case of combinatorial analysis is given by rolling a pair of dice, taking the sum of the two up faces as result. The possible number of outcomes is given by the $6 \times 6 = 36$ different combinations that have a sum ranging from 2 to 12. All possible combinations are enumerated in Table 1.1, and probabilities corresponding to the sum, computed as the number of favorable cases divided by 36, are shown in Fig. 1.1.

Table 1.1 Possible values of the sum of two dice with all possible pair combinations and their corresponding probabilities

Sum	Favorable cases	Probability
2	(1, 1)	1/36
3	(1, 2), (2, 1)	1/18
4	(1, 3), (2, 2), (3, 1)	1/12
5	(1, 4), (2, 3), (3, 2), (4, 1)	1/9
6	(1, 5), (2, 4), (3, 3), (4, 2), (5, 1)	5/36
7	(1, 6), (2, 5), (3, 4), (4, 3), (5, 2), (6, 1)	1/6
8	(2, 6), (3, 5), (4, 4), (5, 3), (6, 2)	5/36
9	(3, 6), (4, 5), (5, 4), (6, 3)	1/9
10	(4, 6), (5, 5), (6, 4)	1/12
11	(5, 6), (6, 5)	1/18
12	(6, 6)	1/36

(continued)

---

[2] Many role-playing games use dice of different shapes than the cube and can have a number of faces different from six. Dice with different number of faces from six dating back to ancient Romans have also been found.

Example 1.1 (continued)

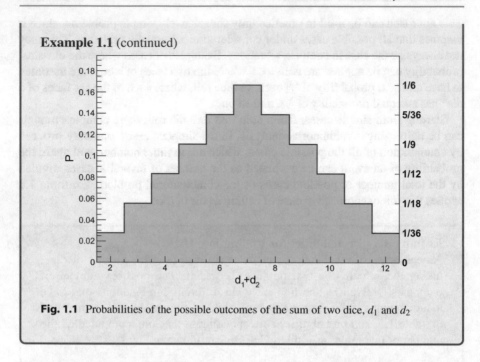

Fig. 1.1 Probabilities of the possible outcomes of the sum of two dice, d_1 and d_2

Many combinatorial problems can be decomposed in all possible *elementary events*. An event corresponds to the occurrence of one in a specific set of possible outcomes. For instance, the event *"sum of dice = 4"* corresponds to the set of possible elementary outcomes $\{(1, 3), (2, 2), (3, 1)\}$. Other events (e.g., *"sum is an odd number"*, or *"sum is greater than 5"*) may be associated with different sets of possible pair combinations. In general, formulating the problem in terms of sets allows to replace logical operators *"and"*, *"or"* and *"not"* in a sentence by the intersection, union and complement of the corresponding sets of elementary outcomes, respectively.

In more complex combinatorial problems, the decomposition into equally probable elementary cases may be a laborious task, in practice. One of the possible approaches to such complex cases is to use of computer simulation performed with Monte Carlo methods, which are introduced in Chap. 4. Numerical methods are also widely used to simulate realistic detector response, which include effects like efficiency, resolution, etc.

1.5   Problems with the Generalization to the Continuum

The generalization of classical probability, introduced for discrete cases in Sect. 1.4 above, to continuous random variables cannot be done in an unambiguous way.

A way to extend the concept of equiprobability, as applied to the six possible outcomes of a die roll, to a continuous variable x consists in partitioning its range of possible values, say $[x_1, x_2]$, into intervals $I_1, \cdots, I_N$, all having the same width equal to $(x_2 - x_1)/N$, for an arbitrarily large number N. The *probability distribution* of the variable x can be considered *uniform* if the probability that a random extraction of x falls within each of the N intervals is the same, i.e., it is equal to $1/N$.

This definition of uniform probability of x in the interval $[x_1, x_2]$ clearly changes under reparameterization. If the variable x is transformed into $y = Y(x)$, assuming for simplicity that the transformation Y is a monotonically increasing function of x, the interval $[x_1, x_2]$ is transformed int $[y_1, y_2] = [Y(x_1), Y(x_2)]$. In this case, the transformed intervals $J_1, \cdots, J_N = Y(I_1), \cdots, Y(I_N)$ do not necessarily have all the same width, unless the transformation Y is a special one, such as a linear function. Therefore, if x has a uniform distribution, $y = Y(x)$ in general does not necessarily have a uniform distribution, according to the previous definition of equiprobability.

1.6   The Bertrand's Paradox

The arbitrariness in the definition of uniformity of a random extraction becomes evident in a famous paradox, called the Bertrand's paradox [2], that can be formulated as follows:

- Consider an equilateral triangle inscribed in a circle. Extract *uniformly* one of the possible chords of the circle. What is the probability that the length of the extracted chord is larger than the side of the inscribed triangle?

The apparent paradox arises because the uniform extraction of a chord in a circle is not a uniquely defined process. Below three possible examples are presented of apparently uniform random extractions that give different results in terms of probability:

1. Let us take the circle's diameter passing by one of the vertices of the triangle; extract *uniformly* a point in this diameter and take the chord perpendicular to the diameter passing by the extracted point. Evidently from Fig. 1.2 (left plot), the basis of the triangle cuts the vertical diameter, leaving a segment at the bottom whose length is half the radius. This corresponds to one half of the extracted chords having a radius less than the triangle basis. Considering all possible diameters of the circle that we assume to extract uniformly with respect to the azimuth angle, all possible chords of the circle are spanned. Hence, the probability in question would result to be $P = 1/2$.
2. Let us take, instead, one of the chords starting from the top vertex of the triangle (Fig. 1.2, center plot) and extract *uniformly* an angle with respect to the tangent to the circle passing by that vertex. The chord is longer than the triangle's side when it intersects the basis of the triangle, and it is shorter otherwise. This occurs in one-third of the cases since the angles of an equilateral triangle measure $\pi/3$ each, and the chords span an angle of π. By uniformly extracting a possible point

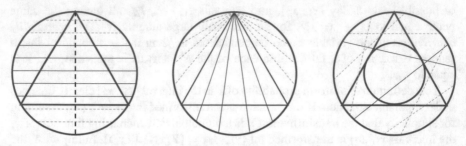

Fig. 1.2 Illustration of Bertrand's paradox: three different choices of *random extraction* of a chord in a circle lead apparently to probabilities that the cord is longer than the inscribed triangle's side of $^1/_2$ (left), $^1/_3$ (center), and $^1/_4$ (right), respectively. Red solid lines and blue dashed lines represent chords longer and shorter than the triangle's side, respectively

on the circumference of the circle, one would conclude that $P = ^1/_3$, which is different from $P = ^1/_2$, as found in the first case.

3. Let us extract *uniformly* a point inside the circle and construct the chord passing by that point perpendicular to the radius that passes through the same point (Fig. 1.2, right plot). With this extraction, it is possible to conclude that $P = ^1/_4$, since the chords starting from a point contained inside (outside) the circle inscribed in the triangle have a length longer (shorter) than the triangle's side, and the ratio of the areas of the circle inscribed in the triangle to the area of the circle that inscribes the triangle is equal to $^1/_4$. $P = ^1/_4$ is different from the values found in both above cases.

The paradox is only apparent because the process of uniform random extraction of a chord in a circle is not uniquely defined, as already discussed in Sect. 1.5.

1.7   Axiomatic Probability Definition

An axiomatic definition of probability, founded on measure theory and valid both in the discrete and the continuous case, is due to Kolmogorov [3]. Let us consider a *measure space*, $(\Omega, F \subseteq 2^{\Omega}, P)$, where P is a function that maps elements of F to real numbers. F is a subset of the power set[3] 2^{Ω} of Ω, i.e., F contains subsets of Ω. Ω is called *sample space*, and F is called *event space*. P is a *probability measure* if the following properties are satisfied:

1. $P(E) \geq 0 \quad \forall E \in F$,
2. $P(\Omega) = 1$,
3. $\forall (E_1, \cdots, E_N) \in F^N : E_i \cap E_j = 0, \quad P\left(\bigcup_{i=1}^{N} E_i\right) = \sum_{i=1}^{N} P(E_i)$.

---

[3] The power set of a set S is the set of all subsets of S, including the empty set.

Both Bayesian probability, defined in Chap. 5, and frequentist probability, defined in Chap. 6, can be demonstrated to obey Kolmogorov's axioms. The following sections present some properties of probability, as defined above, that descend from Kolmogorov's axioms and are valid for different approaches to probability.

1.8 Conditional Probability

Given two events A and B, the *conditional probability* represents the probability of the event A given the condition that the event B has occurred and is defined as:

$$P(A \mid B) = \frac{P(A \cap B)}{P(B)} . \tag{1.2}$$

Conditional probability is illustrated in Fig. 1.3. The probability of A, $P(A)$, corresponds to the area of the set A relative to the area of the whole sample space Ω, which is equal to one. The conditional probability, $P(A \mid B)$, corresponds to the area of the *intersection* of A and B relative to the area of the set B.

1.9 Independent Events

An event A is said to be *independent* on event B if the conditional probability of A, given B, is equal to the probability of A. In other words, the occurrence of B does not change the probability of A:

$$P(A \mid B) = P(A) . \tag{1.3}$$

Given the definition of conditional probability in Eq. (1.2), A is independent of B if, and only if, the probability of the simultaneous occurrence of both events is equal to the product of their probabilities:

$$P(\text{``}A \text{ and } B\text{''}) = P(A \cap B) = P(A) \, P(B) . \tag{1.4}$$

Fig. 1.3 Illustration of conditional probability. Two events, A (blue) and B (red), are represented as sets of elementary outcomes. The conditional probability $P(A \mid B)$ is equal to the area of the intersection, $A \cap B$, divided by the area of B

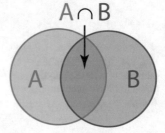

From the symmetry of Eq. (1.4), if A is independent of B, then B is also independent of A. We say that A and B are independent events.

Example 1.2 - Combination of Detector Efficiencies

Consider an experimental apparatus made of two detectors A and B, and a particle traversing both. The detector A produces a signal when crossed by a particle with probability ε_A and the detector B with probability ε_B. ε_A and ε_B are called *efficiencies* of the detectors A and B, respectively.

If the signals are produced independently in the two detectors, the probability ε_{AB} that a particle gives a signal in both detectors, according to Eq. (1.4), is equal to the product of ε_A and ε_B:

$$\varepsilon_{AB} = \varepsilon_A \, \varepsilon_B \,. \tag{1.5}$$

This result does not hold if there are causes of simultaneous inefficiency of both detectors, e.g., a fraction of times where the electronics systems for both A and B are simultaneously switched off for short periods, or geometrical overlap of inactive regions, and so on.

1.10   Law of Total Probability

Let us consider a number of events corresponding to the sets $E_1, \cdots, E_N$, which are in turn subsets of another set E_0 included in the sample space Ω. Assume that the set of all E_i is a *partition* of E_0, as illustrated in Fig. 1.4, i.e.,

$$E_i \cap E_j = 0 \quad \forall i, j \quad \text{and} \quad \bigcup_{i=1}^{N} E_i = E_0. \tag{1.6}$$

Given Kolmogorov's third axiom (Sect. 1.7), the probability corresponding to E_0 is equal to the sum of the probabilities of E_i:

$$P(E_0) = \sum_{i=1}^{N} P(E_i) \,. \tag{1.7}$$

Fig. 1.4 Illustration of the partition of a set E_0 into the union of the disjoint sets E_1, $E_2, E_3, \cdots, E_N$

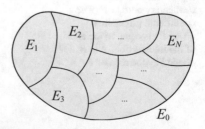

Consider a partition $A_1, \cdots, A_N$ of the entire sample space Ω, i.e., a number disjoint sets,

$$A_i \cap A_j = 0 \,, \tag{1.8}$$

such that:

$$\bigcup_{i=1}^{N} A_i = \Omega \,. \tag{1.9}$$

We have:

$$\sum_{i=1}^{N} P(A_i) = 1 \,, \tag{1.10}$$

and we can build the following sets, as illustrated in Fig. 1.5:

$$E_i = E_0 \cap A_i \,. \tag{1.11}$$

Each set E_i, using Eq. (1.2), corresponds to a probability:

$$P(E_i) = P(E_0 \cap A_i) = P(E_0 \mid A_i) \, P(A_i) \,. \tag{1.12}$$

Using the partition $A_1, \cdots, A_N$, Eq. (1.7) can be rewritten as:

$$P(E_0) = \sum_{i=1}^{N} P(E_0 \mid A_i) \, P(A_i) \,. \tag{1.13}$$

Fig. 1.5 Illustration of the law of total probability. The sample space Ω is partitioned into the sets $A_1, A_2, A_3, \cdots,$ A_N, and the set E_0 is correspondingly partitioned into $E_1, E_2, E_3, \cdots, E_N$, where each E_i is $E_0 \cap A_i$ and has probability equal to $P(E_0 \mid A_i) \, P(A_i)$

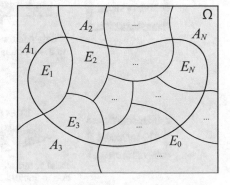

This decomposition is called *law of total probability* and can be interpreted as *weighted average* (see Sect. 7.4) of the probabilities $P(A_i)$ with weights $w_i = P(E_0|A_i)$.

Equation (1.13) has application in the computation of Bayesian probability, as discussed in Chap. 5.

1.11 Inference

Measurements in physics are based on experimental data samples. From the statistical point of view, data are the outcome of random processes. The randomness of the process determines *fluctuations* in data. An experiment at a particle collider that records many collision events for further analysis, for instance, can be seen as a way to sample the PDF of the measurable quantities.

Among the parameters of the PDF that models the distribution of data, there are fundamental constants of Nature that we want to measure. The process of determining an *estimate* $\hat{\theta}$ of some unknown parameter θ from observed data is called *inference*. The inference procedure must also determine the *uncertainty* or *error* $\delta\hat{\theta}$ associated with the estimated value $\hat{\theta}$.

Statistical fluctuations of the data sample, due to the intrinsic theoretical and experimental randomness of our observable quantities, reflect into a finite uncertainty $\delta\hat{\theta}$. The complementary role of data fluctuation and measurement uncertainty is illustrated in the diagram shown in Fig. 1.6. The smaller the amount of data fluctuation, which means that the probability distribution concentrates a large probability into a small interval of possible data values, the smaller the uncertainty in the determination of unknown parameters. Ideally, if the data sample would exhibit no fluctuation at all, and if our detector would have a perfect response, an exact

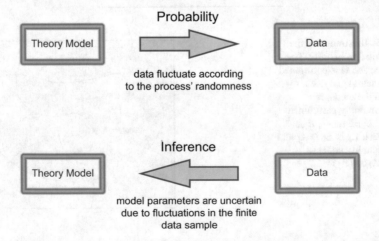

Fig. 1.6 Illustration of the complementary role of data fluctuation and measurement uncertainty

knowledge of the unknown parameters could be possible. This case is never present in real experiments, and every real-life measurement is affected by some level of uncertainty.

An important aspect of inference is to distinguish parameters of interest that are the result of our final measurement, from other parameters that are only relevant to provide a complete description of our PDF, that are called nuisance parameters:

- **Parameters of Interest**
  Theory models predict the probability distribution of observable quantities. Those models depend on unknown parameters, some of which are the goal of our measurement. Such parameters are called *parameters of interest*. There are cases where we are interested to measure parameters that provide a characterization of our detector, like calibration constant, efficiency, resolution, or others. In such cases, our parameters of interest are not strictly related to a fundamental physics theory but are more related to a model that describes a detector response.
- **Nuisance Parameters**
  The distribution of experimental data is the result of the combination of a theory model and the effect of the detector response: detector's finite resolution, miscalibrations, the presence of background, etc. The detector response itself can be described by a probability model that depends on unknown parameters. If we are interested to measure theory parameters, and the modeling of detector response is only an auxiliary tool to precisely describe experimental distributions, the additional unknown parameters that are not part of our measurement called *nuisance parameters*. Nuisance parameters appear in the inference procedure and must be determined together with the parameters of interest. For instance, when determining the yield of a signal peak, often other parameters need to be determined from data, such as the experimental resolution that affects the peak width, and the detector efficiencies that need to be corrected for to determine the signal production yield from the measured area under the peak. Moreover, additional nuisance parameters may be needed to determine the amount and distribution of the possible background sources, and so on.

1.12 Measurements and Their Uncertainties

A *data sample* consist of measured values of the observable quantities. Those values can also be considered as the result of a sampling of the PDF determined by a combination of theory and instrumental effects.

We can determine, or *estimate*, the value of unknown parameters, either parameters of interest or nuisance parameters, using the data collected by our experiment. The estimate is not exactly equal to the true value of the parameter but provides an approximation of the true value, within some *uncertainty*, or *error*.

As result of the measurement of a parameter θ, one quotes the estimated value $\hat{\theta}$ and its uncertainty $\delta\hat{\theta}$:

$$\theta = \hat{\theta} \pm \delta\hat{\theta} \,. \tag{1.14}$$

The estimate $\hat{\theta}$ is often also called *point estimate* or, sometimes in physics, also *central value*.[4] The interval $[\hat{\theta} - \delta\hat{\theta}, \ \hat{\theta} + \delta\hat{\theta}]$ is called *uncertainty interval*.

The meaning of the uncertainty interval is different in the frequentist and in the Bayesian approaches. In both approaches, a probability level needs to be specified to determine the size of the uncertainty. When not otherwise specified, by convention, a 68.27% probability level is assumed, corresponding to the area under a Gaussian distribution in a $\pm 1\sigma$ interval (see Sect. 3.8). Other choices are 90% or 95% probability level, usually adopted when quoting upper or lower limits.

In some cases, *asymmetric errors*, with positive and negative uncertainties, are determined, and the result is quoted as:

$$\theta = \hat{\theta} \, {}^{+\delta\hat{\theta}_+}_{-\delta\hat{\theta}_-} \,, \tag{1.15}$$

corresponding to the uncertainty interval $[\hat{\theta} - \delta\hat{\theta}_+, \ \hat{\theta} + \delta\hat{\theta}_-]$.

1.13   Statistical and Systematic Uncertainties

Nuisance parameters can be often determined from experimental data samples. In some cases, dedicated data samples may be needed, e.g., data from test beams that allow to measure calibration constants of a detector, or cosmic-ray runs for alignment constants, etc. Dedicated simulation programs or complex theoretical calculations may be needed to estimate some otherwise unknown parameter, with their uncertainties.

The uncertainty on the determination of nuisance parameters reflects into uncertainties on the estimate of parameters of interest (see, for instance, Sect. 12.20). Uncertainties due to the propagation of imperfect knowledge of nuisance parameters produces *systematic uncertainties* in the final measurement. Uncertainties related to the determination of the parameters of interest purely reflecting fluctuation in data, regardless of possible uncertainties due to the estimate of nuisance parameters, are called *statistical uncertainties*. A typical notation to quote separately statistical and systematic uncertainty contributions is:

$$\theta = \hat{\theta} \pm \delta\hat{\theta} \,(\text{stat.}) \pm \delta\hat{\theta}' \,(\text{syst.}) \,. \tag{1.16}$$

---

[4] The name "central value" is sometimes used in physics as synonymous of point estimate. This may create some ambiguity because in statistics central value is more frequently used for median, mean, or mode.

Sometimes, separate contributions to systematic uncertainties due to individual *sources* (i.e., individual nuisance parameters) are quoted.

1.14 Frequentist vs Bayesian Inference

As introduced in Sect. 1.2, two main complementary approaches to probability exist in literature and correspond to two different interpretations of uncertainty intervals, point estimates, and corresponding uncertainties:

- **Bayesian approach**
  The point estimate is the value $\hat{\theta}$ that maximizes the posterior probability density (see Chap. 5.3). The uncertainty interval is reported as *credible interval* $[\theta^{lo}, \theta^{hi}]$. One's *degree of belief* that the unknown parameter θ is contained in the quoted interval can be quantified, usually, with a 68.27% probability.
- **Frequentist approach**
  The *point estimate* is the value of a function of the data sample, called *estimator*, whose statistical properties can be determined (see Chap. 6). The uncertainty interval is reported as *confidence interval* $[\theta^{lo}, \theta^{hi}]$. In a large fraction, usually taken as 68.27%, of repeated experiments, the unknown true value of θ is contained in the quoted interval. The fraction is intended in the limit of infinitely large number of repetitions of the experiment. The extremes of the interval θ^{lo} and θ^{hi} may vary from one experiment to another and are themselves random variables. In the frequentist approach, the property that the confidence interval contains the true value in 68.27% of an infinitely large number of experiments is called *coverage*. The probability level, usually taken as 68.27%, is called *confidence level*. Interval estimates that have a larger or smaller probability to contain the true value compared to the desired confidence level are said to *overcover* or *undercover*, respectively.

Inference using the Bayesian and frequentist approaches is discussed in Sect. 5.6 of Chap. 5.3 and in Sect. 6.1 of Chap. 6, respectively. With both the frequentist and the Bayesian approaches, there are some degrees of arbitrariness in the choice of uncertainty intervals (central interval, extreme intervals, smallest-length interval, etc.), as discussed in Sect. 5.10 for the Bayesian approach. Chapter 8 presents the corresponding implications for the frequentist approach and more rigorously defines a procedure to build confidence intervals.

References

1. P. Laplace, *Essai Philosophique Sur les Probabilités*, 3rd edn. (Courcier Imprimeur, Paris, 1816)
2. J. Bertrand, *Calcul des probabilités* (Gauthier-Villars, 1889), pp. 5–6
3. A. Kolmogorov, *Foundations of the Theory of Probability* (Chelsea Publishing, New York, 1956)

Discrete Probability Distributions

2

2.1 Introduction

This chapter presents general definitions and discusses the main properties of probability distributions of discrete random variables and some of the most frequently used discrete probability distributions.

Consider a random variable x which has possible values $x_1, \cdots, x_N$, each occurring with a probability $P(\{x_i\}) = P(x_i)$, $i = 1, \cdots N$. Using Kolmogorov's terminology, the sample space is $\Omega = \{x_1 \cdots, x_N\}$. The function that associates the probability $P(x_i)$ to each possible value x_i of x:

$$P : x_i \mapsto P(x_i) \tag{2.1}$$

is called *probability distribution*. A probability distribution of a discrete random variable is also called *probability mass function*, or PMF.

An *event* E corresponds to a set of K distinct possible elementary events $\{x_{E_1}, \cdots, x_{E_K}\}$, where we used the set of K indices $(E_1, \cdots, E_K) \in \{1, \cdots, N\}^K$. According to the third Kolmogorov's axiom, the probability of the event E is equal to:

$$P(E) = P(\{x_{E_1}, \cdots, x_{E_K}\}) = P\left(\bigcup_{j=1}^{K}\{x_{E_j}\}\right) = \sum_{j=1}^{K} P(\{x_{E_j}\}) = \sum_{j=1}^{K} P(x_{E_j}). \tag{2.2}$$

From the second Kolmogorov's axiom, the probability of the event Ω corresponding to the set of all possible elementary outcomes, $x_1, \cdots, x_N$, must be equal to one.

© The Author(s), under exclusive license to Springer Nature Switzerland AG 2023
L. Lista, *Statistical Methods for Data Analysis*, Lecture Notes in Physics 1010,
https://doi.org/10.1007/978-3-031-19934-9_2

Equivalently, using Eq. (2.2), the sum of the probabilities of all possible elementary outcomes is equal to one:

$$\sum_{i=1}^{N} P(x_i) = 1 .$$ (2.3)

This property, common of all probability distributions, is called *normalization condition*.

2.2 Joint and Marginal Probability Distributions

If we have n random variables, the probability that those variables assume the values given by the n-tuple $x_1, \cdots, x_n$ simultaneously is called *joint probability* and is indicated as $P(x_1, \cdots, x_n)$. For discrete random variables, it is also called *joint probability mass function*.

The probability that a single variable, say the first one, assumes a specific value x_1 is obtained by adding the joint probabilities of all n-tuples of possible values where the value of the first variable is x_1:

$$P_1(x_1) = \sum_{x_2, \cdots, x_n} P(x_1, x_2, \cdots, x_n) .$$ (2.4)

This is in general called *marginal probability distribution* or, specifically for discrete variables, *marginal probability mass function*. For two variables, for instance, we can define the marginal probability distributions for the individual variables as:

$$P_x(x) = \sum_{y} P(x, y) ,$$ (2.5)

$$P_y(y) = \sum_{x} P(x, y) .$$ (2.6)

Marginal probability distributions for continuous variables and more general properties of marginal probability distributions are presented in Sect. 3.6.

2.3 Conditional Distributions and Chain Rule

Consider the events " $x = x^\star$ " and " $y = y^\star$ ". According to Eq. (1.2), we have:

$$P(\text{"} x = x^\star \text{ and } y = y^\star \text{"}) = P(\text{"} x = x^\star \text{"} | \text{"} y = y^\star \text{"}) \, P(\text{"} y = y^\star \text{"}),$$ (2.7)

where $P("x = x^\star$ and $y = y^{\star "})$ is the joint distribution $P(x^\star, y^\star)$ and $P("y = y^{\star "})$ is the marginal distribution $P_y(y^\star)$. With an abuse of notation, marginal distributions like P_y are also denoted as P when the meaning is obvious from the context.

We can define the *conditional probability distribution*, or in particular for a discrete case, the *conditional probability mass function*, as:

$$P(x^\star \mid y^\star) = P("x = x^{\star "} \mid "y = y^{\star "}) , \tag{2.8}$$

and we can write:

$$P(x, y) = P(x \mid y) \, P(y) . \tag{2.9}$$

If we have three variables, we can write, similarly:

$$P(x, y, z) = P(x \mid y, z) \, P(y, z) , \tag{2.10}$$

$$P(x, y) = P(x \mid y) \, P(y) . \tag{2.11}$$

Hence, we have the *chain rule*:

$$P(x, y, z) = P(x \mid y, z) \, P(y \mid z) \, P(z) . \tag{2.12}$$

More in general, for n variables, the chain rule is:

$$P(x_1, \cdots, x_n) = P(x_1) \prod_{i=2}^{n} P(x_i \mid x_1, \cdots, x_{i-1}) . \tag{2.13}$$

2.4 Independent Random Variables

Consider two variables x and y, with probability distribution $P(x, y)$, and let us assume that the events "$x = x^{\star "}$ and "$y = y^{\star "}$ are independent. According to Eq. (1.4):

$$P(x^\star, y^\star) = P("x = x^\star \text{ and } y = y^{\star "}) = P("x = x^{\star "}) \, P("y = y^{\star "}) =$$

$$= P_x(x^\star) \, P_y(y^\star) . \tag{2.14}$$

This means that the probability distribution can be factorized into the product of the marginal distributions. In this case, the variables x and y are said to be independent.

More in general, the variables $x_1, \cdots, x_n$ are independent if:

$$P(x_1, \cdots, x_n) = P_1(x_1) \cdots P_n(x_n) . \tag{2.15}$$

Independence can also be defined for groups of variables: the sets of variables $x_1, \cdots, x_h$ and $y_1, \cdots, y_k$ are independent if the probability distribution can be factorized as:

$$P(x_1, \cdots, x_h, y_1, \cdots, y_k) = P_x(x_1, \cdots, x_h) \, P_y(y_1, \cdots, y_k) \,. \qquad (2.16)$$

2.5 Statistical Indicators: Average, Variance, and Covariance

In this section, several useful quantities related to probability distributions of discrete random variables are defined. The extension to continuous random variables is presented in Sect. 3.3.

Consider a discrete random variable x which can assume N possible values, $x_1, \cdots, x_N$, with probability distribution P. The *average value* of x, also called *expected value* or *mean value*, is defined as:

$$\langle x \rangle = \sum_{i=1}^{N} x_i \, P(x_i) \,. \qquad (2.17)$$

Alternative notations are $\mu_x = \mathbb{E}[x] = \langle x \rangle$. The notation $\mathbb{E}[x]$ is often used in statistical literature to indicate the average value, while the notation $\langle x \rangle$ is more familiar to physicists and recalls the notation used for bra-ket inner product in quantum mechanics. More in general, given a function g of x, the average value of g is defined as:

$$\langle g(x) \rangle = \sum_{i=1}^{N} g(x_i) \, P(x_i) \,. \qquad (2.18)$$

The value of x that corresponds to the largest probability is called *mode*. The mode may not be unique. In that case, the distribution of x is said to be *multimodal*; in particular, it is called *bimodal* in presence of two maxima.

The *variance* of x is defined as:

$$\mathbb{V}\mathrm{ar}[x] = \langle (x - \langle x \rangle)^2 \rangle = \sum_{i=1}^{N} (x_i - \langle x \rangle)^2 \, P(x_i) \,, \qquad (2.19)$$

and the *standard deviation* is the square root of the variance:

$$\sigma_x = \sqrt{\mathbb{V}\mathrm{ar}[x]} = \sqrt{\sum_{i=1}^{N} (x_i - \langle x \rangle)^2 P(x_i)} \,. \qquad (2.20)$$

Sometimes the variance is indicated as the square of the standard deviation: $\sigma_x^2 = \mathbb{V}\text{ar}[x]$. The *root mean square* of x, abbreviated as RMS or r.m.s., is defined as:[1]

$$x_{\text{rms}} = \sqrt{\sum_{i=1}^{N} x_i^2 \, P(x_i)} = \sqrt{\langle x^2 \rangle} \,. \tag{2.21}$$

It is easy to demonstrate that the product of a constant a times x has average:

$$\langle a\,x \rangle = a \, \langle x \rangle \tag{2.22}$$

and variance:

$$\mathbb{V}\text{ar}[ax] = a^2 \, \mathbb{V}\text{ar}[x] \,. \tag{2.23}$$

The average of the sum of two random variables x and y is equal to the sum of their averages:

$$\langle x + y \rangle = \langle x \rangle + \langle y \rangle \,. \tag{2.24}$$

A demonstration for the continuous case, also valid for the discrete case, is given in Sect. 3.6.

The variance in Eq. (2.19) is also equal to:

$$\mathbb{V}\text{ar}[x] = \langle x^2 \rangle - \langle x \rangle^2 \tag{2.25}$$

since, from Eqs. (2.22) and (2.24):

$$\mathbb{V}\text{ar}[x] = \langle (x - \langle x \rangle)^2 \rangle = \langle x^2 - 2x\langle x \rangle + \langle x \rangle^2 \rangle = \langle x^2 \rangle - \langle x \rangle^2 \,. \tag{2.26}$$

Given two random variables x and y, their *covariance* is defined as:

$$\boxed{\mathbb{C}\text{ov}(x,\,y) = \langle x\,y \rangle - \langle x \rangle \, \langle y \rangle \,,} \tag{2.27}$$

and the *correlation coefficient* of x and y is defined as:

$$\boxed{\rho_{xy} = \frac{\mathbb{C}\text{ov}(x,\,y)}{\sigma_x \, \sigma_y} \,.} \tag{2.28}$$

---

[1] In some physics literature, the standard deviation is sometimes also called root mean square or r.m.s. This causes some ambiguity.

In particular,

$$\mathbb{V}\mathrm{ar}[x] = \mathbb{C}\mathrm{ov}(x, x) \,. \tag{2.29}$$

Note that the covariance is symmetric:

$$\mathbb{C}\mathrm{ov}(x, y) = \mathbb{C}\mathrm{ov}(y, x) \,. \tag{2.30}$$

Two variables that have null covariance are said to be *uncorrelated*. It can be demonstrated that:

$$\mathbb{V}\mathrm{ar}[x + y] = \mathbb{V}\mathrm{ar}[x] + \mathbb{V}\mathrm{ar}[y] + 2\,\mathbb{C}\mathrm{ov}(x, y) \,. \tag{2.31}$$

From Eq. (2.31), the variance of the sum of uncorrelated variables is equal to the sum of their variances. Given n random variables, $x_1, \cdots, x_n$, the symmetric matrix:

$$C_{ij} = \mathbb{C}\mathrm{ov}(x_i, y_j) \tag{2.32}$$

is called *covariance matrix*. The diagonal elements C_{ii} are always positive or null and correspond to the variances of the n variables. A covariance matrix is diagonal if and only if all variables are uncorrelated.

Moments m'_k of the distribution of a random variable x are defined as:

$$m'_k = \left\langle x^k \right\rangle \,, \tag{2.33}$$

while *central moments m_k* of the distribution are defined as:

$$m_k = \left\langle (x - \langle x \rangle)^k \right\rangle \,. \tag{2.34}$$

m'_1 is equal to the expected value, and m_2 is the variance.

Another useful quantity is the *skewness* that measures the asymmetry of a distribution and is defined as:

$$\boxed{\gamma_1[x] = \left\langle \left(\frac{x - \langle x \rangle}{\sigma_x} \right)^3 \right\rangle = \frac{\gamma[x]}{\sigma_x^3} \,,} \tag{2.35}$$

where the quantity:

$$\boxed{\gamma[x] = \langle x^3 \rangle - 3\,\langle x \rangle\,\langle x^2 \rangle + 2\,\langle x \rangle^3} \tag{2.36}$$

is called *unnormalized skewness* and is equal to m_3. Symmetric distributions have skewness equal to zero, while negative (positive) skewness corresponds to a mode greater then (less than) the average value $\langle x \rangle$. It is possible to demonstrate the following properties:

$$\gamma[x + y] = \gamma[x] + \gamma[y] \,, \tag{2.37}$$

$$\gamma[ax] = a^3 \gamma[x] \,. \tag{2.38}$$

The *kurtosis*, finally, is defined as:

$$\beta_2[x] = \left\langle \left(\frac{x - \langle x \rangle}{\sigma_x} \right)^4 \right\rangle = \frac{m_4}{\sigma_x^4} \,. \tag{2.39}$$

Usually, the *kurtosis coefficient*, γ_2, is defined as:

$$\gamma_2[x] = \beta_2[x] - 3 \,. \tag{2.40}$$

This definition gives $\gamma_2 = 0$ for a normal distribution (defined in Sect. 3.8). γ_2 is also called *excess*.

2.6   Statistical Indicators for Finite Samples

For finite samples, statistical indicators have definitions similar to the ones introduced in the previous section for probability distributions. The *mean*, or *arithmetic mean* of a finite sample $x_1, \cdots, x_N$, is defined as:[2]

$$\bar{x} = \frac{1}{N} \sum_{i=1}^{N} x_i \,. \tag{2.41}$$

The *median* is defined, assuming that the sample $x_1, \cdots, x_N$ is an ordered sequence, as:

$$\tilde{x} = \begin{cases} x_{(N+1)/2} & \text{if } N \text{ is odd} \,, \\ \frac{1}{2}(x_{N/2} + x_{N/2+1}) & \text{if } N \text{ is even} \,. \end{cases} \tag{2.42}$$

The notation $\mathrm{Med}[x]$ is also sometimes used instead of $\tilde{x}$.

---

[2] The notation $\bar{x}$ is sometimes also used for the mean of a distribution in some literature.

For a finite sample $x_1, \cdots, x_N$, the variance and standard deviation are defined as:

$$s_x^2 = \frac{1}{N-1} \sum_{i=1}^{N} (x_i - \bar{x})^2 , \qquad (2.43)$$

$$s_x = \sqrt{s_x^2} . \qquad (2.44)$$

The presence of the term $N-1$ in the denominator of Eq. (2.43) is discussed in Exercise 6.5 and is motivated by the removal of the *bias* (see Sect. 6.3) in the estimate of the standard deviation of a normal distribution from which the random sample $x_1 \cdots, x_N$ is extracted.

Moments m_k' of the sample are defined as:

$$m_k' = \frac{1}{N} \sum_{i=1}^{N} x_i^k , \qquad (2.45)$$

while *central moments* m_k of the sample are defined as:

$$m_k = \frac{1}{N} \sum_{i=1}^{N} (x_i - \bar{x})^k . \qquad (2.46)$$

m_1' is the mean, and m_2, multiplied by the correction factor $N/(N-1)$, is the variance of the sample. The *skewness* of the sample can defined from the moments m_k as:

$$g_1 = \frac{m_3}{m_2^{3/2}} \qquad (2.47)$$

or, more frequently, according to the following unbiased version [3]:

$$G_1 = \frac{\sqrt{N(N-1)}}{N-2} \frac{m_3}{m_2^{3/2}} . \qquad (2.48)$$

The presence of the term $\sqrt{N(N-1)}/(N-2)$ has a motivation similar to the term $N-1$ at the denominator of Eq. (2.43), again to remove the bias for a normally distributed sample. The *kurtosis coefficient* of the sample is defined similarly to Eq. 2.40 as:

$$g_2 = \frac{m_4}{m_2^2} - 3 . \qquad (2.49)$$

The unbiased version, for a normally distributed sample, has a more complex expression [4] which depend on N:

$$G_2 = \frac{N-1}{(N-2)(N-3)} \left[(N+1) \frac{m_4}{m_2^2} - 3(N-1) \right] . \qquad (2.50)$$

Statistical indicators for finite samples have many properties similar to the corresponding indicators defined for probability distributions.

2.7   Transformations of Variables

Given a random variable x, we can transform it into another variable y with a function $Y : x \mapsto y = Y(x)$. If x can assume one of the values $\{x_1, \cdots, x_N\}$, then y can assume one of the values $\{y_1, \cdots, y_M\} = \{Y(x_1), \cdots, Y(y_N)\}$. N is equal to M only if all the values $Y(x_1), \cdots, Y(x_N)$ are different from each other, i.e., the function Y is injective. The probability corresponding to each possible value y is given by the sum of the probabilities of all values x that are transformed into y by Y:

$$P(y) = \sum_{x:\, y=Y(x)} P(x) . \qquad (2.51)$$

In case, for a given y, there is a single value of x for which $Y(x) = y$, we have $P(y) = P(x)$. Equation (2.51) can also be written as:

$$P(y) = \sum_{x \in X} \delta_{y\, Y(x)} P(x) , \qquad (2.52)$$

where δ_{ab} is the Kronecker's delta and is not null and equal to one only if $a = b$, i.e., only if $y = Y(x)$ in Eq. (2.52).

The generalization to more variables is straightforward: assume that a variable z is determined from two random variables x and y as $z = Z(x, y)$. If $\{z_1, \cdots, z_M\}$ is the set of all possible values that z can assume, the probability corresponding to each possible value z is given by:

$$P(z) = \sum_{x,\, z:\, z=Z(x,\, y)} P(x, y) . \qquad (2.53)$$

This expression is consistent with the results obtained for the sum of two dice considered in Sect. 1.4, as discussed in Exercise 2.1 below.

Equation (2.53) can be equivalently written, using the Kronecker's delta, as:

$$P(z) = \sum_{x \in X} \sum_{y \in Y} \delta_{z\, Z(x,\, y)} P(x, y) . \qquad (2.54)$$

Example 2.1 - Variable Transformation Applied to Dice Rolls

The computation of the probabilities that the sum of two dice rolls is even or odd is straightforward. We take this simple case in order to demonstrate the validity of the general method expressed by Eq. (2.53). The probability of the sum of two dice can be evaluated using Eq. (2.53), where x and y are the outcome of each dice, and $z = x + y$.

All cases in Table 1.1 can be considered, and the probabilities for all even and odd values can be added. Even values and their corresponding probabilities are $2 : \frac{1}{36}, 4 : \frac{1}{12}, 6 : \frac{5}{36}, 8 : \frac{5}{36}, 10 : \frac{1}{12}, 12 : \frac{1}{36}$. Therefore, the probability of an even result is: $(1 + 3 + 5 + 5 + 3 + 1)/36 = \frac{1}{2}$, and the probability of an odd result is $1 - \frac{1}{2} = \frac{1}{2}$.

Another way to proceed is the following: each dice has probability $\frac{1}{2}$ to give an even or an odd result. The sum of two dice is even if either two even or two odd results are added. Each case has probability $\frac{1}{2} \times \frac{1}{2} = \frac{1}{4}$, since the two dice extractions are independent. Hence, the probability to have either two odd or two even results is $\frac{1}{4} + \frac{1}{4} = \frac{1}{2}$, since the two events have no intersection.

2.8　The Bernoulli Distribution

Let us consider a basket containing a number of balls each having one of two possible colors, say red and white. Assume we know the number R of red balls in the basket and the number W of white balls (Fig. 2.1). The probability to randomly extract a red ball in basket is $p = R/(R + W)$, according to Eq. (1.1).

A variable x equal to the number of red balls in one extraction is called *Bernoulli variable* and can assume only the values 0 or 1. The probability distribution of x, called *Bernoulli distribution*, is simply given by:

$$\begin{cases} P(1) = p \,, \\ P(0) = 1 - p \,. \end{cases} \tag{2.55}$$

Fig. 2.1　A set of $R = 3$ red balls plus $W = 7$ white balls considered in a Bernoulli process. The probability to randomly extract a red ball is $p = R/(R + W) = \frac{3}{10} = 30\%$

The average of a Bernoulli variable is easy to compute from the definition in Eq. (2.17):

$$\langle x \rangle = 0 \times P(0) + 1 \times P(1) = P(1) = p \, . \tag{2.56}$$

Similarly, the average of x^2 is:

$$\langle x^2 \rangle = 0^2 \times P(0) + 1^2 \times P(1) = P(1) = p \, . \tag{2.57}$$

Hence, the variance of x, using Eq. (2.25), is:

$$\mathbb{V}\mathrm{ar}[x] = \langle x^2 \rangle - \langle x \rangle^2 = p\,(1-p) \, . \tag{2.58}$$

For a Bernoulli process, the ratio:

$$o = \frac{p}{1-p} \tag{2.59}$$

is called *odds*.

2.9   The Binomial Distribution

A *binomial process* consists of a given number N of independent Bernoulli extractions, each with probability p. This could be implemented, for instance, by randomly extracting a ball from a basket containing a fraction p of red balls; after each extraction, the extracted ball is placed again in the basket and then the extraction is repeated, for a total of N extraction. Figure 2.2 shows the possible outcomes of a binomial process as subsequent random extractions of a single ball.

The number n of positive outcomes (red balls) is called *binomial variable* and is equal to the sum of the N Bernoulli random variables. Its probability distribution can be simply determined from Fig. 2.2, considering how many red and white balls are present in each extraction, assigning each extraction a probability p or $1 - p$, respectively, and considering the number of possible paths leading to a given combination of red/white extractions. The latter term is called *binomial coefficient* (Fig. 2.3) and can be demonstrated by recursion to be equal to [1]:

$$\boxed{\binom{n}{N} = \frac{N!}{n!\,(N-n)!} \, .} \tag{2.60}$$

The probability distribution of a binomial variable n for given N and p can be obtained considering that the N extractions are independent, hence the corresponding probabilities, p for a red extraction, $1 - p$ for a white extraction, can be multiplied, according to Eq. (1.4). This product has to be multiplied by the binomial coefficient from Eq. (2.60) in order to take into account all possible

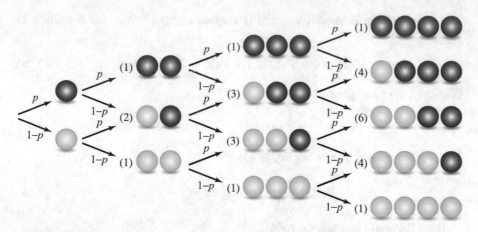

Fig. 2.2 Binomial process represented as subsequent random extractions of a single red or white ball (Bernoulli process). The tree shows all the possible combinations at each extraction step. Each branching has a corresponding probability equal to p or $1 - p$ for a red or white ball, respectively. The number of paths corresponding to each possible combination is shown in parentheses and is equal to the binomial coefficient in Eq. (2.60)

extraction paths leading to the same outcome. The probability to obtain n red and $N - n$ white extractions, called *binomial distribution*, can be written, in this way, as:

$$P(n;\ N,\ p) = \frac{N!}{n!\,(N-n)!} p^n (1 - p)^{N-n}\,. \tag{2.61}$$

A more rigorous demonstration by recurrence leads to the same result. Binomial distribution are shown in Fig. 2.4 for $N = 15$ and for $p = 0.2, 0.5$, and 0.8.

Since a binomial variable n is equal to the sum of N independent Bernoulli variables with probability p, the average and variance of n are equal to N times the average and variance of a Bernoulli variable, respectively (Eqs. (2.56) and (2.58)):

$$\langle n \rangle = Np\,, \tag{2.62}$$

$$\mathbb{V}\mathrm{ar}[n] = Np\,(1 - p)\,. \tag{2.63}$$

Those formulae can also be obtained directly from Eq. (2.61).

2.10 The Multinomial Distribution

The binomial distribution introduced in Sect. 2.9 can be generalized to the case in which, out of N extractions, there are more than two outcome categories. Consider k categories ($k = 2$ for the binomial case: success and failure), let the possible numbers of outcomes be equal to $n_1, \cdots, n_k$ for each of the k categories, with

Fig. 2.3 The Yang Hui triangle showing the construction of binomial coefficients. Reprinted from [2], public domain

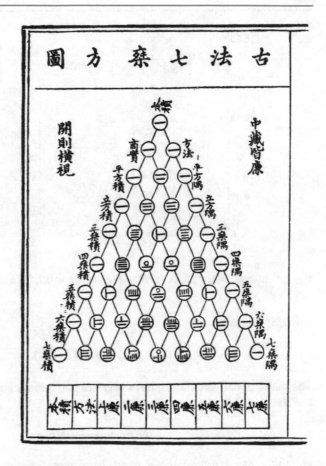

$\sum_{i=1}^{k} n_i = N$. Each category has a probability p_i, $i = 1, \cdots, k$ of the individual extraction, with $\sum_{i=1}^{k} p_i = 1$. The joint probability for the values of the k random variables $n_1, \cdots, n_k$ is given by:

$$P(n_1, \cdots, n_k; N, p_1, \cdots, p_k) = \frac{N!}{n_1! \cdots n_k!} p_1^{n_1} \cdots p_k^{n_k} \qquad (2.64)$$

and is called *multinomial distribution*. Equation (2.61) is equivalent to Eq. (2.64) for $k = 2, n = n_1, n_2 = N - n_1, p_1 = p$, and $p_2 = 1 - p$.

The average values of the multinomial variables n_i are:

$$\langle n_i \rangle = N p_i , \qquad (2.65)$$

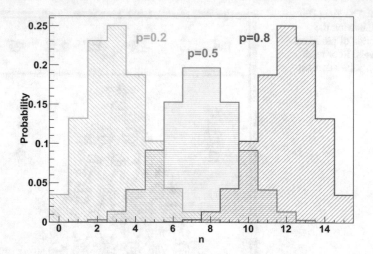

Fig. 2.4 Binomial distributions for $N = 15$ and for $p = 0.2, 0.5$ and 0.8

and their variances are:

$$\mathbb{V}\mathrm{ar}[n_i] = Np_i(1 - p_i) . \tag{2.66}$$

The multinomial variables n_i have negative correlation, and their covariance is, for $i \neq j$:

$$\mathbb{C}\mathrm{ov}(n_i, n_j) = -Np_i p_j . \tag{2.67}$$

For a binomial distribution, Eq. (2.67) leads to the obvious conclusion that $n_1 = n$ and $n_2 = N - n$ are 100% anti-correlated.

2.11   The Poisson Distribution

The binomial distribution has an interesting limit, when N is very large and p is very small, so that the product $\nu = pN$, called the *rate* parameter, is finite. Assume that n has a binomial distribution (Eq. (2.61)). We can reparameterize its distribution using ν instead of p:

$$P(n; \, N, \nu) = \frac{N!}{n! \, (N - n)!} \left(\frac{\nu}{N}\right)^n \left(1 - \frac{\nu}{N}\right)^{N-n} . \tag{2.68}$$

Equation (2.68) can also be written as:

$$P(n; \, N, \nu) = \left(\frac{\nu^n}{n!}\right) \frac{N(N - 1) \cdots (N - n + 1)}{N^n} \left(1 - \frac{\nu}{N}\right)^N \left(1 - \frac{\nu}{N}\right)^{-n} . \tag{2.69}$$

The first term, $v^n/n!$, does not depend on N. The limit for $N \to \infty$ of the remaining three terms are, respectively:

- $\lim\limits_{N \to \infty} \dfrac{N(N-1)\cdots(N-n+1)}{N^n} = 1$: the numerator has n terms, all tend to N, and simplifies with the denominator;
- $\lim\limits_{N \to \infty} \left(1 - \dfrac{v}{N}\right)^N = \lim\limits_{N \to \infty} e^{N \ln(1 - v/N)} = e^{-v}$;
- $\lim\limits_{N \to \infty} \left(1 - \dfrac{v}{N}\right)^{-n} = 1$ because $v/N \to 0$.

The distribution of n from Eq. (2.69), in this limit, becomes the Poisson distribution:

$$P(n;\ v) = \frac{v^n e^{-v}}{n!} \ . \tag{2.70}$$

A non-negative integer random variable n is called *Poissonian random variable* if it is distributed, for a given value of the parameter v, according to the distribution in Eq. (2.70). A Poissonian random variable has expected value and variance both equal to v. This descends directly from the limit of a binomial distribution. Figure 2.5 shows examples of Poisson distributions for different values of the parameter v.

Poisson distributions have several interesting properties, some of which are listed below:

- A binomial distribution with a number of extractions N and probability $p \ll 1$ can be approximated with a Poisson distribution with average $v = pN$, as discussed above.

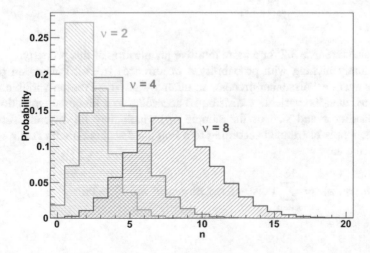

Fig. 2.5 Poisson distributions with different value of the rate parameter v

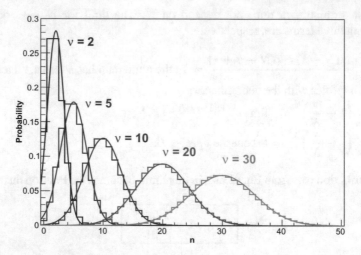

Fig. 2.6 Poisson distributions with different value of the parameter ν compared with Gaussian distributions with $\mu = \nu$ and $\sigma = \sqrt{\nu}$

- For large ν, a Poisson distribution can be approximated with a Gaussian distribution (see Sect. 3.25) having average ν and standard deviation $\sqrt{\nu}$. See Fig. 2.6 for a visual comparison.
- If two variables n_1 and n_2 follow Poisson distributions with averages ν_1 and ν_2, respectively, it is easy to demonstrate, using Eq. (2.53), that their sum $n = n_1 + n_2$ follows again a Poisson distribution with average $\nu_1 + \nu_2$. In formulae:

$$P(n;\ \nu_1,\ \nu_2) = \sum_{\substack{n_1 = 0 \\ n_2 = n - n_1}}^{n} \text{Pois}(n_1;\ \nu_1)\, \text{Pois}(n_2;\ \nu_2) = \text{Pois}(n;\ \nu_1 + \nu_2)\ .$$

(2.71)

See also Exercise 3.2 for a more intuitive justification of this property.

- Randomly picking with probability ε occurrences from a Poissonian process gives again a Poissonian process. In other words, if a Poisson variable n_0 has rate ν_0, then the variable n distributed according to a binomial extraction with probability ε and size of the sample n_0, which may be different for every extraction, is distributed according to a Poisson distribution with rate $\nu = \varepsilon\,\nu_0$. In formulae:

$$P(n;\ \nu_0,\ \varepsilon) = \sum_{n_0=0}^{\infty} \text{Pois}(n_0;\ \nu_0)\, \text{Binom}(n;\ n_0,\ \varepsilon) = \text{Pois}(n;\ \varepsilon\,\nu_0)\ . \quad (2.72)$$

This is the case, for instance, of the distribution of the number of counting n of cosmic rays with rate ν_0 recorded by a detector whose efficiency ε is not ideal ($\varepsilon < 1$).

2.12 The Law of Large Numbers

Assume to repeat N times an experiment whose outcome is a random variable x having a given probability distribution. The mean of all results is:

$$\bar{x} = \frac{x_1 + \cdots + x_N}{N} .$$ (2.73)

$\bar{x}$ is itself a random variable, and its expected value, from Eq. (2.24), is equal to the expected value of x. The distribution of $\bar{x}$, in general, tends to have smaller fluctuations compared with the variable x. Its probability distribution has in general a shape that favors values in the central part of the allowed range.

A formal demonstration is omitted here, but this law can be demonstrated numerically, using classical probability and combinatorial analysis, in the simplest cases. A case with $N = 2$ is, for instance, the distribution of the sum of two dice $d_1 + d_2$ shown in Fig. 1.1, where $\bar{x}$ is just given by $(d_1 + d_2)/2$. The distribution has the largest probability value for $(d_1 + d_2)/2 = 3.5$, which is the expected value of a single dice roll:

$$\langle x \rangle = \frac{1 + 2 + 3 + 4 + 5 + 6}{6} = 3.5 .$$ (2.74)

Repeating the combinatorial exercise for the mean of three or more dice gives an even more peaked distribution.

In general, it is possible to demonstrate that, under some conditions about the distribution of x, as N increases, most of the probability for possible values of $\bar{x}$ gets concentrated in a region around the expected value $\langle \bar{x} \rangle = \langle x \rangle$, and that region, which contains most of the probability, becomes smaller and smaller as N tends to infinity. In other words, the probability distribution of $\bar{x}$ becomes a narrow peak around the value $\langle x \rangle$, and the interval of values around $\langle x \rangle$ that correspond to a large fraction of the total probability, say, 90% or 95%, becomes smaller and smaller. Eventually, for $N \to \infty$, the distribution becomes a Dirac's delta centered at $\langle x \rangle$.

This convergence of the probability distribution towards a Dirac's delta is called *law of large numbers* and can be illustrated in a simulated experiment consisting of repeated dice roll, as shown in Fig. 2.7, where $\bar{x}$ is plotted as a function of N for two independent random extractions. Larger values of N correspond to smaller fluctuations of the result and to a visible convergence towards the value of 3.5. If we would ideally increase *to infinity* the total number of trials N, the mean value $\bar{x}$ of the infinite random extractions would no longer be a random variable but would take a single possible value, equal to $\langle x \rangle = 3.5$.

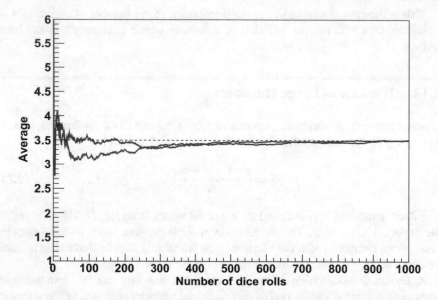

Fig. 2.7 An illustration of the law of large numbers using a computer simulation of die rolls. The average of the first N out of 1000 random extraction is reported as a function of N. 1000 extractions have been repeated twice (red and blue lines) with independent random extractions in order to show that two different lines converge to the same value

Note that the value 6, which is the most distant from the expected average of 3.5, is a *possible* value, even for a very large number of extractions N. The probability to have all extractions equal to 6, which is the only possibility that would give an average of 6, is equal to $(1/6)^N$, which may be extremely small, but not null. Nonetheless, even if 6 is a possible value, it is not *plausible* that it may really occur if N is extremely large. The confusion between possible and plausible may lead to believe in possible but not plausible hypothesis and is one of the reasons for debates between rational people and supporters of unreasonable theories, like flat-Earth, anti-vax, conspiracies, etc.

The *law of large numbers* has many empirical verifications for most of random experiments and has a broad validity range. It is valid whatever approach to probability is taken, classical, Bayesian, or frequentist. See also Sect. 3.20, where the central limit theorem is discussed, which is also somewhat related to the law of large numbers.

2.13   Law of Large Numbers and Frequentist Probability

Consider now an experiment repeated N times, and an event E of our interest with probability p. E may occur or not in each experiment; therefore, the number of times E occurs in an experiment may be zero or one. This number is a Bernoulli variable. According to the law of large numbers, the mean number of occurrences in N experiments, which is equal to the frequency $N(E)/N$, converges in probability, in the limit $N \to \infty$, to the expected value, which, for a Bernoulli variable, is equal to p. In the extreme limit $N \to \infty$, therefore, the frequency $N(E)/N$ becomes equal to the probability p in Eq. (6.1). This justifies the frequentist approach to probability, as discussed in Sect. 6.1.

Anyway, to demonstrate that the law of large numbers applies in this case, an underlying probability definition must be introduced before justifying the frequentist definition of probability introduced above. For this reason, either frequentist probability is considered as a consequence of another more fundamental probability definition, e.g., Bayesian or classic, or the frequentist definition of probability is somewhat affected by circular logic.

References

1. The coefficients present in the binomial distribution are the same that appear in the expansion a binomial raised to the nth power, $(a + b)^n$. A simple iterative way to compute those coefficients is known as Pascal's triangle. In different countries this triangle is named after different authors, e.g., the Tartaglia's triangle in Italy, Yang Hui's triangle in China, and so on. In particular, the following publications of the triangle are present in literature:

   * India: published in the tenth century, referring to the work of Pingala, dating back to fifth–second century BC.
   * Persia: Al-Karaju (953–1029) and Omar Jayyám (1048–1131)
   * China: Yang Hui (1238–1298), see Fig. 2.3
   * Germany: Petrus Apianus (1495–1552)
   * Italy: Nicolò Fontana Tartaglia (1545)
   * France: Blaise Pascal (1655)

2. Yang Hui (杨辉) triangle as published by Zhu Shijie (朱世杰) in *Siyuan yujian*, (四元玉鉴) *Jade Mirror of the four unknowns* (1303), public domain image.
3. D. Zwillinger, S. Kokoska, *CRC Standard Probability and Statistics Tables and Formulae* (Chapman & Hall, New York, 2000)
4. D.N. Joanes, C.A. Gill, Comparing measures of sample skewness and kurtosis. J. R. Stat. Soc. D **47**(1), 183–189 (1998). https://doi.org/10.1111/1467-9884.00122

Probability Density Functions

3

3.1 Introduction

In Sect. 1.6 we discussed the difficulty to define the concept of equal probability for continuous cases using Bertrand's paradox as example. While for discrete random variables we can reduce a composite event into elementary events, each having a well-defined and possibly equal probabilities, motivated by symmetries of the system, for a continuous variable x there is no unique way to decompose the range of possible values into equally probable elementary intervals.

In Sect. 1.5, we considered a continuous random variable x with possible values in an interval $[x_1, x_2]$, and we saw that, if x is uniformly distributed in $[x_1, x_2]$, a transformed variable $y = Y(x)$ is not in general uniformly distributed in $[y_1, y_2] = [Y(x_1), Y(x_2)]$. Y was here taken as a monotonically increasing function of x, for simplicity. The choice of the continuous variable on which equally probable intervals are defined is therefore an arbitrary choice.

The following sections show how to overcome this difficulty using a definition consistent with the axiomatic approach to probability due to Kolmogorov, introduced in Sect. 1.7.

3.2 Definition of Probability Density Function

The concept of probability distribution introduced in Sect. 2.1 can be generalized to the continuous case. Let us consider a *sample space* $\Omega \subseteq \mathbb{R}^n$. Each random extraction gives an outcome, we would say a *measurement*, in case of an experiment, corresponding to one point $\vec{x} = (x_1, \cdots, x_n)$ in the sample space Ω. We can associate to any point $\vec{x} \in \Omega$ a real value greater or equal to zero, $f(\vec{x}) = f(x_1, \cdots, x_n)$. The function f is called *probability density function* or PDF.

© The Author(s), under exclusive license to Springer Nature Switzerland AG 2023
L. Lista, *Statistical Methods for Data Analysis*, Lecture Notes in Physics 1010,
https://doi.org/10.1007/978-3-031-19934-9_3

The probability of an *event* A, where $A \subseteq \Omega$, i.e., the probability that a random extraction $\vec{x}$ belongs to the set A, is given by:

$$P(A) = \int_A f(x_1, \cdots, x_n) \, \mathrm{d}^n x \, . \tag{3.1}$$

The product $f(\vec{x}) \, \mathrm{d}^n x$ can be interpreted as *differential probability*, and $f(\vec{x})$ is equal to the probability $\mathrm{d}P$ corresponding to the infinitesimal hypervolume $\mathrm{d}x_1 \cdots \mathrm{d}x_n$ divided by the infinitesimal hypervolume:

$$f(x_1, \cdots, x_n) = \frac{\mathrm{d}P}{\mathrm{d}x_1 \cdots \mathrm{d}x_n} \, . \tag{3.2}$$

The normalization condition for discrete probability distributions (Eq. (2.3)) can be generalized to continuous case as follows:

$$\int_\Omega f(x_1, \cdots, x_n) \, \mathrm{d}^n x = 1 \, . \tag{3.3}$$

In one dimension, one can write:

$$\int_{-\infty}^{+\infty} f(x) \, \mathrm{d}x = 1 \, . \tag{3.4}$$

Note that the probability corresponding to a set containing a single point is rigorously zero if f is a real function, i.e., $P(\{x_0\}) = 0$ for any x_0, since the set $\{x_0\}$ has a null measure. It is possible to treat discrete variables in one dimension using the same formalism of PDFs, extending the definition of PDF to Dirac's delta function $\delta(x - x_0)$, which is normalized by definition:

$$\int_{-\infty}^{+\infty} \delta(x - x_0) \, \mathrm{d}x = 1 \, . \tag{3.5}$$

Linear combinations of Dirac's delta functions model discrete probabilities if the weights in the linear combination are equal to the probabilities of the corresponding discrete values. A distribution representing a discrete random variable x that can only take the values $x_1, \cdots, x_N$ with probabilities $P_1, \cdots, P_N$, respectively, can be written, using the continuous PDF formalism, as:

$$f(x) = \sum_{i=1}^{N} P_i \, \delta(x - x_i) \, . \tag{3.6}$$

The normalization condition in this case becomes:

$$\int_{-\infty}^{+\infty} f(x)\, dx = \sum_{i=1}^{N} P_i \int_{-\infty}^{+\infty} \delta(x - x_i)\, dx = \sum_{i=1}^{N} P_i = 1\,, \tag{3.7}$$

which gives again the normalization condition for a discrete variable, already shown in Eq. (2.3).

Discrete and continuous distributions can be combined. For instance, the PDF:

$$f(x) = \frac{1}{2}\delta(x) + \frac{1}{2}f(x) \tag{3.8}$$

describes a random variable x that has a 50% probability to give $x = 0$ and a 50% probability to give a value of distributed according to $f(x)$. Discrete peaks may appear if a distribution is truncated by a transformation of variable. For instance, if the variable y is distributed according to a distribution $g(y)$, the variable $x = \max(0, y)$ is distributed according to:

$$f(x) = \alpha\, \delta(x) + (1 - \alpha)\, \theta(x)\, g(x), \tag{3.9}$$

where θ is the step function:

$$\theta(x) = \begin{cases} 0 & \text{if } x < 0\,, \\ 1 & \text{if } x \geq 0\,, \end{cases} \tag{3.10}$$

and α is the probability that y is negative:

$$\alpha = \int_{-\infty}^{0} g(y)\, dy\,. \tag{3.11}$$

3.3 Statistical Indicators in the Continuous Case

This section generalizes the definitions of statistical indicators introduced in Sect. 2.5 for continuous variables.

The average value of a continuous variable x whose PDF is f is:

$$\boxed{\langle x \rangle = \int x\, f(x)\, dx\,.} \tag{3.12}$$

The integration range is from $-\infty$ to $+\infty$, or the entire validity range of the variable x if the range is not $\mathbb{R}$. More in general, the average value of $g(x)$ is:

$$\boxed{\langle g(x) \rangle = \int g(x)\, f(x)\, dx\,.} \tag{3.13}$$

The variance of x is:

$$\mathbb{V}\text{ar}[x] = \int (x - \langle x \rangle)^2 f(x) \, \mathrm{d}x = \langle (x - \langle x \rangle)^2 \rangle = \langle x^2 \rangle - \langle x \rangle^2 . \tag{3.14}$$

It is also indicated with σ_x^2. The standard deviation is defined as the square root of the variance:

$$\sigma_x = \sqrt{\mathbb{V}\text{ar}[x]}. \tag{3.15}$$

The kth moment of x is defined as:

$$m_k'(x) = \langle x^k \rangle = \int x^k f(x) \, \mathrm{d}x , \tag{3.16}$$

while the kth central moment of x is defined as:

$$m_k(x) = \langle (x - \langle x \rangle)^k \rangle = \int (x - \mu_x)^k f(x) \, \mathrm{d}x . \tag{3.17}$$

Covariance, correlation coefficient, and covariance matrix can be defined for the continuous case in the same way as done in Sect. 2.5, as well as skewness and kurtosis.

The *mode* of a PDF f is the value $\hat{x}$ corresponding to the maximum of $f(x)$:

$$f(\hat{x}) = \max_x f(x) \tag{3.18}$$

or:

$$\hat{x} = \arg \max_x f(x) . \tag{3.19}$$

As for the discrete case, a continuous PDF may have more than one mode, and in that case, it is called *multimodal distribution*.

The *median* of a PDF $f(x)$ is the value $\tilde{x}$ such that:

$$P(x < \tilde{x}) = P(x > \tilde{x}), \tag{3.20}$$

or equivalently:

$$\int_{-\infty}^{\tilde{x}} f(x) \, \mathrm{d}x = \int_{\tilde{x}}^{+\infty} f(x) \, \mathrm{d}x . \tag{3.21}$$

This definition may clearly have problems in case f has some Dirac's delta components, as we have considered in the previous section, because the probability of the single value $\tilde{x}$, in that case, might not be null. More in general, the quantity q_α such that:

$$\int_{-\infty}^{q_\alpha} f(x) \, dx = \alpha = 1 - \int_{q_\alpha}^{+\infty} f(x) \, dx \tag{3.22}$$

is called *quantile*, or α-quantile. The median is the quantile corresponding to a probability $\alpha = {}^1\!/_2$. The 100 quantiles corresponding probabilities of 1%, 2%, $\cdots$, 99% are called 1st, 2nd, $\cdots$, 99th *percentile*, respectively.

3.4   Cumulative Distribution

Given a PDF $f(x)$, its *cumulative distribution* is defined as:

$$F(x) = \int_{-\infty}^{x} f(x') \, dx' \,. \tag{3.23}$$

The cumulative distribution $F(x)$ is a monotonically increasing function of x, and, from the normalization of $f(x)$ (Eq. (3.4)), its values range from 0 to 1. In particular, if the range of x is $\mathbb{R}$:

$$\lim_{x \to -\infty} F(x) = 0 \,, \tag{3.24}$$

$$\lim_{x \to +\infty} F(x) = 1 \,. \tag{3.25}$$

If the variable x follows the PDF $f(x)$, the PDF of the transformed variable $y = F(x)$, where F is the cumulative distribution of x, is uniformly distributed between 0 and 1, as can be easily demonstrated:

$$\frac{dP}{dy} = \frac{dP}{dx}\frac{dx}{dy} = f(x)\frac{dx}{dF(x)} = \frac{f(x)}{f(x)} = 1 \,. \tag{3.26}$$

This property is very useful to generate pseudorandom numbers with the desired PDF with computer algorithms, as discussed in Sect. 4.4.

3.5   Continuous Transformations of Variables

The evaluation of probability distributions under transformation of variables was discussed in Sect. 2.7 for the discrete case and can be generalized to the continuum. Consider a transformation of variable $y = Y(x)$, where x follows a PDF $f_x(x)$. The PDF of the transformed variable y is:

$$f_y(y) = \int \delta(y - Y(x)) \, f_x(x) \, dx \,. \tag{3.27}$$

This generalizes Eq. (2.52) which has a very similar formal structure, where the sum is replaced by the integral, and the Kronecker Delta is replaced by Dirac's delta. The equivalence of the two equations can be demonstrated if we write a discrete distribution using the PDF formalism as in Eq. (3.6).

Assuming a monotonically increasing function Y, we can demonstrate the validity of Eq. (3.27) as follows:

$$f_y(y) = \int \delta(y - Y(x)) f_x(x) \, dx = \int \delta(y - Y(x)) \frac{dP(x)}{dx} \frac{dx}{dy} \, dy =$$

$$= \int \delta(y - Y(x)) \frac{dP(y)}{dy} \, dy = \frac{dP(Y(x))}{dy} = \frac{dP(y)}{dy} . \tag{3.28}$$

If the function Y is not monotonic, we can use the following property of Dirac's delta to perform the integration:

$$\delta(y - Y(x)) = \sum_{i=1}^{n} \frac{\delta(x - x_i)}{|Y'(x_i)|} , \tag{3.29}$$

where x_i are solutions of the equation $y = Y(x)$.

Similarly to Eq. (3.27), considering a transformation $z = Z(x, y)$, the PDF of the transformed variable z can be generalized from Eq. (2.54) as:

$$f_z(z) = \int \delta(z - Z(x, y)) f_{xy}(x, y) \, dx \, dy , \tag{3.30}$$

where $f_{xy}(x, y)$ is the PDF for the variables x and y. The case of transformations into more than one variable can be easily generalized. If we have, for instance: $x' = X'(x, y), y' = Y'(x, y)$, the transformed two-dimensional PDF can be written as:

$$f'(x', y') = \int \delta(x' - X'(x, y)) \delta(y' - Y'(x, y)) f(x, y) \, dx \, dy . \tag{3.31}$$

If the transformation is invertible, the PDF transforms according to the determinant of the Jacobean of the transformation, which appears in the transformation of the n-dimensional volume element $d^n x = dx_1 \cdots dx_n$:

$$f(x_1, \cdots, x_n) = \frac{dP}{d^n x} = \frac{dP'}{d^n x'} \left| \det\left(\frac{\partial x_i'}{\partial x_j}\right) \right| = f'(x_1', \cdots, x_n') \left| \det\left(\frac{\partial x_i'}{\partial x_j}\right) \right| . \tag{3.32}$$

For the simplest case of a single variable:

$$f_x(x) = f_y(y) \left| \frac{dy}{dx} \right| \tag{3.33}$$

which corresponds to:

$$\frac{dP}{dx} = \frac{dP}{dy} \left| \frac{dy}{dx} \right| . \tag{3.34}$$

The absolute value is required in case the transformation from x to y locally decreasing.

3.6 Marginal Distributions

Given a two-dimensional PDF $f(x, y)$, the probability distributions of the two individual variables x and y, called *marginal distributions* or *marginal PDFs*, can be determined by integrating $f(x, y)$ over the other coordinate, y and x, respectively:

$$f_x(x) = \int f(x, y) \, dy , \tag{3.35}$$

$$f_y(y) = \int f(x, y) \, dx . \tag{3.36}$$

The above expressions are similar to the discrete case, defined in Sect. 2.2. They are also a special case of continuous transformation of variables, as described in Sect. 3.5, where the applied transformation maps the two variables into one of the two:

$$X : (x, y) \longmapsto x , \tag{3.37}$$

$$Y : (x, y) \longmapsto y . \tag{3.38}$$

More in general, if we have a PDF in $n = h+k$ variables $(\vec{x}, \vec{y}) = (x_1, \cdots, x_h, y_1, \cdots, y_k)$, the marginal PDF of the subset of n variables $(x_1, \cdots, x_h)$ can be determined by integrating the PDF $f(\vec{x}, \vec{y})$ over the remaining set of variables $(y_1, \cdots, y_k)$:

$$\boxed{f_{x_1, \cdots, x_h}(\vec{x}) = \int f(\vec{x}, \vec{y}) \, d^k y .} \tag{3.39}$$

The definition of marginal distributions allows to demonstrate that the average of the sum of two random variables is equal to the sum of their average values:

$$\boxed{\langle x + y \rangle = \langle x \rangle + \langle y \rangle .} \tag{3.40}$$

The sum $x + y$ can be considered a special case of transformation:

$$Z : (x, y) \mapsto z = x + y .$$ (3.41)

The average value of $z = x + y$ is:

$$\langle x + y \rangle = \int (x + y) f(x, y) \, dx \, dy = \int x f(x, y) \, dx \, dy + \int y f(x, y) \, dx \, dy =$$

$$= \int x \, dx \int f(x, y) \, dy + \int y \, dy \int f(x, y) \, dx =$$

$$= \int x f_x(x) \, dx + \int y f_y(y) \, dy = \langle x \rangle + \langle y \rangle .$$ (3.42)

3.7   Uniform Distribution

A variable x is *uniformly distributed* in the interval $[a, b[$ if the PDF is constant in a range $[a, b[$. This condition was discussed in Sect. 1.6, before formally introducing the concept of PDF. Imposing the normalization condition, a uniform PDF can be written as:

$$p(x) = \begin{cases} \dfrac{1}{b - a} & \text{if } a \le x < b , \\ 0 & \text{if } x < a \text{ or } x \ge b . \end{cases}$$ (3.43)

Examples of uniform distributions are shown in Fig. 3.1 for different extreme values a and b.

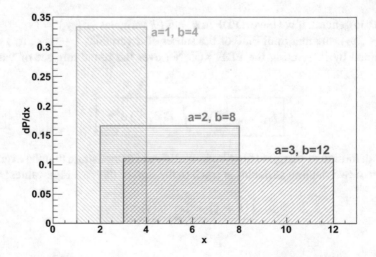

Fig. 3.1 Uniform distributions with different values of the extreme values a and b

The average of a uniformly distributed variable x is:

$$\langle x \rangle = \frac{a+b}{2}, \tag{3.44}$$

and its standard deviation is:

$$\sigma_x = \frac{b-a}{\sqrt{12}}. \tag{3.45}$$

Example 3.1 - Strip Detectors

A detector instrumented with strips of a given pitch l receives particles at positions uniformly distributed along each strip. The standard deviation of the distribution of the position of the impact points of the particles on the strip along the direction transverse to the strips is given by $l/\sqrt{12}$, according to Eq. (3.45).

3.8 Gaussian Distribution

A *Gaussian distribution*, or *normal distribution*, is defined by the following PDF, where μ and σ are fixed parameters:

$$p(x; \mu, \sigma) = \frac{1}{\sigma\sqrt{2\pi}} \exp\left[-\frac{(x-\mu)^2}{2\sigma^2}\right]. \tag{3.46}$$

A random variable distributed according to normal distribution is called *Gaussian random variable*, or *normal random variable*. The function has a single maximum at $x = \mu$ and two inflection points at $x = \mu \pm \sigma$. Examples of Gaussian distributions are shown in Fig. 3.2 for different values of μ and σ.

The average value and standard deviation of a normal variable are μ and σ, respectively. The *full width at half maximum* (FWHM) of a Gaussian distribution is equal to $2\sigma\sqrt{2\log 2} \simeq 2.3548\,\sigma$.

For $\mu = 0$ and $\sigma = 1$, a normal distribution is called *standard normal* or *standard Gaussian* distribution. A *standard normal random variable* is often indicated with z, and its distribution is:

$$\phi(z) = \frac{1}{\sqrt{2\pi}} e^{-z^2/2}. \tag{3.47}$$

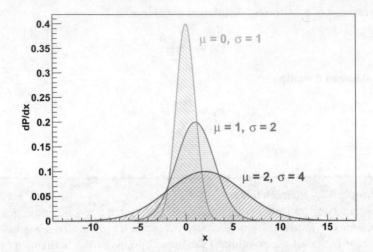

Fig. 3.2 Gaussian distributions with different values of average and standard deviation parameters μ and σ

The cumulative distribution of a standard normal distribution is:

$$\Phi(z) = \frac{1}{\sqrt{2\pi}} \int_{-\infty}^{x} e^{-z'^2/2}\, \mathrm{d}z' = \frac{1}{2}\left[\mathrm{erf}\left(\frac{z}{\sqrt{2}}\right) + 1\right]. \qquad (3.48)$$

If we have a normal random variable x with average μ and standard deviation σ, a standard normal variable can be built with the following transformation:

$$z = \frac{x - \mu}{\sigma}. \qquad (3.49)$$

The probability for a Gaussian distribution corresponding to the symmetric interval around μ [$\mu - Z\sigma$, $\mu + Z\sigma$], frequently used in many applications, can be computed as:

$$P(Z\sigma) = \frac{1}{\sqrt{2\pi}} \int_{-Z}^{Z} e^{-x^2/2}\, \mathrm{d}x = \Phi(Z) - \Phi(-Z) = \mathrm{erf}\left(\frac{Z}{\sqrt{2}}\right). \qquad (3.50)$$

The most frequently used values are the ones corresponding to 1σ, 2σ, and 3σ ($Z = 1, 2, 3$) and have probabilities of 68.27%, 95.45% and 99.73%, respectively.

The importance of the Gaussian distribution resides in the central limit theorem (see Sect. 3.20), which allows to approximate to Gaussian distributions may realistic distributions resulting from the superposition of more random effects.

3.9 χ^2 **Distribution**

A χ^2 *random variable* with k *degrees of freedom* is the sum of the squares of k independent standard normal variables $z_1, \cdots, z_k$, defined above in Sect. 3.8:

$$\chi^2 = \sum_{j=1}^{k} z_i^2 . \tag{3.51}$$

If we have k independent normal random variables $x_1, \cdots, x_k$, each with mean μ_j and standard deviation σ_j, a χ^2 variable can be built by transforming each x_j in a standard normal variable using the transformation in Eq. (3.49):

$$\chi^2 = \sum_{j=1}^{k} \frac{(x_j - \mu_j)^2}{\sigma_j^2} . \tag{3.52}$$

The distribution of a χ^2 variable with k degrees of freedom is given by:

$$\boxed{p(\chi^2; k) = \frac{2^{-k/2}}{\Gamma(k/2)} \chi^{k-2} e^{-\chi^2/2} .} \tag{3.53}$$

Γ is the so-called *gamma function* and is the analytical extension of the factorial.[1] Sometimes, the notation χ_k^2 is used to explicitate the number of degrees of freedom in the subscript k. χ^2 distributions are shown in Fig. 3.3 for different number of degrees of freedom k.

The expected value of a χ^2 variable is equal to the number of degrees of freedom k and the variance is equal to $2k$. The cumulative χ^2 distribution is given by:

$$P(\chi^2 < v) = \int_0^v p(\chi^2; k) \, \mathrm{d}\chi^2 = P\left(\frac{k}{2}, \frac{v}{2}\right) , \tag{3.54}$$

where $P(a, k) = \gamma(a, k)/\Gamma(a)$ is the so-called *regularized Gamma function*. It is related to the Poisson distribution by the following formula:

$$\sum_{j=0}^{k-1} \frac{e^{-v} v^j}{j!} = 1 - P(k, v) , \tag{3.55}$$

so that the following relation holds:

$$\sum_{j=0}^{k-1} \frac{e^{-v} v^j}{j!} = \int_{2v}^{+\infty} p(\chi^2; 2k) \, \mathrm{d}\chi^2 . \tag{3.56}$$

---

[1] $\Gamma(k) = (k-1)!$ if k is an integer value.

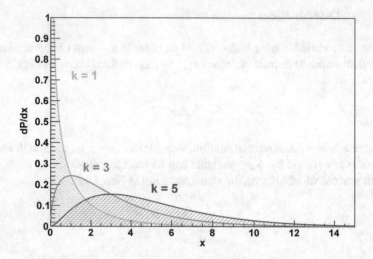

Fig. 3.3 χ^2 distributions with different numbers of degrees of freedom k

Typical applications of the χ^2 distribution are goodness-of-fit tests (see Sect. 6.17), where the cumulative χ^2 distribution is used.

3.10   Log Normal Distribution

If a random variable y is distributed according to a normal distribution with average μ and standard deviation σ, the variable $x = e^y$ is distributed according to the following distribution, called *log normal distribution*:

$$p(x; \mu, \sigma) = \frac{1}{x \sigma \sqrt{2\pi}} \exp\left[-\frac{(\log x - \mu)^2}{2\sigma^2}\right] . \tag{3.57}$$

Note that x must be greater than zero. The PDF in Eq. (3.57) can be determined by applying Eq. (3.27) to the case of a normal distribution. A log normal variable has the following average and standard deviation:

$$\langle x \rangle = e^{\mu + \sigma^2/2} , \tag{3.58}$$

$$\sigma_x = e^{\mu + \sigma^2/2} \sqrt{e^{\sigma^2} - 1} . \tag{3.59}$$

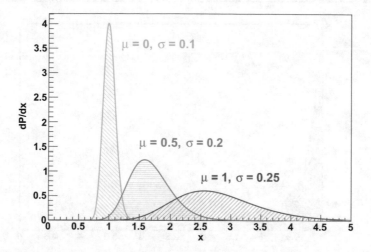

Fig. 3.4 Log normal distributions with different values of the parameters μ and σ

Note that Eq. (3.58) implies that $\langle e^y \rangle > e^{\langle y \rangle}$ for a normal random variable y:

$$\langle e^y \rangle = \langle x \rangle = e^\mu e^{\sigma^2/2} > e^\mu = e^{\langle y \rangle} . \tag{3.60}$$

Examples of log normal distributions are shown in Fig. 3.4 for different values of μ and σ.

3.11 Exponential Distribution

An *exponential distribution* of a variable $x \geq 0$ is characterized by a PDF proportional to $e^{-\lambda x}$, where λ is a constant. The expression of an exponential PDF, including the normalization factor λ, is given by:

$$\boxed{p(x; \lambda) = \lambda e^{-\lambda x} .} \tag{3.61}$$

Examples of exponential distributions are shown in Fig. 3.5 for different values of the parameter λ.

The mean value and the standard deviation of an exponential distribution are equal to $1/\lambda$. The distribution is asymmetric with a skewness equal to 2.

Exponential distributions are widely used in physics to model the distribution of particle lifetimes. In those cases, the lifetime parameter τ is defined as the inverse

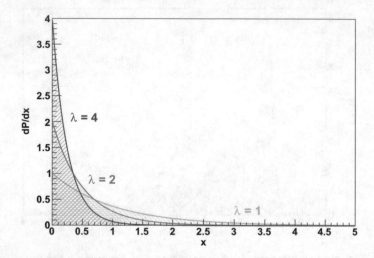

Fig. 3.5 Exponential distributions with different values of the parameter λ

of λ, and the PDF may also be parameterized as:

$$p(t;\ \tau) = \frac{1}{\tau}\, e^{-t/\tau}\,. \tag{3.62}$$

3.12   Gamma Distribution

A gamma distribution is an asymmetric distribution defined for positive x as:

$$p(x;\alpha,\beta) = \frac{\beta^{\alpha} x^{\alpha-1} e^{-\beta x}}{\Gamma(\alpha)}\,. \tag{3.63}$$

The parameters α and β must be positive. An alternative parameterization found in literature is:

$$p(x;\kappa,\theta) = \frac{x^{\kappa-1} e^{-x/\theta}}{\theta^{\kappa}\,\Gamma(\kappa)}\,, \tag{3.64}$$

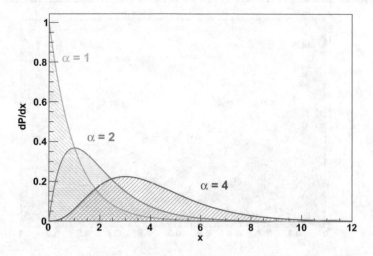

Fig. 3.6 Examples of gamma distributions for parameter values $\alpha = \kappa = 1, 2$ and 4 and for $\beta = 1/\theta = 1$

where $\kappa = \alpha$ is called shape parameter, and $\theta = 1/\beta$ is called scale parameter. The mean of the distribution is $\kappa\theta = \alpha/\beta$ and the variance is $\kappa\theta^2 = \alpha\beta^2$. The mode is $(\kappa - 1)\theta = (\alpha - 1)/\beta$. Figure 3.6 shows different examples of gamma distributions.

3.13 Beta Distribution

A beta distribution is defined for $0 < x < 1$ as:

$$p(x; \alpha, \beta) = \frac{x^{\alpha-1}(1 - x)^{\beta-1}}{B(\alpha, \beta)}, \tag{3.65}$$

where $B(\alpha, \beta)$ at the denominator is a normalization factor:

$$B(\alpha, \beta) = \int_0^1 x^{\alpha-1}(1 - x)^{\beta-1}\,\mathrm{d}x = \frac{\Gamma(\alpha)\Gamma(\beta)}{\Gamma(\alpha + \beta)}. \tag{3.66}$$

The parameters α and β must be positive. The mean of the distribution is $\alpha/(\alpha + \beta)$ and the mode is $(\alpha - 1)/(\alpha + \beta - 2)$ only for $\alpha, \beta > 1$. For lower values of α and/or β, the distribution diverges at $x = 0$ and/or $x = 1$. The distribution is symmetric for $\alpha = \beta$. Figure 3.7 shows different examples of beta distributions with fixed sum of $\alpha + \beta$.

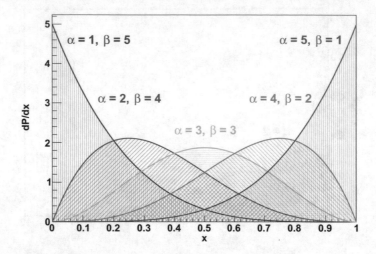

Fig. 3.7 Examples of beta distributions for different values of α and β. The sum $\alpha + \beta$ is equal to 6 for all distributions. The distribution with $\alpha = \beta$ is symmetric

Example 3.2 - Poisson and Binomial Distributions from a Uniform Process

Consider a random variable ξ uniformly distributed over an interval $[0, X [$. ξ could be either a time or space variable, in a concrete case. For instance, ξ could be the arrival position of a rain drop on the ground, or the impact position of a particle on a detector along one direction, or the arrival time of a cosmic ray.

Consider N independent uniform random extractions of ξ in the range $[0, X [$. We introduce the rate r, equal to N/X, which represents the expected number of extractions per unit of ξ.

Let us now consider only the values of ξ contained in a shorter interval $[0, x [$ (Fig. 3.8). The extraction of n occurrences out of N in the interval $[0, x [$, while the remaining $N - n$ occurrences are in $[x, X [$, is a binomial process (see Sect. 2.9).

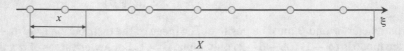

Fig. 3.8 Example of uniform random extractions of a variable ξ. Two intervals are shown of sizes x and X. We consider $x \ll X$

Consider N and X as constants, i.e., not subject to random fluctuations, in the limit $N \rightarrow \infty$, $X \rightarrow \infty$ obtained while keeping the ratio $N/X = r$ as a constant.

(continued)

Example 3.2 (continued)

The expected value ν of the number of extracted values of ξ in the interval $[0, x[$ can be determined with a simple proportion:

$$\nu = \langle n \rangle = \frac{Nx}{X} = rx \,, \tag{3.67}$$

where n follows a binomial distribution. Those conditions describe the limit of a binomial distribution already discussed in Sect. 2.11, which give a Poisson distribution with rate parameter $\nu = rx$.

Note that the superposition of two uniform processes, like the one considered here, is again a uniform process, whose total rate is equal to the sum of the two individual rates. This is consistent with the property that the sum of two Poissonian variables follow again a Poisson distribution having a rate equal to the sum of the two rates, as discussed in Sect. 2.11.

Example 3.3 - Exponential Distributions from a Uniform Process
Consider a sequence of events characterized by a time variable t that is uniformly distributed over an indefinitely large time interval. The time t could be the arrival time of a cosmic ray, for instance, or the arrival time of a car at a highway gate, if we assume the arrival time to be uniformly distributed.

Let t_1 be the time of the first occurrence after an arbitrary time t_0 that we can take as origin of times. The distribution of the time difference $t_1 - t_0$ can be demonstrated to follow an exponential distribution. The situation is sketched in Fig. 3.9.

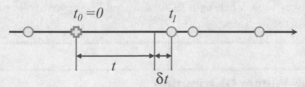

Fig. 3.9 Occurrence times (dots) of events uniformly distributed in time, represented along a horizontal axis. The time origin ($t_0 = 0$) is marked as a cross. The occurrence time of the first event is marked as t_1

If $t = t_1$ is the first occurrence after t_0, there is no other occurrence between t_0 and t. The number of occurrences in the interval $[t_0, t[$ and in the consecutive interval $[t, t+dt[$ are independent random variables. Therefore,

(continued)

Example 3.3 (continued)

it is possible to write the probability to have the first occurrence t_1 in the interval $[t, t + dt [$ as the product of the probability to have *zero* occurrences in $[t_0, t [$ times the probability to have *one* occurrence in $[t, t + dt [$:

$$dP(t) = P(0; [t_0, t [) P(1; [t, t + dt [) . \tag{3.68}$$

The probability to have n occurrences in a generic interval $[t_a, t_b[$ is given by the Poisson distribution with a rate equal to $\nu = r (t_b - t_a)$:

$$P(n; \nu) = \frac{\nu^n e^{-\nu}}{n!} . \tag{3.69}$$

Therefore, setting $t_0 = 0$, we have:

$$P(0; [0, t[) = P(0; rt) = e^{-rt} , \tag{3.70}$$

$$P(1; [t, t + dt [) = P(1; r \, dt) = r \, dt \, e^{-r \, dt} \simeq r \, dt , \tag{3.71}$$

and finally:

$$\frac{dP(t)}{dt} = re^{-rt} . \tag{3.72}$$

The term r in front of the exponential ensures the proper normalization of the PDF. Note that it is possible to apply the same demonstration to the difference of two consecutive occurrence times considering $t_0 = t_i$ and $t = t_{i+i}$.

The exponential distribution is characteristic of particles lifetimes. The possibility to measure the decay parameter of an exponential distribution, independently on the initial time t_0, allows to measure particle lifetimes even if the particle's creation time is not known. For instance, the lifetime of cosmic-ray muons can be measured at sea level even if the muon was produced in the high atmosphere.

3.14   Breit–Wigner Distribution

This section and the followings, up to Sect. 3.18, present some of the most used PDF in physics. The list is of course not exhaustive.

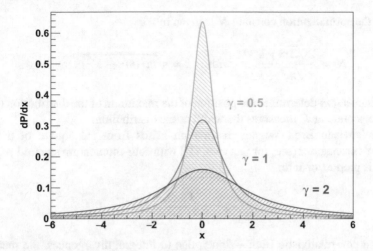

Fig. 3.10 Breit–Wigner distributions centered around zero for different values of the width parameter γ

A (non-relativistic) *Breit–Wigner distribution*, also known as *Lorentz distribution* or *Cauchy distribution*, has the following expression:

$$p(x;\, m,\, \gamma) = \frac{1}{\pi}\left[\frac{\gamma}{(x-m)^2 + \gamma^2}\right]. \tag{3.73}$$

The parameter m determines the position of the maximum of the distribution (mode) and twice the parameter γ is equal to the full width at half maximum of the distribution. Breit–Wigner distributions arise in many resonance problems in physics. Figure 3.10 shows examples of Breit–Wigner distributions for different values of the width parameter γ and for fixed $m = 0$.

Since the integrals of both $x\,p(x)$ and $x^2 p(x)$ are divergent, the mean and variance of a Breit–Wigner distribution are undefined.

3.15   Relativistic Breit–Wigner Distribution

A *relativistic Breit–Wigner distribution* has the following expression:

$$p(x;\, m,\, \Gamma) = \frac{N}{(x^2 - m^2)^2 + m^2\Gamma^2}, \tag{3.74}$$

where the normalization constant N is given by:

$$N = \frac{2\sqrt{2}\,m\,\gamma\,k}{\pi\sqrt{m^2 + k}}\,, \quad \text{with:} \quad k = \sqrt{m^2(m^2 + \Gamma^2)}\,. \tag{3.75}$$

The parameter m determines the position of the maximum of the distribution (mode) and the parameter Γ measures the width of the distribution.

A relativistic Breit–Wigner distribution arises from the square of a virtual particle's propagator (see, for instance, [1]) with four-momentum squared $p^2 = x^2$, which is proportional to:

$$\frac{1}{(x^2 - m^2) + i\,m\,\Gamma}\,. \tag{3.76}$$

As for a non-relativistic Breit–Wigner, due to integral divergences, the mean and variance of a relativistic Breit–Wigner distribution are undefined. Figure 3.11 shows examples of relativistic Breit–Wigner distribution for different values of Γ and fixed m.

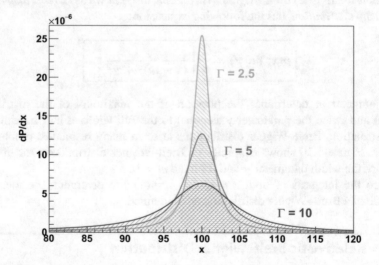

Fig. 3.11 Relativistic Breit–Wigner distributions with mass parameter $m = 100$ and different values of the width parameter Γ

3.16 Argus Distribution

The Argus collaboration introduced a function that models many cases of combinatorial backgrounds where kinematical bounds produce a sharp edge [2]. The *Argus distribution* or *Argus function* is given by:

$$p(x; \theta, \xi) = Nx\sqrt{1 - \left(\frac{x}{\theta}\right)^2}\, e^{-\xi^2[1-(x/\theta)^2]/2}, \tag{3.77}$$

where N is a normalization coefficient which depends on the parameters θ and ξ. Examples of Argus distributions are shown in Fig. 3.12 for different values of the parameters θ and ξ. The primitive function of Eq. (3.77) can be computed analytically to save computing time in the evaluation of the normalization coefficient N. Assuming $\xi^2 \geq 0$, the normalization condition for an Argus distribution can be written as follows:

$$\frac{1}{N} \int p(x; \theta, \xi)\, dx =$$

$$= \frac{\theta^2}{\xi^2} \left\{ e^{-\xi^2[1-(1-x/\theta)^2]/2} \sqrt{1 - \frac{x^2}{\theta^2}} - \sqrt{\frac{\pi}{2\xi^2}}\, \mathrm{erf}\sqrt{\frac{1}{2}\xi^2 \left(1 - \frac{x^2}{\theta^2}\right)} \right\}. \tag{3.78}$$

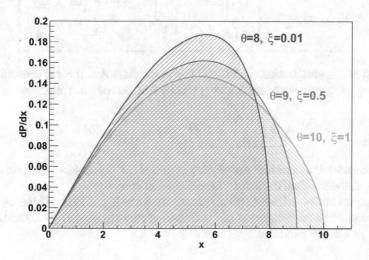

Fig. 3.12 Argus distributions with different values of the parameters θ and ξ

The normalized expression of the Argus function can also be written as:

$$p(x; \theta, \xi) = \frac{\xi^3}{\sqrt{2\pi}\,\Psi(\xi)} \frac{x}{\theta^2} \sqrt{1 - \left(\frac{x}{\theta}\right)^2}\, e^{-\xi^2[1-(x/\theta)^2]/2}\,, \tag{3.79}$$

where $\Psi(\xi) = \Phi(\xi) - \xi\phi(\xi) - 1/2$, $\phi(\xi)$ is a standard normal distribution (Eq. (3.47)) and $\Phi(\xi)$ is its cumulative distribution (Eq. (3.48)).

3.17   Crystal Ball Function

Several realistic detector responses and various physical processes only approximately follow a Gaussian distribution, and exhibit an asymmetric tail on one of the two sides. In most of the cases, a broader tail appears on the left side of the Gaussian core distribution, for instance, when there is a leakage in the energy deposited in a calorimeter, or energy lost by a particle due to radiation.

A description of such a distribution was provided by the collaboration working on the Crystal ball experiment at SLAC [3]. They introduced the following a PDF where a power-law distribution is used in place of one of the two Gaussian tail. The continuity of the PDF and its derivative is ensured with a proper choice of the parameters that model the tail.

The *Crystal ball distribution* is defined as follows:

$$p(x; \alpha, n, \bar{x}, \sigma) = N \cdot \begin{cases} \exp\left[-\dfrac{(x-\bar{x})^2}{2\sigma^2}\right] & \text{for} \quad \dfrac{x-\bar{x}}{\sigma} > -\alpha\,, \\ A\left(B - \dfrac{x-\bar{x}}{\sigma}\right)^{-n} & \text{for} \quad \dfrac{x-\bar{x}}{\sigma} \leq -\alpha\,, \end{cases} \tag{3.80}$$

where N is a normalization coefficient. The parameters A and B are determined by imposing the continuity of the function and its first derivative. This gives:

$$A = \left(\frac{n}{|\alpha|}\right)^n e^{-\alpha^2/2}\,, \qquad B = \frac{n}{|\alpha|} - |\alpha|\,. \tag{3.81}$$

The parameter α determines the starting point of the power-law tail, measured in units of σ, the standard deviation of the Gaussian core.

Examples of Crystal ball distributions are shown in Fig. 3.13 where the parameter α is varied, while the parameters of the Gaussian core are fixed at $\mu = 0$ and $\sigma = 1$, and the power-law exponent is set to $n = 2$.

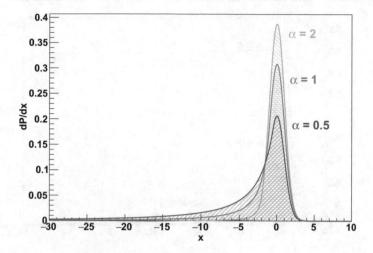

Fig. 3.13 Crystal ball distributions with $\mu = 0$, $\sigma = 1$, $n = 2$ and different values of the parameter α

3.18   Landau Distribution

The fluctuations of energy loss of particles traversing a thin layers of matter have been modeled by Lev Landau [4, 5]. The distribution of the energy loss x depends on the following integral expression called *Landau distribution*:

$$L(x) = \frac{1}{\pi} \int_0^\infty e^{-t \log t - xt} \sin(\pi t)\, dt \;.$$

(3.82)

More frequently, the distribution is shifted by a constant μ and scaled by a constant σ, and a transformed PDF is defined according to the following expression:

$$p(x; \mu, \sigma) = L\left(\frac{x - \mu}{\sigma}\right) \;.$$

(3.83)

Examples of Landau distributions are shown in Fig. 3.14 for different values of σ and fixed $\mu = 0$. This distribution is also used as empiric model for several asymmetric distributions.

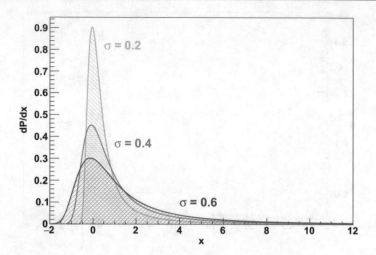

Fig. 3.14 Landau distributions with $\mu = 0$, and different values of σ

3.19 Mixture of PDFs

Given a number of PDFs, $f_1, \cdots f_n$, the combination:

$$f(x) = \sum_{i=1}^{n} w_i f_i(x) \tag{3.84}$$

is a properly normalized PDF if:

$$\sum_{i=1}^{n} w_i = 1 . \tag{3.85}$$

This PDF represent a mixture of random processes, whose outcome is x, each occurring with probability $P(i) = w_i$ and described by the PDF $f_i(x)$. Example of such a distribution are energy or mass spectra in a data sample containing a mixture of different physical processes. See Sect. 6.14 for a concrete example. In this perspective, f can also be written with the following notation, similar to conditional probability:

$$f(x) = \sum_{i=1}^{n} P(i) f(x \mid i) . \tag{3.86}$$

$f(x \mid i)$ denotes the PDF for x assuming that i has been the process extracted with probability $P(i)$.

3.20 Central Limit Theorem

Consider N random variables, $x_1, \cdots, x_N$, all independently extracted from the same PDF. Such variables are an example of *independent and identically distributed* random variables, or IID, or also i.i.d. It can be demonstrated that, if the original PDF has finite variance, the average of those N variables is distributed, approximately, in the limit of $N \rightarrow \infty$, according to a Gaussian distribution, regardless of the underlying PDFs. The demonstration is not reported here. Anyway, it is easy to produce approximate numerical demonstrations of the *central limit theorem* for specific cases using computer generated sequences of pseudorandom numbers with algorithms that are discussed in Chap. 4.

Two examples of such numerical exercise are shown in Figs. 3.15 and 3.16, where multiple random extractions, using two different PDFs, are summed. The sum is

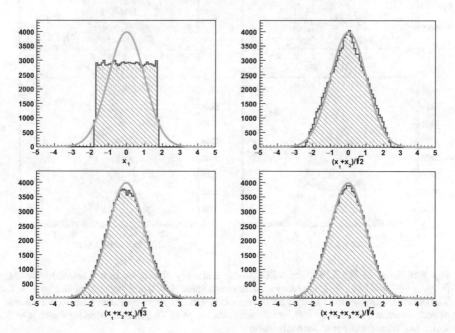

Fig. 3.15 Illustration of the central limit theorem using a Monte Carlo extraction. Random variables x_i are generated uniformly in the interval $[-\sqrt{3}, \sqrt{3}[$ to have average value $\mu = 0$ and standard deviation $\sigma = 1$. All plots correspond to 10^5 random extractions. The top-left plot shows the distribution of a single variable x_1; the other plots show the distribution of $(x_1 + x_2)/\sqrt{2}$, $(x_1 + x_2 + x_3)/\sqrt{3}$ and $(x_1 + x_2 + x_3 + x_4)/\sqrt{4}$, respectively, where all x_i are extracted from the same uniform distribution. A Gaussian curve with $\mu = 0$ and $\sigma = 1$, with proper normalization, in order to match the sample size, is superimposed for comparison. The Gaussian approximation is visually more and more stringent as a larger number of variables is added

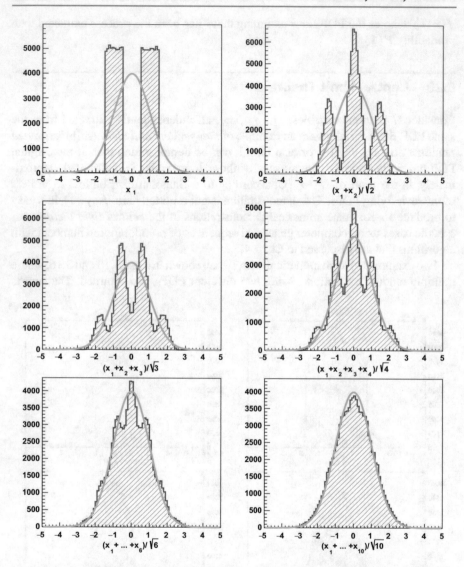

Fig. 3.16 Same as Fig. 3.15, using a PDF that is uniformly distributed in two disjoint intervals, $[-3/2, -1/2[$ and $[1/2, 3/2[$, in order to have average value $\mu = 0$ and standard deviation $\sigma = 1$. The individual distribution and the sum of $N = 2, 3, 4, 6$ and 10 independent random extractions of such a variable, divided by $\sqrt{N}$, are shown in the six plots. Gaussian distributions with $\mu = 0$ and $\sigma = 1$ are superimposed for comparison

divided by the square root of the number of generated variables to preserve in the combination the same variance of the original distribution. The variance of the sum of independent variable, in fact, is equal to the sum of variances (see Sect. 2.5). The distributions obtained with a sample of 10^5 random extractions are plotted, and Gaussian distributions are superimposed for comparison, in two PDF cases.

3.21 Probability Distributions in Multiple Dimension

Probability densities can be defined in more than one dimension as introduced in Sect. 3.2. In the case of two dimensions, a PDF $f(x, y)$ measures the probability density per unit area, i.e., the ratio of the differential probability dP corresponding to an infinitesimal interval around a point (x, y) and the differential area $dx\, dy$:

$$f(x, y) = \frac{dP}{dx\, dy}. \tag{3.87}$$

In three dimensions the PDF measures the probability density per volume area:

$$f(x, y, z) = \frac{dP}{dx\, dy\, dz}, \tag{3.88}$$

and so on in more dimensions.

A probability distribution in more dimensions that describes the distribution of more than one random variable is called *joint probability density function* and is the continuous case of a joint probability distribution.

3.22 Independent Variables

A visual illustration of the interplay between the joint distribution $f(x, y)$ and the marginal distributions $f_x(x)$ and $f_y(y)$ is shown in Fig. 3.17. The events A and B shown in the figure correspond to two values $x^\star$ and $y^\star$ extracted in the intervals $[x, x + \delta x[$ and $[y, y + \delta y[$, respectively:

$$A = \{x^\star : x \le x^\star < x + \delta x\} \quad \text{and} \quad B = \{y^\star : y \le y^\star < y + \delta y\}. \tag{3.89}$$

The probability of their intersection is:

$$P(A \cap B) = P(\text{``}x \le x^\star < x + x \text{ and } y \le y^\star < y + \delta y\text{''}) = f(x, y)\, \delta x\, \delta y. \tag{3.90}$$

By definition of marginal PDF:

$$P(A) = \delta P(x) = f_x(x)\, \delta x \quad \text{and} \quad P(B) = \delta P(y) = f_y(y)\, \delta y, \tag{3.91}$$

Fig. 3.17 Illustration of relation between marginal distributions and joint distribution for independent variables. In the xy plane, a slice in x of width δx corresponds to a probability $\delta P(x) = f_x(x)\,\delta x$, and a slice in y with width δy corresponds to a probability $\delta P(y) = f_y(y)\,\delta y$. Their intersection corresponds to a probability $\delta P(x, y) = f(x, y)\,\delta x\,\delta y$

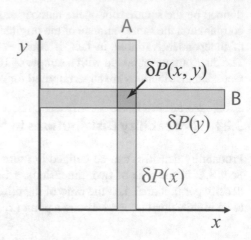

hence, the product of the two probabilities is:

$$P(A)\,P(B) = \delta P(x, y) = f_x(x)\,f_y(y)\,\delta x\,\delta y\,. \tag{3.92}$$

According to Eq. (1.4), two events A and B are independent if $P(A \cap B) = P(A)\,P(B)$. Therefore, the equality $P(A \cap B) = P(A)\,P(B)$ holds, given Eq. (3.92), if and only if $f(x, y)$ can be factorized into the product of the two marginal PDFs:

$$f(x, y) = f_x(x)\,f_y(y)\,. \tag{3.93}$$

From this result, x and y can be defined as *independent random variables* if their joint PDF is factorized into the product of a PDF of the variable x times a PDF of the variable y.

More in general, n variables $x_1, \cdots, x_n$ are said to be independent if their n-dimensional PDF can be factorized into the product of n one-dimensional PDF in each of the variables:

$$f(x_1, \cdots, x_n) = f_1(x_1) \cdots f_n(x_n)\,. \tag{3.94}$$

In a weaker sense, the variables sets $\vec{x} = (x_1, \cdots, x_h)$ and $\vec{y} = (y_1, \cdots, y_k)$ are independent if:

$$f(\vec{x}, \vec{y}) = f_x(\vec{x})\,f_y(\vec{y})\,. \tag{3.95}$$

3.23 Covariance, Correlation, and Independence

As for discrete variables, the covariance of two variables x and y is defined as:

$$\boxed{\mathbb{C}\text{ov}(x, y) = \langle x\, y \rangle - \langle x \rangle \langle y \rangle ,}$$
(3.96)

and the correlation coefficient as:

$$\boxed{\rho_{xy} = \frac{\mathbb{C}\text{ov}(x, y)}{\sigma_x\, \sigma_y} .}$$
(3.97)

If x and y are independent, it can be easily demonstrated that they are also uncorrelated. Conversely, if two variables are uncorrelated, they are not necessarily independent, as shown in Example 3.4 below. If x and y are uncorrelated the variance of the sum of the two variables is equal to the sum of the variances:

$$\mathbb{V}\text{ar}[x + y] = \mathbb{V}\text{ar}[x] + \mathbb{V}\text{ar}[y] .$$
(3.98)

Example 3.4 - Uncorrelated Variables May Not Be Independent
Two independent variables are uncorrelated, but two uncorrelated variables are not necessarily independent. An example of PDF that describes uncorrelated variables that are not independent is given by the sum of four two-dimensional Gaussian PDFs as specified below:

$$f(x, y) = \frac{1}{4} [g(x;\, \mu,\, \sigma)\, g(y;\, 0,\, \sigma) + g(x;\, -\mu,\, \sigma)\, g(y;\, 0,\, \sigma) +$$

$$g(x;\, 0,\, \sigma)\, g(y;\, \mu,\, \sigma) + g(x;\, 0,\, \sigma)\, g(y;\, -\mu,\, \sigma)] ,$$
(3.99)

where g is a one-dimensional Gaussian distribution. This example is illustrated in Fig. 3.18 which plots the PDF in Eq. (3.99) with numerical values $\mu = 2.5$ and $\sigma = 0.7$.

Considering that, for a variable z distributed according to $g(z;\, \mu,\, \sigma)$, the following relations hold:

$$\langle z \rangle = \mu ,$$

$$\langle z^2 \rangle = \mu^2 + \sigma^2 ,$$

(continued)

Example 3.4 (continued)

it is easy to demonstrate that, for x and y distributed according to $f(x, y)$, the following relations also hold:

$$\langle x \rangle = \langle y \rangle = 0 \,,$$
$$\langle x^2 \rangle = \langle y^2 \rangle = \sigma^2 + \frac{\mu^2}{2} \,,$$
$$\langle xy \rangle = 0 \,.$$

Applying the definition of covariance in Eq. (2.27) gives $\mathbb{C}\mathrm{cov}(x, y) = 0$, and for this reason x and y are uncorrelated. Anyway, x and y are clearly not independent because $f(x, y)$ cannot be factorized into the product of two PDF, i.e., there is no pair of functions $f_x(x)$ and $f_y(y)$ such that $f(x, y) = f_x(x) f_y(y)$. Consider, for instance, three slices, $f(x_0, y)$, at $x_0 = 0$ and $x_0 = \pm \mu$. The function of y, $f(x_0, y)$, for fixed x_0, has two maxima for $x_0 = 0$ and a single maximum for $x_0 = \pm \mu$. For a factorized PDF, instead, the shape of $f(x_0, y) = f_x(x_0) f_y(y)$ should be the same for all values of x_0, up to the scale factor $f_x(x_0)$.

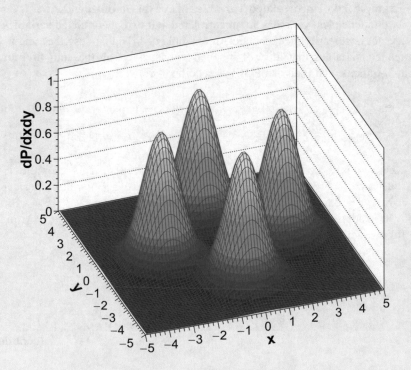

Fig. 3.18 Example of a PDF of two variables x and y that are uncorrelated but not independent

3.24 Conditional Distributions

Given a two-dimensional PDF $f(x, y)$ and a fixed value x_0 of the variable x, the *conditional probability density function* of y given x_0 is defined as:

$$f(y \mid x_0) = \frac{f(x_0, y)}{\int f(x_0, y') \, dy'}.$$

(3.100)

It is the continuous case of a *conditional probability distribution*, which may also be introduced for discrete random variables. The conditional distribution is obtained by slicing $f(x, y)$ at $x = x_0$, and then applying a normalization factor to the sliced one-dimensional distribution. An illustration of conditional PDF, ignoring the normalization factor for display purposes, is shown in Fig. 3.19.

Reminding Eq. (1.2), and considering again the example in Fig. 3.17, the definition of conditional distribution in Eq. (3.100) is consistent with the definition of conditional probability: $P(B \mid A) = P(A \cap B)/P(A)$, where $B = \{y^\star : y \leq$

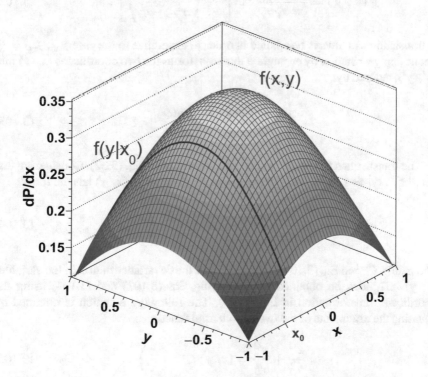

Fig. 3.19 Illustration of conditional PDF in two dimensions. The PDF $f(y|x_0)$ can be obtained by rescaling the red curve that shows $f(x_0, y)$ by a proper factor, in order to ensure normalization

$y^* < y + \delta y$}, and $A = \{x^* : x_0 \leq x^* < x_0 + \delta x\}$, x^* and y^* being extracted value of x and y, respectively.

In more than two dimensions, Eq. (3.100) can be generalized for a PDF of $h + k$ variables $(\vec{x}, \vec{y}) = (x_1, \cdots, x_h, y_1, \cdots, y_k)$ as:

$$f(\vec{y} \mid \vec{x}_0) = \frac{f(\vec{x}_0, \vec{y})}{\int f(\vec{x}_0, \vec{y}) \, dy'_1 \cdots dy'_k} . \tag{3.101}$$

3.25   Gaussian Distributions in Two or More Dimensions

Let us consider the product of two Gaussian distributions for the variables x' and y' having standard deviations $\sigma_{x'}$ and $\sigma_{y'}$, respectively, and for simplicity having both averages $\mu_{x'} = \mu_{y'} = 0$:

$$g'(x', y') = \frac{1}{2\pi \, \sigma_{x'}\sigma_{y'}} \exp\left[-\frac{1}{2}\left(\frac{x'^2}{\sigma_{x'}^2} + \frac{y'^2}{\sigma_{y'}^2} \right) \right] . \tag{3.102}$$

A translation can always be applied in order to generalize to the case $\mu_{x'}$, $\mu_{y'} \neq 0$. Let us apply a rotation by an angle ϕ that transforms the two coordinates (x, y) into (x', y'), defined by:

$$\begin{cases} x' = x \cos\phi - y \sin\phi , \\ y' = x \sin\phi + y \cos\phi . \end{cases} \tag{3.103}$$

The transformed PDF $g(x, y)$ can be obtained from Eq. (3.32). Considering that $\det\left| \partial x'_i / \partial x_j \right| = 1$, we have $g'(x', y') = g(x, y)$, where $g(x, y)$ has the form:

$$g(x, y) = \frac{1}{2\pi \, |C|^{\frac{1}{2}}} \exp\left[-\frac{1}{2}(x, y)\, C^{-1} \begin{pmatrix} x \\ y \end{pmatrix} \right] . \tag{3.104}$$

The matrix C^{-1} in Eq. (3.104) is the inverse of the covariance matrix of the variables (x, y). C^{-1} can be obtained by comparing Eqs. (3.102) and (3.104) using the coordinate transformation in Eq. (3.103). The following equation is obtained by equating the arguments of the two exponential functions:

$$\frac{x'^2}{\sigma_{x'}^2} + \frac{y'^2}{\sigma_{y'}^2} = (x, y)\, C^{-1} \begin{pmatrix} x \\ y \end{pmatrix} , \tag{3.105}$$

and, substituting the transformed coordinates, we obtain:

$$
C^{-1} = \begin{pmatrix} \dfrac{\cos^2 \phi}{\sigma_{x'}^2} + \dfrac{\sin^2 \phi}{\sigma_{y'}^2} & \sin \phi \cos \phi \left(\dfrac{1}{\sigma_{y'}^2} - \dfrac{1}{\sigma_{x'}^2} \right) \\ \sin \phi \cos \phi \left(\dfrac{1}{\sigma_{y'}^2} - \dfrac{1}{\sigma_{x'}^2} \right) & \dfrac{\sin^2 \phi}{\sigma_{x'}^2} + \dfrac{\cos^2 \phi}{\sigma_{y'}^2} \end{pmatrix} .
\tag{3.106}
$$

Considering that the covariance matrix should have the form:

$$
C = \begin{pmatrix} \sigma_x^2 & \rho_{xy}\,\sigma_x \sigma_y \\ \rho_{xy}\,\sigma_x \sigma_y & \sigma_y^2 \end{pmatrix} ,
\tag{3.107}
$$

where ρ_{xy} is the correlation coefficient defined in Eq. (2.28), the determinant of C^{-1} must be equal to:

$$
\left| C^{-1} \right| = \frac{1}{\sigma_{x'}^2 \sigma_{y'}^2} = \frac{1}{\sigma_x^2 \sigma_y^2 \left(1 - \rho_{xy}^2 \right)} .
\tag{3.108}
$$

Inverting the matrix C^{-1} in Eq. (3.106), the covariance matrix in the rotated variables (x, y) is:

$$
C = \begin{pmatrix} \cos^2 \phi\, \sigma_{x'}^2 + \sin^2 \phi\, \sigma_{y'}^2 & \sin \phi \cos \phi \left(\sigma_{y'}^2 - \sigma_{x'}^2 \right) \\ \sin \phi \cos \phi \left(\sigma_{y'}^2 - \sigma_{x'}^2 \right) & \sin^2 \phi\, \sigma_{x'}^2 + \cos^2 \phi\, \sigma_{y'}^2 \end{pmatrix} .
\tag{3.109}
$$

The variances of x and y and their correlation coefficient can be determined by comparing Eq. (3.109) to Eq. (3.107):

$$
\sigma_x^2 = \cos^2 \phi\, \sigma_{x'}^2 + \sin^2 \phi\, \sigma_{y'}^2 ,
\tag{3.110}
$$

$$
\sigma_y^2 = \sin^2 \phi\, \sigma_{x'}^2 + \cos^2 \phi\, \sigma_{y'}^2 ,
\tag{3.111}
$$

$$
\rho_{xy} = \frac{\mathrm{cov}(x, y)}{\sigma_x \sigma_y} = \frac{\sin 2\phi \left(\sigma_{y'}^2 - \sigma_{x'}^2 \right)}{\sqrt{\sin^2 2\phi \left(\sigma_{x'}^2 - \sigma_{y'}^2 \right)^2 + 4 \sigma_{x'}^2 \sigma_{y'}^2}} .
\tag{3.112}
$$

The invariance of the area of the ellipses is expressed by the equation:

$$
\sigma_{x'}^2 + \sigma_{y'}^2 = \sigma_x^2 + \sigma_y^2
\tag{3.113}
$$

that is obtained by adding Eqs. (3.110) and (3.111). Equation (3.112) implies that the correlation coefficient is equal to zero if either $\sigma_{y'} = \sigma_{x'}$ or if ϕ is a multiple of

$\pi/2$. If we subtract Eqs. (3.111) and (3.110), we obtain:

$$\sigma_y^2 - \sigma_x^2 = \left(\sigma_{x'}^2 - \sigma_{y'}^2\right)\cos 2\phi .$$

(3.114)

Using the above equation together with Eq. (3.112) and (3.108), we have the following relation gives $\tan 2\phi$ in terms of the elements of the covariance matrix:

$$\boxed{\tan 2\phi = \frac{2\,\rho_{xy}\,\sigma_x\sigma_y}{\sigma_y^2 - \sigma_x^2} .}$$

(3.115)

The transformed PDF can be finally written in terms of all the results obtained so far:

$$\boxed{g(x,\,y) = \frac{1}{2\pi\,\sigma_x\sigma_y\sqrt{1 - \rho_{xy}^2}}\exp\left[-\frac{1}{2(1 - \rho_{xy}^2)}\left(\frac{x^2}{\sigma_x^2} + \frac{y^2}{\sigma_y^2} - \frac{2x\,y\,\rho_{xy}}{\sigma_x\sigma_y}\right)\right] .}$$

(3.116)

The geometric interpretation of σ_x and σ_y in the rotated coordinate system is shown in Fig. 3.20, where the one-sigma ellipse determined by the following equation is drawn:

$$\frac{x^2}{\sigma_x^2} + \frac{y^2}{\sigma_y^2} - \frac{2x\,y\,\rho_{xy}}{\sigma_x\sigma_y} = 1 .$$

(3.117)

It is possible to demonstrate that the distance of the horizontal and vertical tangent lines to the ellipse defined by Eq. (3.117) have a distance with respect to their corresponding axes equal to σ_y and σ_x, respectively.

Similarly to the 1σ contour, defined in Eq. (3.117), a $Z\sigma$ contour is defined by:

$$\frac{x^2}{\sigma_x^2} + \frac{y^2}{\sigma_y^2} - \frac{2x\,y\,\rho_{xy}}{\sigma_x\sigma_y} = Z^2 .$$

(3.118)

The projection of a two-dimensional Gaussian in Eq. (3.116) along the two axes gives the following marginal PDFs, which correspond to the expected one-dimensional Gaussian distributions with standard deviations σ_x and σ_y, respectively:

$$g_x(x) = \int_{-\infty}^{+\infty} g(x,\,y)\,\mathrm{d}y = \frac{1}{\sigma_x\sqrt{2\pi}}\,e^{-x^2/2\sigma_x^2} ,$$

(3.119)

$$g_y(y) = \int_{-\infty}^{+\infty} g(x,\,y)\,\mathrm{d}x = \frac{1}{\sigma_y\sqrt{2\pi}}\,e^{-y^2/2\sigma_y^2} .$$

(3.120)

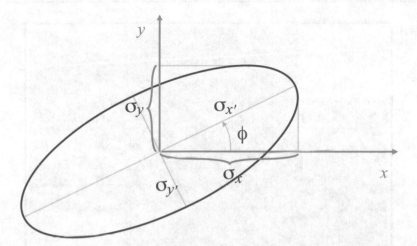

Fig. 3.20 One-sigma contour for a two-dimensional Gaussian PDF. The two ellipse axes have length equal to $\sigma_{x'}$ and $\sigma_{y'}$. The x' axis is rotated by an angle ϕ with respect to the x axis. The lines tangent to the ellipse parallel to the x and y axes, shown in gray, have a distance with respect to the corresponding axes equal to σ_y and σ_x. In this example, the correlation coefficient ρ_{xy} is positive

In general, projecting a two-dimensional Gaussian PDF in any direction gives a one-dimensional Gaussian whose standard deviation is equal to the distance of the tangent line to the 1σ ellipse perpendicular to the axis along which the two-dimensional Gaussian is projected. This is illustrated in Fig. 3.21 where 1σ and 2σ contours are shown for a two-dimensional Gaussian. Figure 3.21 shows three possible choices of 1σ and 2σ bands: one along the x axis, one along the y axis, and one along an oblique direction.

Note that the probability corresponding to area inside the 1σ ellipse defined by Eq. (3.117) is smaller than 68.27%, which corresponds to a one-dimensional 1σ interval defined by the one-dimensional version of the ellipse equation:

$$\frac{x^2}{\sigma^2} = 1 \quad \Longrightarrow \quad x = \pm\sigma . \tag{3.121}$$

The probability values corresponding to $Z\sigma$ one-dimensional intervals for a Gaussian distribution are determined by Eq. (3.50). The corresponding probabilities of an area inside a two-dimensional $Z\sigma$ ellipse can be performed by integrating $g(x,\,y)$ in two dimensions over the area E_Z contained inside the $Z\sigma$ ellipse:

$$P_{2D}(Z\sigma) = \int_{E_Z} g(x,\,y)\,\mathrm{d}x\,\mathrm{d}y , \tag{3.122}$$

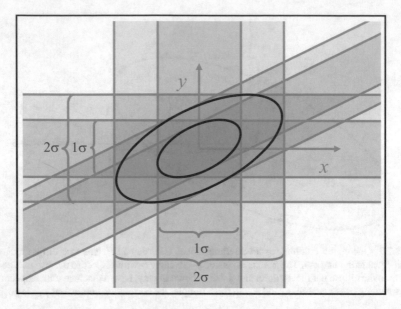

Fig. 3.21 Illustration of two-dimensional 1σ and 2σ Gaussian contours. Each one-dimensional projection of the 1σ or 2σ contour corresponds to a band which has a 68.27% or 95.45% probability content, respectively. As an example, three possible projections are shown: a vertical, a horizontal and an oblique one. The probability content of the areas inside the ellipses are smaller than the corresponding one-dimensional projected intervals

where

$$E_Z = \left\{ (x, y) : \frac{x^2}{\sigma_x^2} + \frac{y^2}{\sigma_y^2} - \frac{2x\,y\,\rho_{xy}}{\sigma_x\sigma_y} \le Z^2 \right\} . \tag{3.123}$$

The integral in Eq. (3.122), written in polar coordinates, simplifies to:

$$P_{2D}(Z\sigma) = \int_0^Z e^{-r^2/2} r \, \mathrm{d}r = 1 - e^{-Z^2/2} , \tag{3.124}$$

which can be compared to the one-dimensional case:

$$P_{1D}(Z\sigma) = \sqrt{\frac{2}{\pi}} \int_0^Z e^{-x^2/2} \, \mathrm{d}x = \mathrm{erf}\left(\frac{Z}{\sqrt{2}} \right) . \tag{3.125}$$

The probability corresponding to 1σ, 2σ, and 3σ for the one- and two-dimensional cases are reported in Table 3.1. The two-dimensional integrals are, in all cases, smaller than in one dimension for a given Z. In particular, to recover the same probability content as the corresponding one-dimensional interval, one would need to artificially enlarge a two-dimensional ellipse from 1σ to 1.515σ, from 2σ

Table 3.1 Probabilities corresponding to $Z\sigma$ one-dimensional intervals and corresponding two-dimensional areas inside the elliptic contours for different values of Z

	P_{1D}	P_{2D}
1σ	**0.6827**	0.3934
2σ	**0.9545**	0.8647
3σ	**0.9973**	0.9889
1.515σ	0.8702	**0.6827**
2.486σ	0.9871	**0.9545**
3.439σ	0.9994	**0.9973**

Bold values show identical probability values for 1D and 2D intervals

to 2.486σ, and from 3σ to 3.439σ. Usually, results are reported in literature as 1σ and 2σ contours where the probability content of 68.27% or 95.45% refers to any one-dimensional projection of those contours.

The generalization to n dimensions of Eq. (3.116) is:

$$
g(x_1, \cdots, x_n) = \frac{1}{(2\pi)^{\frac{n}{2}} |C|^{-\frac{1}{2}}} \exp\left[-\frac{1}{2}(x_i - \mu_i)\, C_{ij}^{-1}(x_j - \mu_j) \right], \quad (3.126)
$$

where we introduced the parameters μ_i, which are the average of the variable x_i, and the $n \times n$ covariance matrix C_{ij} of the variables $x_1, \cdots, x_n$.

References

1. J. Bjorken, S. Drell *Relativistic Quantum Fields* (McGraw-Hill, New York, 1965)
2. ARGUS collaboration, H. Albrecht et al., Search for hadronic b → u decays. Phys. Lett. **B241**, 278–282 (1990)
3. J. Gaiser, *Charmonium Spectroscopy from Radiative Decays of the J/ψ and ψ'*. Ph.D. Thesis, Stanford University (1982), Appendix F
4. L. Landau, On the energy loss of fast particles by ionization. J. Phys. (USSR) **8**, 201 (1944)
5. W. Allison, J. Cobb, Relativistic charged particle identification by energy loss. Annu. Rev. Nucl. Part. Sci. **30**, 253–298 (1980)

Random Numbers and Monte Carlo Methods 4

4.1 Pseudorandom Numbers

Many applications, ranging from simulations to video games and 3D-graphics, take advantage of computer-generated numeric sequences that have properties very similar to truly random variables. Sequences of numbers generated by computers are not truly random because algorithms are necessarily deterministic and reproducible. The apparent randomness arises from the strong nonlinearity of the algorithms, as detailed in the following. For this reason, such variables are called *pseudorandom* numbers. The possibility to reproduce the same sequence of computer-generated numbers, nonetheless, is often a good feature for many applications and for software testing.

A sequence of pseudorandom number should ideally have, in the limit of large numbers, the desired statistical properties of a real random variable. Considering that computers have finite floating-point machine precision, pseudorandom numbers, in practice, have a finite number of discrete possible values, depending on the number of bits used to store a floating-point value. Moreover, at some point, a sequence of pseudorandom numbers becomes periodic and the same subsequence repeats over and over. To avoid the problems that may be related to the periodicity, the *period* of a random sequence, i.e., the number of extractions after which the sequence repeats itself, should be as large as possible, and anyway larger than the number of random numbers required by our specific application.

Numerical methods involving the repeated use of computer-generated pseudo-random numbers are also known as *Monte Carlo* methods, abbreviated as MC, from the name of the city hosting the famous casino, where the properties of (truly) random numbers resulting from roulette and other games are exploited to generate profit. In the following, we sometimes refer to pseudorandom numbers simply as random numbers, when the context creates no ambiguity.

© The Author(s), under exclusive license to Springer Nature Switzerland AG 2023
L. Lista, *Statistical Methods for Data Analysis*, Lecture Notes in Physics 1010,
https://doi.org/10.1007/978-3-031-19934-9_4

4.2   Properties of Pseudorandom Generators

Good pseudorandom number generators should produce sequences of numbers that are statistically independent on previous extractions, though unavoidably each number is determined mathematically, via the generator's algorithm, from the previously extractions.

All numbers in a sequence should be independent and distributed according to the same PDF. Such variables are called *independent and identically distributed* random variables, or IID, or also i.i.d., as already introduced in Sect. 3.20. Those properties can be written as follows, where $f(x)$ is the desired PDF:

$$f(x_i) = f(x_j), \quad \forall i, j, \tag{4.1}$$

$$f(x_n \mid x_{n-m}) = f(x_n), \quad \forall n, m. \tag{4.2}$$

Example 4.1 - Transition From Regular to "Unpredictable" Sequences
There are several examples of mathematical algorithms that lead to sequences that are poorly predictable. One example of transition from a regular to a chaotic regime is given by the *logistic map* [1]. The sequence is defined by recursion, starting from an initial value x_0, as:

$$x_{n+1} = \lambda x_n (1 - x_n). \tag{4.3}$$

Depending on the value of λ, the sequence may have very different possible behaviors. If the sequence converges to a single asymptotic value x for $n \to \infty$, we have:

$$\lim_{n \to \infty} x_n = x, \tag{4.4}$$

where x must satisfy:

$$x = \lambda x (1 - x). \tag{4.5}$$

Excluding the trivial solution $x = 0$, Eq. (4.5) leads to:

$$x = (\lambda - 1)/\lambda. \tag{4.6}$$

This solution is stable for values of λ from 1 to 3. Above $\lambda = 3$, the sequence stably approaches a state where it oscillates between two values x_1 and x_2 that satisfy the following system of two equations:

$$\begin{cases} x_1 = \lambda x_2 (1 - x_2), \\ x_2 = \lambda x_1 (1 - x_1). \end{cases} \tag{4.7}$$

(continued)

Example 4.1 (continued)

For larger values, up to $1 + \sqrt{6}$, the sequences oscillate among four values, and further *bifurcations* occur for even larger values of λ, until it achieves a very complex and poorly predictable behavior. For $\lambda = 4$, the sequence finally densely covers the interval $]0, 1[$. The density of the first n values of the sequence with $\lambda = 4$, in the limit for $n \to \infty$, can be demonstrated to be equal to:

$$f(x) = \lim_{\Delta x \to 0} \frac{\Delta N}{\Delta x} = \frac{1}{\pi \sqrt{x(1-x)}}. \tag{4.8}$$

This is a special case of a *beta distribution* (Sect. 3.13) with parameters $\alpha = \beta = 0.5$. A comparison between the density of values in the logistic sequence for $\lambda = 4$ and the function in Eq. (4.8) is shown in Fig. 4.1. The behavior of the logistic map for different values of λ is shown in Fig. 4.2.

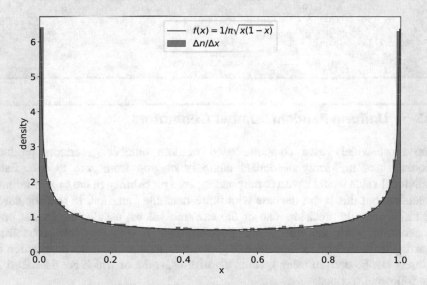

Fig. 4.1 Density of 100,000 consecutive values in the logistic sequence for $\lambda = 4$ (blue histogram) compared to the function $f(x) = 1/\pi\sqrt{x(1-x)}$ (red line)

(continued)

Example 4.1 (continued)

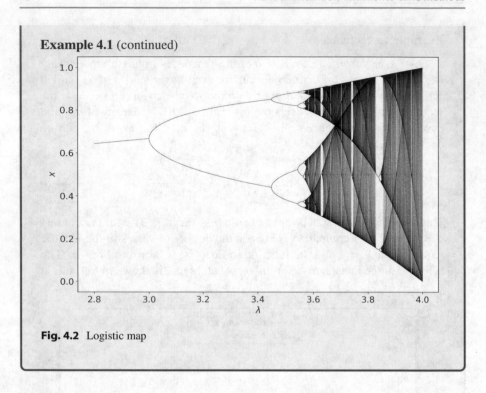

Fig. 4.2 Logistic map

4.3   Uniform Random Number Generators

The most widely used computer-based random number generators produce
sequences of uniformly distributed numbers ranging from zero to one. Each
individual value would have a corresponding zero probability, in the case of infinite
precision, but this is not the case with finite machine precision. In realistic cases
of finite numeric precision, one of the extreme values, usually the largest one,
is excluded, while the other is still possible, though with very low probability.
Starting from uniform random number generators, most of the other distribution of
interest can be derived using specific algorithms, some of which are described in
the following sections.

One example of uniform random number generator is the function `lrand48`
[2], which is a standard of C programming language. Given an initial value x_0, the
sequence is defined by recursion according to the following algorithm:

$$x_{n+1} = (ax_n + c) \bmod m \,, \tag{4.9}$$

where the values of m, a, and c are:

$$m = 2^{48} ,$$

$$a = 25214903917 = 5DEECE66D_{hex} ,$$

$$c = 11 = B_{hex} .$$

The sequence of integer numbers obtained from Eq. (4.9) for a given initial value x_0 is distributed uniformly, to a good approximation, between 0 and $2^{48} - 1$. A pseudorandom sequence of bits generated in this way can also be used to define floating-point pseudorandom numbers uniformly distributed from zero to one that are more useful for mathematical applications. This is done with the function drand48.

4.4 Nonuniform Distribution: Inversion of the Cumulative Distribution

Random numbers distributed according to a nonuniform PDF can be generated starting from a uniform random number generator using various algorithms. Examples are provided in this and in the following sections.

To generate a pseudorandom number x distributed according to a given function $f(x)$, first its cumulative distribution can be built (Eq. (3.23)):

$$F(x) = \int_{-\infty}^{X} f(x') \, dx' . \tag{4.10}$$

By inverting the cumulative distribution $F(x)$, and computing the value of its inverse for a random number r uniformly extracted in $[0, \ 1 \ [$, the transformed variable:

$$x = F^{-1}(r) \tag{4.11}$$

is distributed according to $f(x)$. The demonstration proceeds as follows. If we write:

$$r = F(x) , \tag{4.12}$$

we have:

$$dr = \frac{dF}{dx} \, dx = f(x) \, dx . \tag{4.13}$$

Introducing the differential probability dP, we have:

$$\frac{dP}{dx} = f(x) \frac{dP}{dr} . \tag{4.14}$$

Since r is uniformly distributed, $dP/dr = 1$, hence:

$$\frac{dP}{dx} = f(x) \, , \tag{4.15}$$

which demonstrates that x is distributed according to the desired PDF.

The simplest application of the inversion of the cumulative distribution is the extraction of a random number x uniformly distributed between the extreme values a and b. If r is uniformly extracted between 0 and 1, x can be obtained by transforming r according to:

$$x = a + r\,(b - a) \, . \tag{4.16}$$

This method only works conveniently if the cumulative distribution $F(x)$ can be easily computed and inverted using either analytical or fast numerical methods. If not, usually this algorithm may be slow, and alternative implementations often provide better CPU performances.

4.5   Random Numbers Following a Finite Discrete Distribution

A variation of the inversion of the cumulative distribution, described in Sect. 4.4, can be used to extract discrete random values j from 1 to n, each with probability P_j. One can first build the discrete cumulative distribution that can be stored into an array:[1]

$$C_j = P(i \le j) = \sum_{i=1}^{j} P_i \, . \tag{4.17}$$

Then, if a random number r is uniformly generated between 0 and 1, i can be chosen as the smallest j such that $r < C_j$. A binary search or other optimizations may speed up the algorithm that extracts j in case of the number of possible values is very large. Optimized implementations exist for discrete random extractions and are described in [3] and [4].

Example 4.2 - Extraction of an Exponential Random Variable
The inversion of the cumulative distribution presented in Sect. 4.4 allows to extract random numbers x distributed according to an exponential PDF:

$$f(x) = \lambda\, e^{-\lambda x} \, . \tag{4.18}$$

(continued)

---

[1] Note that $C_n = 1$ is a trivial value, so there is no need to store the last value of the array in memory.

Example 4.2 (continued)
The cumulative distribution of $f(x)$ is:

$$F(x) = \int_0^x f(x') \, dx' = 1 - e^{-\lambda x} . \tag{4.19}$$

Setting $F(x) = r$ gives:

$$1 - e^{-\lambda x} = r , \tag{4.20}$$

which can be inverted in order to obtain x as:

$$x = -\frac{1}{\lambda} \log(1 - r) . \tag{4.21}$$

If the extraction of r gives values in the interval $[0, \, 1 \, [$, like with `drand48`, $r = 1$ is never be extracted, hence the argument of the logarithm is never null, ensuring the numerical validity of Eq. (4.21).

Example 4.3 - Extraction of a Uniform Point on a Sphere
Assume we want to generate two variables, θ and ϕ, distributed in such a way that they correspond in spherical coordinates to a point uniformly distributed on a sphere. The probability density per unit of solid angle Ω is uniform:

$$\frac{dP}{d\Omega} = \frac{dP}{\sin\theta \, d\theta \, d\phi} = k , \tag{4.22}$$

where k is a normalization constant such that the PDF integrates to unity over the entire solid angle. From Eq. (4.22), the joint two-dimensional PDF can be factorized into the product of two PDFs, as functions of θ and ϕ:

$$\frac{dP}{d\theta \, d\phi} = f(\theta) \, g(\phi) = k \sin\theta , \tag{4.23}$$

where

$$f(\theta) = \frac{dP}{d\theta} = c_\theta \sin\theta , \tag{4.24}$$

$$g(\phi) = \frac{dP}{d\phi} = c_\phi . \tag{4.25}$$

The constants c_θ and c_ϕ ensure the normalization of $f(\theta)$ and $g(\phi)$ individually. This factorization implies that θ and ϕ are independent.

(continued)

Example 4.3 (continued)

θ can be extracted by inverting the cumulative distribution of $f(\theta)$ (Eq. (4.11)). ϕ is uniformly distributed, since $g(\phi)$ is a constant, so it can be extracted by remapping the interval $[0, 1[$ into $[0, 2\pi[$ (Eq. (4.16)).

The generation of θ proceeds by inverting the cumulative distribution of $f(\theta)$:

$$F(\theta) = \int_0^\theta c_\theta \sin\theta' \, d\theta' = c_\theta \left[-\cos\theta'\right]_0^\theta = c_\theta (1 - \cos\theta) . \qquad (4.26)$$

c_θ can be determined by imposing $F(\pi) = 1$:

$$c_\theta (1 - \cos\pi) = 1 , \qquad (4.27)$$

which gives $c_\theta = 1/2$, and:

$$F(\theta) = \frac{1 - \cos\theta}{2} . \qquad (4.28)$$

By equating $F(\theta)$ to a value r_θ uniformly extracted from zero to one, one obtains:

$$2r_\theta = 1 - \cos\theta , \qquad (4.29)$$

hence:

$$\cos\theta = 1 - 2r_\theta . \qquad (4.30)$$

In summary, the generation may proceed as follows:

$$\theta = \arccos(1 - 2r_\theta) \in \,]0, \pi] , \qquad (4.31)$$

$$\phi = 2\pi r_\phi \in [0, 2\pi[, \qquad (4.32)$$

where r_θ and r_ϕ are extracted in $[0, 1[$ with a standard uniform generator.

Example 4.4 - A Simple Cosmic-Ray Generator

Cosmic rays reach the ground with an angular distribution that has been measured at sea level to be, to a good approximation, equal to:

$$\frac{dP(\theta, \phi)}{d\Omega} = k \cos^\alpha\theta , \qquad (4.33)$$

(continued)

Example 4.4 (continued)

with $0 \leq \theta < \pi/2$ and k a normalization constant. The exponent α is equal to two within a good approximation. Arrival positions at the ground are uniformly distributed in two dimensions. We call the two coordinates x and y, which we can generate uniformly within the surface of a detector that we assume to be placed on a horizontal plane.

As in Example 4.3, we can generate the arrival direction by extracting θ and ϕ distributed according to:

$$\frac{dP}{d\Omega} = \frac{dP}{\sin\theta \, d\theta \, d\phi} = k\cos^2\theta . \tag{4.34}$$

The PDF is factorized as:

$$\frac{dP}{d\theta \, d\phi} = f(\theta) \, g(\phi) = k \sin\theta \cos^2\theta, \tag{4.35}$$

where

$$f(\theta) = c_\theta \sin\theta \cos^2\theta , \tag{4.36}$$

$$g(\phi) = c_\phi . \tag{4.37}$$

The normalization constant is given by $k = c_\theta \, c_\phi$.

The cumulative distribution for $f(\theta)$ is:

$$F(\theta) = c_\theta \int_0^\theta \sin\theta' \cos^2\theta' \, d\theta' = -\frac{c_\theta}{3} \left[\cos^3\theta'\right]_{\theta'=0}^{\theta'=\theta} = \frac{c_\theta}{3}(1 - \cos^3\theta) . \tag{4.38}$$

The normalization coefficient c_θ is determined by imposing $F(\pi/2) = 1$:

$$1 = \frac{c_\theta}{3}\left(1 - \cos^3\frac{\pi}{2}\right) = \frac{c_\theta}{3} . \tag{4.39}$$

Therefore, $c_\theta = 3$, and:

$$F(\theta) = 1 - \cos^3\theta . \tag{4.40}$$

$F(\theta)$ must be inverted in order to obtain θ from the equation $F(\theta) = r_\theta$, with r_θ uniformly extracted in $[0, 1[$:

$$\left(1 - \cos^3\theta\right) = r_\theta . \tag{4.41}$$

(continued)

Example 4.4 (continued)
This allows to generate the polar angle as:

$$\theta = \arccos(\sqrt[3]{1 - r_\theta}) \in \left[0, \frac{\pi}{2}\right[. \tag{4.42}$$

The azimuth angle, as usual, can be uniformly generated from 0 to 2π as in the previous example.

A sample of randomly extracted cosmic rays with the described algorithm is shown in Fig. 4.3, where intersections with two parallel detector planes are also shown. Such a simulation setup may help determine the detector acceptance and the expected angular distribution of cosmic rays intersecting the detector setup with a given geometry.

Figure 4.4 shows the angular distribution of all generated cosmic rays, and the angular distribution of the ones that cross a detector made by two square 1m × m horizontal detector planes, vertically separated by a 1m distance. Distributions like this one are usually very hard to compute with analytical methods or other alternatives to Monte Carlo.

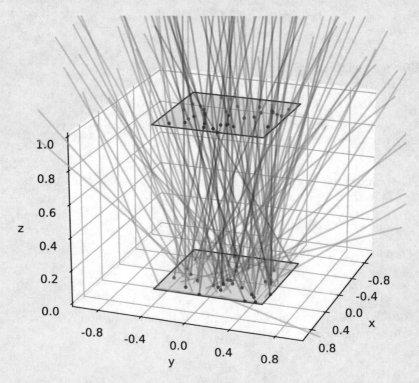

Fig. 4.3 Simulated cosmic rays generated in all direction (blue) intersecting a pair of detector planes (red)

(continued)

Example 4.4 (continued)

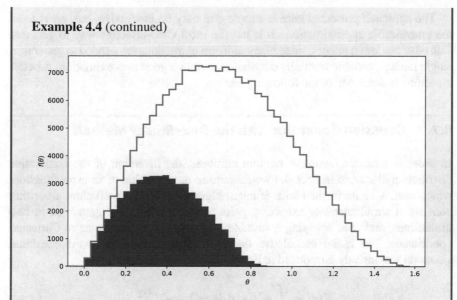

Fig. 4.4 Angular distribution of simulated cosmic rays. The blue histogram represents all generated cosmic rays, the red histogram represents only the ones that cross the detector geometry. The histograms are normalized to the total number of generated cosmic rays

4.6 Gaussian Generator Using the Central Limit Theorem

By virtue of the central limit theorem (see Sect. 3.20), the sum of N random variables, each having a finite variance, is distributed, in the limit $N \to \infty$, according to a Gaussian distribution. A finite but sufficiently large number of random numbers, $x_1, \cdots, x_N$, can be extracted using a uniform random generator. Each value can be remapped from $[0, 1[$ to $[-\sqrt{3}, \sqrt{3}[$ using Eq. (4.16), so that the average and variance of each x_i are equal to zero and one, respectively. Then, their average, times $\sqrt{N}$, is computed:

$$x_c = \frac{x_1 + \cdots + x_N}{\sqrt{N}}. \tag{4.43}$$

The variable x_c has again average equal to zero and variance equal to one. Figure 3.15 shows the distribution of x_c for N from one to four, compared with a Gaussian distribution. The distribution of x_c, as determined from Eq. (4.43), is necessarily truncated, and x_c lies in the range $[-\sqrt{3N}, \sqrt{3N}[$ by construction. Conversely, the range of a truly Gaussian random variable has no upper nor lower bound.

The approach presented here is simple and may be instructive, but, apart from the unavoidable approximations, it is not the most CPU-effective way to generate Gaussian random numbers, since many uniform extractions are needed to generate a single random variable normally distributed, within a good approximation. A better algorithm is described in the following section.

4.7   Gaussian Generator with the Box–Muller Method

In order to generate Gaussian random numbers, the inversion of the cumulative distribution discussed in Sect. 4.4 would require the inversion of an error function, which cannot be performed with efficient algorithms. A more efficient algorithm consists of simultaneously extracting pairs of random numbers generated in two dimensions, and then applying a suitable transformation from polar to Cartesian coordinates. The radial cumulative distribution of a standard two-dimensional Gaussian was already introduced in Eq. (3.124):

$$F(r) = \int_0^r e^{-\rho^2/2}\rho \, dr = 1 - e^{-r^2/2} \, . \tag{4.44}$$

The inversion of $F(r)$ gives the Box–Muller transformation [5] that takes two variables r_1 and r_2 uniformly distributed in $[0, 1[$ and maps them into two variables z_1 and z_2 distributed according to a standard normal. The transformation is the following:

$$\begin{cases} z_1 = r \cos \phi \, , \\ z_2 = r \sin \phi \, , \end{cases} \tag{4.45}$$

with:

$$\begin{cases} r = \sqrt{-2 \log(1 - r_1)} \, , \\ \phi = 2\pi \, r_2 \, . \end{cases} \tag{4.46}$$

A standard normal random number z can be easily transformed into a Gaussian random number x with average μ and standard deviation σ using the following linear transformation, which is the inverse of Eq. (3.49):

$$x = \mu + \sigma z \, . \tag{4.47}$$

More efficient generators for Gaussian random numbers exist. For instance, the so-called Ziggurat algorithm is described in [6].

4.8   Hit-or-Miss Monte Carlo

The generation of random numbers according to a desired distribution can be performed with computer algorithms also for cases where the cumulative distribution cannot be easily computed or inverted. A rather general-purpose and simple random number generator is the *hit-or-miss Monte Carlo*. Consider a PDF $f(x)$ defined in an interval $x \in [a, b[$. The function $f(x)$ must not be necessarily normalized,[2] but we assume to know either the maximum value m of $f(x)$, or at least a value m that is known to be greater or equal to the maximum of $f(x)$. The algorithm proceeds according to the following steps:

- A uniform random number x is extracted in the interval $[a, b[$ and $f = f(x)$ is computed.
- A random number r is extracted uniformly in the interval $[0, m[$.
- If $r > f$, the extraction of x is repeated until $r \le f$. In this case, x is accepted as randomly extracted value.

The method is illustrated in Fig. 4.5.

The probability distribution of the accepted values of x is, by construction, equal to $f(x)$, or $f(x)/\int_a^b f(x)\,dx$, if f is not normalized.

The method accepts a fraction of extractions equal to the ratio of area under the curve $f(x)$ to the area of the rectangle $[a, b[\times [0, m[$ that contains the curve f. This may be inefficient if the ratio is particularly small.

We can define *efficiency* of the method the fraction of accepted values of x, equal to:

$$\varepsilon = \frac{\int_a^b f(x)\,dx}{m(b-a)}. \tag{4.48}$$

The algorithm may be suboptimal if ε is very low, as in the cases where the shape of $f(x)$ is very peaked.

Hit-or-miss Monte Carlo can also be applied to multidimensional cases with no conceptual difference: first a point in n dimensions $\vec{x} = (x_1, \cdots, x_n)$ is extracted; then, $\vec{x}$ is accepted or rejected according to a random extraction $r \in [0, m[$, compared with $f(\vec{x})$, where m is greater or equal to the maximum of f.

---

[2] That is, the integral $\int_a^b f(x)\,dx$ may be different from one.

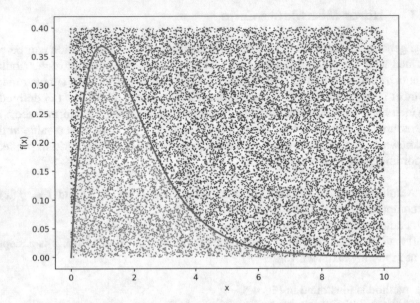

Fig. 4.5 Illustration of the hit-or-miss Monte Carlo method. Points are uniformly generated within a rectangle that contains the PDF curve $f(x)$. Green points are accepted, while red points are rejected. The x values of the accepted points are distributed according to $f(x)$

4.9   Importance Sampling

If the function $f(x)$ is particularly peaked, the efficiency of the hit-or-miss method (Eq. (4.48)) may be very low. The algorithm may be adapted in order to improve the efficiency by identifying, in a preliminary stage of the algorithm, a partition of the interval $[a, b[$ such that in each subinterval the function $f(x)$ has a smaller variation than in the overall range. In each subinterval, the maximum of $f(x)$ is estimated, as sketched in Fig. 4.6. The modified algorithm proceeds as follows: first, a subinterval of the partition is randomly chosen with a probability proportional to its area (see Sect. 4.5); then, the hit-or-miss approach is followed in the corresponding rectangle. This approach is called *importance sampling*.

A possible variation of this method is to use, instead of the rectangular partition, an envelope for the function $f(x)$, i.e., a function $g(x)$ that is always greater than or equal to $f(x)$:

$$g(x) \geq f(x) , \quad \forall x \in [a, b[, \qquad (4.49)$$

and for which a convenient method to extract x according to the normalized distribution $g(x)$ is known. A concrete case of this method is presented in Example 4.5.

It is evident that the efficiency of the importance sampling may be significantly larger compared to the basic hit-or-miss Monte Carlo if the partition and the corresponding maxima in each subinterval are properly chosen.

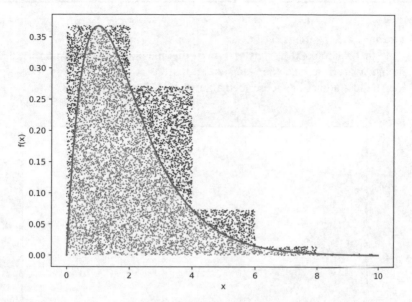

Fig. 4.6 Variation of the hit-or-miss Monte Carlo using the importance sampling. Points are uniformly distributed within rectangles whose union contains $f(x)$. This extraction has a greater efficiency compared to Fig. 4.5

Example 4.5 - Combining Different Monte Carlo Techniques

We want to generate a random variable x distributed according to a PDF:

$$f(x) = Ce^{-\lambda x} \cos^2 kx \,, \tag{4.50}$$

where C is a normalization constant, while λ and k are two known parameters. $f(x)$ is the product of an oscillating term $\cos^2 kx$ times an exponential term $e^{-\lambda x}$ that damps the oscillation.

As envelope, the function $Ce^{-\lambda x}$ can be taken, and, as first step, a random number x is generated according to this exponential distribution (see Example 4.2). Then, a hit-or-miss technique is applied, accepting or rejecting x with probability equal to $\cos^2 kx$. The probability distribution, given the two independent processes, is the product of the exponential envelope times the cosine-squared oscillating term. In summary, the algorithm may proceed as follows:

1. generate r uniformly in $[0, 1[$;
2. compute $x = -\log(1 - r)/\lambda$, which is distributed according to an exponential;
3. generate s uniformly in $[0, 1[$;
4. if $s > \cos^2 kx$ repeat the extraction at the point 1, else return x.

(continued)

Example 4.5 (continued)

Note that random extractions of x with this method do not have an upper bound, apart from machine precision limitations.

An illustration of the described algorithm is shown in Fig. 4.7.

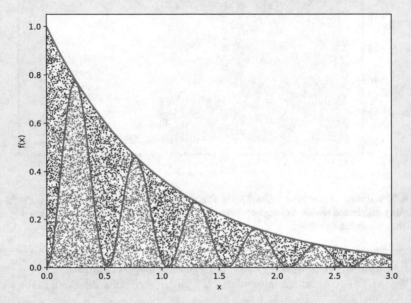

Fig. 4.7 Hit-or-miss Monte Carlo using the importance sampling with an exponential envelope, shown as the orange curve

4.10 Numerical Integration with Monte Carlo Methods

Monte Carlo methods are often used as numerical integration algorithms. The *hit-or-miss* method described in Sect. 4.8, for instance, allows to estimate the integral $\int_a^b f(x)\,dx$ as the fraction of the number of accepted hits n over the total number of extractions N:

$$I = \int_a^b f(x)\,dx \simeq \hat{I} = (b-a)\frac{n}{N} . \tag{4.51}$$

With this approach, n follows a binomial distribution. If n is not too close to either 0 or N, the error on $\hat{I}$ can be approximately estimated (see Example 6.3; a more rigorous approach is discussed in Sect. 8.5) as:

$$\sigma_{\hat{I}} = (b - a)\sqrt{\frac{\hat{I}(1 - \hat{I})}{N}}. \tag{4.52}$$

The error on $\hat{I}$ from Eq. (4.52) decreases as $\sqrt{N}$. This result is true also if the hit-or-miss method is applied to a multidimensional integration, regardless of the number of dimensions d of the problem. This makes Monte Carlo methods advantageous in cases the number of dimensions is large, since other numerical methods, not based on random number extractions, may suffer from severe computing time penalties for large number of dimensions d.

Hit-or-miss Monte Carlo may suffer, anyway, from issues related to the algorithm that finds the maximum value of the PDF, which may be challenging for large number of dimensions. Also, partitioning the multidimensional integration range in an optimal way, in the case of importance sampling, may be a non-trivial task.

4.11 Markov Chain Monte Carlo

The Monte Carlo methods considered in the previous sections are based on sequences of uncorrelated pseudorandom numbers that follow a given probability distribution. There are classes of algorithms that sample more efficiency a probability distribution by producing sequences of *correlated* pseudorandom numbers, where each element in the sequence depends on previous ones.

A sequence of random variables $\vec{x}_0, \cdots, \vec{x}_n$ is called a *Markov chain* if the probability distributions of each variable, given the values of the previous ones, obey:

$$f_n(\vec{x}_{n+1}; \vec{x}_0, \cdots, \vec{x}_n) = f_n(\vec{x}_{n+1}; \vec{x}_n), \tag{4.53}$$

i.e., if $f_n(\vec{x}_{n+1})$ only depends on the previously extracted value $\vec{x}_n$. The dependence on the last value only may be considered a sort of loss of memory. A Markov chain is said *homogeneous* if the functions f_n do not depend on n, i.e., if:

$$f_n(\vec{x}_{n+1}; \vec{x}_n) = f(\vec{x}_{n+1}; \vec{x}_n). \tag{4.54}$$

One example of Markov chain is implemented with the Metropolis–Hastings algorithm [7,8] described below. Our goal is to sample a PDF $f(\vec{x})$ starting from an initial point $\vec{x} = \vec{x}_0$. We do not assume here that f is properly normalized, similarly to what we did for the hit-or-miss Monte Carlo algorithm. Given a point x_i, a point $\vec{x}$ is randomly generated according to a PDF $q(\vec{x}; \vec{x}_i)$, called *proposal distribution*, that depends on the previous point $\vec{x}_i$. The generated point $\vec{x}$ is accepted or not based on the Hastings test ratio:

$$r = \min \left(1, \frac{f(\vec{x}) \, q(\vec{x}_i; \, \vec{x})}{f(\vec{x}_i) \, q(\vec{x}; \, \vec{x}_i)} \right) . \tag{4.55}$$

If a uniformly generated value u is less or equal to r, the generated point $\vec{x}$ is accepted as new point $\vec{x}_{i+1}$, otherwise, the generation of $\vec{x}$ is repeated. Once the proposed point is accepted, a new generation restarts from $\vec{x}_{i+1}$, as above, in order to generate $\vec{x}_{i+2}$, and so on. The process is repeated until the desired number of points is generated.

Each value with non-null probability is eventually reached, within an arbitrary precision, after a sufficiently large number of extractions (*ergodicity*), and the sequence can be proven to follow, in the limit of infinite number of generations, the desired PDF $f(\vec{x})$, provided that f and q satisfy some conditions.

Consider a PDF f such that its domain has two areas A and B where the probability density f is not null, separated by a large region where the probability density is zero or very small. q should guarantee a non-negligible probability to jump from A to B and vice versa. This is the case, for instance, with a Gaussian q, which has tails extending to infinity. A uniform distribution within a limited interval may be problematic if the range is smaller than the minimum distance from A to B. Even with a Gaussian, anyway, if the standard deviation is very small, the probability to explore all areas may be very small. For this reason, the choice of the proposal distribution should be done with some care.

The first generated values may be placed in an area where the probability density is very low and may distort the distribution of the generated sequence if the total number of generated points is not very large. For this reason, usually a certain number of the first extracted values is discarded. This number is usually a tunable parameter of the algorithm. The number of initial extractions that needs to be discarded is not easy to predict and may depend on the choice of q. Empiric tests may be needed to check that the sequence reaches stably a region where the probability density is not too small, after several initial extractions.

It is convenient to choose a symmetric proposal distribution such that $q(\vec{x}, \vec{x}_i) = q(\vec{x}_i, \vec{x})$. In this way, Eq. (4.55) simplifies to the so-called Metropolis–Hastings ratio, which does not depend on q:

$$r = \min \left(1, \frac{f(\vec{x})}{f(\vec{x}_i)} \right) . \tag{4.56}$$

A typical proposal choice may be a multidimensional Gaussian centered around $\vec{x}_i$ with a fixed standard deviation.

The Metropolis–Hastings method falls under the category of Markov chain Monte Carlo (MCMC) techniques. Such techniques are very powerful to compute posterior probability densities for Bayesian inference (see Sect. 5.6). Figure 4.8 shows an example of application of the Metropolis–Hastings Monte Carlo.

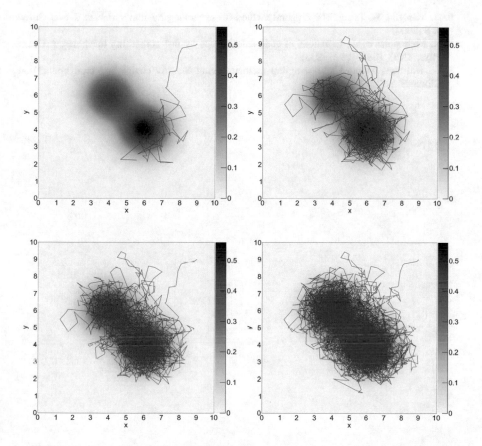

Fig. 4.8 Example of random generation with the Metropolis–Hastings method. The sampled PDF is the sum of two two-dimensional Gaussian distributions with relative weights equal to 0.45 and 0.55. The PDF is plotted as a white-to-red-to-black color map in the background. The generated points are connected by a line in order to show the sequence. The first 20 generated points are shown in purple, all the subsequent ones are in blue. The proposal function is a two-dimensional Gaussian with $\sigma = 0.5$. The total number of generated points is 100 (top, left), 1000 (top, right), 2000 (bottom, left), and 5000 (bottom, right)

References

1. R. May, Simple mathematical models with very complicated dynamics. Nature **621**, 459 (1976)
2. T. O. Group, *The Single UNIX® Specification*, Version 2 (1997). http://www.unix.org
3. G. Marsaglia, W.W. Tsang, J. Wang, Fast generation of discrete random variables. J. Stat. Softw. **11**, 3 (2004)
4. M.D. Vose, A linear algorithm for generating random numbers with a given distribution. IEEE Trans. Softw. Eng. **17**(9), 972–975 (1991). https://doi.org/10.1109/32.92917
5. G.E.P. Box, M. Muller, A note on the generation of random normal deviates. Ann. Math. Stat. **29**, 610–611 (1958)

6. G. Marsglia, W. Tsang, The Ziggurat method for generating random variables. J. Stat. Softw. **5**, 8 (2000)
7. N. Metropolis et al., Equations of state calculations by fast computing machines. J. Chem. Phys. **21(6)**, 1087–1092 (1953)
8. W. Hastings, Monte Carlo sampling methods using Markov chains and their application. Biometrika **57**, 97–109 (1970)

Bayesian Probability and Inference

5

5.1 Introduction

The Bayesian approach to probability is a method that allows to assign probability values to statements which one does not know with certainty whether they are true or false. The probability is determined based of known evidence that is used to update and improve prior subjective knowledge.

Bayesian probability has a wider range of applicability compared to frequentist probability (see Sect. 6.1). The latter can only be applied to repeatable experiments. For instance, under the frequentist approach one can determine the probability that the value of a random variable lies within a certain interval. Under the Bayesian approach it is also possible to determine the probability that the value of an unknown parameter lies within a certain interval, which would not have any frequentist meaning, since an unknown parameter is not a random variable.

Bayes' theorem is presented in Sect. 5.2; then, Sect. 5.3 introduces the definition of Bayesian probability as an extension of Bayes' theorem. Bayesian inference is then discussed in the following sections.

5.2 Bayes' Theorem

The conditional probability, introduced in Eq. (1.2), defines the probability of an event A with the condition that the event B has occurred:

$$P(A \mid B) = \frac{P(A \cap B)}{P(B)} . \qquad (5.1)$$

L. Lista, *Statistical Methods for Data Analysis*, Lecture Notes in Physics 1010, https://doi.org/10.1007/978-3-031-19934-9_5

Fig. 5.1 Illustration of the conditional probabilities, $P(A \mid B)$ and $P(B \mid A)$. The events A and B are represented as subsets of a sample space Ω. This Illustration was originally presented by Robert Cousins, courtesy of the author. All rights reserved

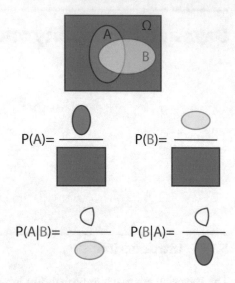

The probability of the event B given the event A, vice versa, can be written as:

$$P(B \mid A) = \frac{P(A \cap B)}{P(A)}.$$
(5.2)

The two above definitions are illustrated in Fig. 5.1 using sets of a sample space Ω that represent the events A and B. By extracting from Eqs. (5.1) and (5.2) the common term $P(A \cap B)$, the following relation is obtained:

$$P(A \mid B)\, P(B) = P(B \mid A)\, P(A) \,,$$
(5.3)

from which the Bayes' theorem can be derived in the following form:

$$\boxed{P(A \mid B) = \frac{P(B \mid A)\, P(A)}{P(B)}.}$$
(5.4)

The probability $P(A)$ can be interpreted as the probability of the event A *before* the knowledge that the event B has occurred, called *prior* probability, while $P(A \mid B)$ is the probability of the same event A considering, as further information, the knowledge that the event B has occurred, called *posterior* probability.

Bayes' theorem can be seen as a rule that allows the inversion of conditional probability. It has general validity for any approach to probability, including frequentist probability. An illustration of Bayes' theorem is presented in Fig. 5.2, using the same set representation of Fig. 5.1. Bayes' theorem has many applications. A couple of examples are illustrated in Exercises 5.1 and 5.2 that follow. In those examples, the denominator $P(B)$ is conveniently computed using the law of total probability from Eq. (1.13).

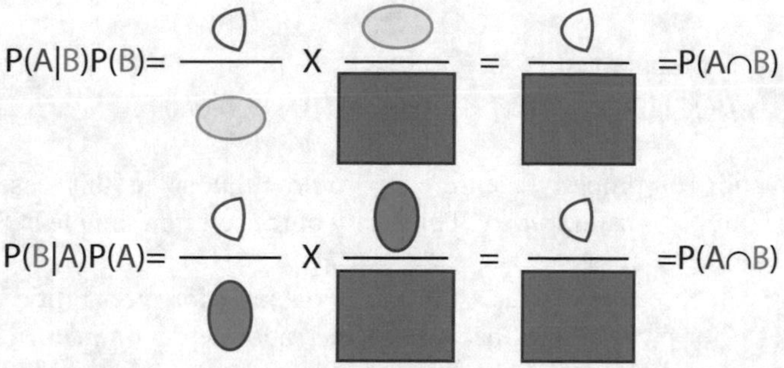

$$P(A|B)P(B)= \frac{\text{⬤}}{\text{⬤}} \; X \; \frac{\text{⬤}}{\text{⬛}} = \frac{\text{⬤}}{\text{⬛}} = P(A\cap B)$$

$$P(B|A)P(A)= \frac{\text{⬤}}{\text{⬤}} \; X \; \frac{\text{⬤}}{\text{⬛}} = \frac{\text{⬤}}{\text{⬛}} = P(A\cap B)$$

Fig. 5.2 Illustration of the Bayes' theorem. The areas of events A and B, equal to $P(A)$ and $P(B)$, respectively, simplify when $P(A \mid B)\,P(B)$ and $P(B \mid A)\,P(A)$ are multiplied. This illustration was originally presented by Robert Cousins, courtesy of the author.

Example 5.1 - Purity and Contamination

A proton beam hits a target and produces a sample of particles that is constituted by 8% of muons, 10% of kaons, while the remaining 82% of the particles are pions. A particle identification detector correctly identifies 95% of muons. It also incorrectly identifies as muons a fraction of 7% of all kaons and 4% of all pions. We want to compute the expected fraction of real muons in the sample of particles identified as such by the detector.

In the notation introduce above, we can write our input data as:

$$P(\mu) = 0.08\,, \qquad\qquad P(\text{ID} \mid \mu) = 0.95\,,$$
$$P(\pi) = 0.82\,, \qquad\qquad P(\text{ID} \mid \pi) = 0.04\,,$$
$$P(K) = 0.10\,, \qquad\qquad P(\text{ID} \mid K) = 0.07\,.$$

The denominator in Bayes' formula can be decomposed using the law of total probability. The three sets of muons, pions, and kaons, in fact, are a partition of the entire sample space, which is the set of all particles. Using Eq. (1.13), we can write:

$$P(\text{ID}) = P(\mu)P(\text{ID} \mid \mu) + P(K)P(\text{ID} \mid K) + P(\pi)P(\text{ID} \mid \pi) = 0.1158\,. \tag{5.5}$$

This means that 11.58% of all particles in the sample are identified as muons.

Applying Bayes' theorem, we have the desired result. More in detail:

$$P(\mu \mid \text{ID}) = P(\text{ID} \mid \mu)P(\mu)/P(\text{ID}) = 0.6563\,, \tag{5.6}$$
$$P(\pi \mid \text{ID}) = P(\text{ID} \mid \pi)P(\mu)/P(\text{ID}) = 0.2832\,, \tag{5.7}$$

(continued)

Example 5.1 (continued)

$$P(K \mid \mathrm{ID}) = P(\mathrm{ID} \mid K) P(\mu) / P(\mathrm{ID}) = 0.0604 \ . \tag{5.8}$$

The fraction of properly identified particles, muons in this case, in a selected sample is called *purity*. The purity of the selected sample is 65.63% (Eq. 5.6).

Bayes' theorem also relates the ratios of posterior probabilities to the ratios of prior probabilities, because in the ratio the denominator cancels. The ratio of posterior probabilities, also called *posterior odds*, can be written as:

$$\frac{P(\mu \mid \mathrm{ID})}{P(\pi \mid \mathrm{ID})} = \frac{P(\mathrm{ID} \mid \mu)}{P(\mathrm{ID} \mid \pi)} \cdot \frac{P(\mu)}{P(\pi)} \ . \tag{5.9}$$

The above expression also holds if more than two possible particle types are present in the sample (muons, pions, and kaons, in this example) and does not require to compute the denominator, as done in Eq. (5.5). Indeed, one does not even need to know the total number of particle types present in the sample. The calculation of the denominator is necessary, instead, to compute individual probabilities related to all possible particle cases, as seen above. Section 5.11 further discusses posterior odds and their use.

Example 5.2 - An Epidemiology Example

The following is a classic example, also reported in several books and lecture series, for instance, in [1] and [2]. Consider a person who received a positive diagnosis of some illness. We want to compute the probability that the tested person is really ill, assuming we know the probability that the test gives a false positive outcome. Assume we know that the probability of a positive test is 100% for an ill person. We also know that the test also has a small probability, say 0.2%, to give a false positive result on a healthy person. A common mistake is to conclude that the probability that the person who tested positive is really ill is is equal to 100%−0.2% = 99.8%. The following demonstrates why this naive answer is wrong.

The problem can be formulated more precisely using the following notation, where "+" and "−" indicate positive and negative test results:

$$P(+ \mid \mathrm{ill}) \simeq 100\% \ , \qquad\qquad P(+ \mid \mathrm{healthy}) = 0.2 \ , \%$$

$$P(- \mid \mathrm{ill}) \simeq 0\% \ , \qquad\qquad P(- \mid \mathrm{healthy}) = 99.8\% \ .$$

(continued)

Example 5.2 (continued)

To answer our question we need to compute, in this notation, $P(\text{ill} \mid +)$, which can be obtained using Bayes' theorem as:

$$P(\text{ill} \mid +) = \frac{P(+ \mid \text{ill})\, P(\text{ill})}{P(+)} \; . \qquad (5.10)$$

Since $P(+ \mid \text{ill}) \simeq 1$, Eq. (5.10) gives approximately:

$$P(\text{ill} \mid +) \simeq \frac{P(\text{ill})}{P(+)} \; . \qquad (5.11)$$

A missing ingredient of the problem appears clearly in Eq. (5.11): from our input data, we do not know $P(\text{ill})$, the probability that a random person in the population under consideration is really ill, regardless of any possibly performed test. In a normal situation where mostly of the population is healthy, we can expect $P(\text{ill}) \ll P(\text{healthy})$. Considering that:

$$P(\text{ill}) + P(\text{healthy}) = 1 \; , \qquad (5.12)$$

and:

$$P(\text{ill } and \text{ healthy}) = 0 \; , \qquad (5.13)$$

the two events "ill" and "healthy" are a partition of the entire sample space, made by the entire population. Therefore, $P(+)$ can be decomposed according to the law of total probability using Eq. (1.13):

$$P(+) = P(+ \mid \text{ill})\, P(\text{ill}) + P(+ \mid \text{healthy})\, P(\text{healthy}) \simeq$$
$$\simeq P(\text{ill}) + P(+ \mid \text{healthy}) \; . \qquad (5.14)$$

The probability $P(\text{ill} \mid +)$ can then be written, using Eq. (5.14) for the denominator, as:

$$P(\text{ill} \mid +) = \frac{P(\text{ill})}{P(+)} \simeq \frac{P(\text{ill})}{P(\text{ill}) + P(+ \mid \text{healthy})} \; . \qquad (5.15)$$

If we assume that $P(\text{ill})$ is smaller than the probability $P(+ \mid \text{healthy})$, $P(\text{ill} \mid +)$ is smaller than 50%, i.e., it is more likely that the positively tested

(continued)

Example 5.2 (continued)

person is healthy than hill. For instance, if $P(\text{ill}) = 0.15\%$, compared with
our assumption of $P(+ \mid \text{healthy}) = 0.2\%$, then:

$$P(\text{ill} \mid +) = \frac{0.15}{0.15 + 0.20} = 43\% . \tag{5.16}$$

This shows that the naive conclusion presented at the beginning,
$P(\text{ill} \mid +) = 99.8\%$, is grossly wrong. The result is illustrated, changing
the proportions to have a better visualization, in Fig. 5.3.

If the probability of a positive diagnosis in case of illness, $P(+ \mid \text{ill})$, is
large, this does not imply that a positive diagnosis corresponds to a large
probability of being really ill. The correct answer also depends, strongly
in this case, on the *prior* probability $P(\text{ill})$ that a random person in the
population is ill.

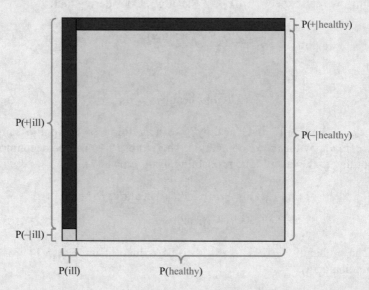

Fig. 5.3 Illustration Bayes' theorem applied to the computation of the probability that
a person with a positive test is really ill. The two red rectangular areas correspond to
the cases of a true positive diagnosis for an ill person ($P(+ \mid \text{ill})$, vertical red area) and
a false positive diagnosis for a healthy person ($P(+ \mid \text{healthy})$, horizontal red area). The
probability that a person with a positive test is really ill, $P(\text{ill} \mid +)$, is equal to the ratio of
the vertical red area to the total red area

5.3   Bayesian Probability Definition

In the above Exercises 5.1 and 5.2, Bayes' theorem was applied to cases where probabilities can also be defined using the frequentist approach. The formulation of Bayes' theorem in Eq. (5.4) can also be used to introduce a new definition of probability. We can interpret the equation:

$$P(A \mid E) = \frac{P(E \mid A)\, P(A)}{P(E)} \tag{5.17}$$

as follows:

- *before* we know that E is true, our *degree of belief* in the event A is equal to the *prior* probability $P(A)$;
- *after* we know that E is true, our degree of belief in the event A changes, and becomes equal to the *posterior* probability $P(A \mid E)$.

Note that this consideration about the update of prior probability of the event A into posterior probabilities reflect the state of our knowledge that E is true, not the fact that the probability of the event A changes before and after the event E has happened, in a chronological sense.

With this interpretation, we can extend the definition of probability, in the new Bayesian sense, to events that are not associated with random outcomes of repeatable experiments but represent statements about facts that are not known with certainty, like *"my football team will win next match,"* or *"the mass of a dark-matter candidate particle is between 1000 and 1500 GeV."*

We first assign a prior probability $P(A)$ to such an unknown statement A that represents a measurement of our prejudice, or preliminary knowledge, about the statement, before the acquisition of any new information that could modify our knowledge. After we know that E is true, our knowledge of A must change, and our degree of belief has to be modified accordingly. The new degree of belief must be equal, according to Bayes' theorem, to the posterior probability $P(A \mid E)$. In other words, Bayes' theorem gives us a quantitative prescription about how to rationally change our subjective degree of belief from an initial prejudice considering newly available information, or *evidence*, E.

The following terminology applies to the elements that appear in Eq. (5.17) in Bayesian probability

- $P(A|E)$ is called *posterior probability*.
- $P(E|A)$ is called *likelihood*.
- $P(A)$ is called *prior probability*.
- $P(E)$ is called *evidence*.

Starting from different subjective priors, i.e., different prejudices, different posteriors are determined. Therefore, an intrinsic and unavoidable feature of Bayesian probability is that the probability associated with an event cannot be defined without depending on a prior probability of that event. This makes Bayesian probability intrinsically subjective. This feature is somewhat unpleasant when the Bayesian approach is used to achieve scientific results that should be objective, and not affected by any subjective assumption. For this reason, many scientists prefer frequentist statistic to the Bayesian approach. This issue and the possibilities to mitigate it are discussed in Sect. 5.12.

5.4   Decomposing the Denominator in Bayes' Formula

The term $P(E)$ that appears in the denominator of Eq. (5.4) can be considered as a normalization factor. Let us assume that the sample space Ω can be decomposed into a partition $A_1, \cdots , A_N$, where

$$\bigcup_{i=1}^{N} A_i = \Omega \quad \text{and} \quad A_i \cap A_j = 0 \quad \forall i, j . \tag{5.18}$$

We can call each of the A_i an *hypothesis*. The law of total probability in Eq. (1.13) can be applied, as already done in Exercises 5.1 and 5.2, previously discussed. We can therefore write:

$$P(E) = \sum_{i=1}^{N} P(E \mid A_i) \, P(A_i) , \tag{5.19}$$

and Bayes' theorem, for one of the hypotheses A_i, can be written as:

$$P(A_i \mid E) = \frac{P(E \mid A_i) P(A_i)}{\sum_{i=j}^{n} P(E \mid A_j) P(A_j)} . \tag{5.20}$$

Written as in Eq. (5.20), it is easy to demonstrate that the Bayesian definition of probability obeys Kolmogorov's axioms of probability, as defined in Sect. 1.7. Hence, all properties of probability discussed in Chap. 1 also apply to Bayesian probability.

Example 5.3 - Extreme Case of Prior Belief: a Dogma

Consider n events, or hypotheses, $A_0, \cdots, A_{n-1}$ that constitute a non-intersecting partition of the sample space Ω. Let us assume that we assign *prior* probabilities to each hypothesis A_i all equal to zero, except for $i = 0$:

$$P(A_i) = \begin{cases} 1 & \text{if } i = 0, \\ 0 & \text{if } i \neq 0. \end{cases} \tag{5.21}$$

This corresponds to the prior belief that the hypothesis A_0 is absolutely true, and all other alternative A_i are absolutely false, for $i \neq 0$. It is possible to demonstrate that, given whatever evidence E, the posterior probabilities of all A_i are identical to the prior probabilities, i.e.,

$$P(A_i \mid E) = P(A_i), \ \forall E. \tag{5.22}$$

From Bayes' theorem, in fact:

$$P(A_i \mid E) = \frac{P(E \mid A_i)\, P(A_i)}{P(E)}. \tag{5.23}$$

If $i \neq 0$, clearly:

$$P(A_i \mid E) = \frac{P(E \mid A_i) \times 0}{P(E)} = 0 = P(A_i). \tag{5.24}$$

If $i = 0$, instead, assuming that $P(E \mid A_0) \neq 0$, even if it is extremely small:

$$P(A_0 \mid E) = \frac{P(E \mid A_0) \times 1}{\sum_{i=1}^{n} P(E \mid A_i)\, P(A_i)} = \frac{P(E \mid A_0)}{P(E \mid A_0) \times 1} = 1 = P(A_0). \tag{5.25}$$

If $P(E \mid A_0) = 0$, $P(A_0 \mid E)$ assumes an indeterminate form $0/0$ but the limit for $P(E \mid A_0) \to 0$ exists:

$$\lim_{P(E \mid A_0) \to 0} P(A_0 \mid E) = P(A_0) = 1. \tag{5.26}$$

This situation reflects the case that we may call *dogma*, or *religious belief*. It means that someone has such a strong prejudice that no evidence, i.e., no new knowledge, can change his/her prior belief.

The scientific method allowed to evolve mankind's knowledge of Nature during history by progressively adding more knowledge based on the observation of evidence. The history of science is full of examples in which

(continued)

Example 5.3 (continued)

theories known to be true have been falsified by new or more precise observations and new better theories have replaced the old ones. This has been possible because even proven theories are never considered absolutely true. According to Eq. (5.22), instead, scientific progress is not possible in the presence of religious beliefs about theories that are falsifiable by the observation of new evidence.

5.5 Bayesian Probability Density and Likelihood Functions

Bayesian probability can be generalized to continuous probability distributions. An observation of an evidence E is replaced by the observation of the value of a randomly extracted variable x, whose PDF $f(x \mid \theta)^1$ depends on the value of an unknown parameter θ. Continuous values of θ replace the discrete events, or hypotheses, A_j. The prior probabilities $P(A_j)$ are replaced by a continuous prior PDF $\pi(\theta)$. Sums are replaced by integrals, and probabilities are replaced by probability densities, according to the following correspondence rules:

- $E \mapsto x$
- $A_i \mapsto \theta$
- $P(E \mid A_i) \mapsto dP(x \mid \theta) = f(x \mid \theta) \, dx$
- $P(A_i \mid E) \mapsto dP(\theta \mid x) = f(\theta \mid x) \, d\theta$
- $P(A_j \mid E) \mapsto dP(\theta' \mid x) = f(\theta' \mid x) \, d\theta'$
- $P(A_i) \mapsto \pi(\theta) \, d\theta$
- $\sum_{j=1}^{n} \mapsto \int d\theta'$

Therefore Bayes' theorem, for differential probabilities, becomes:

$$f(\theta \mid x) \, d\theta = \frac{f(x \mid \theta) \, dx \, \pi(\theta) \, d\theta}{\int f(x \mid \theta') \, dx \, \pi(\theta') \, d\theta'} \tag{5.27}$$

from which we can define the Bayesian posterior probability density:

$$f(\theta \mid x) = \frac{f(x \mid \theta) \, \pi(\theta)}{\int f(x \mid \theta') \, \pi(\theta') \, d\theta'} . \tag{5.28}$$

We now considering a more general number of variables and parameters, given a sample $(x_1, \cdots, x_n)$ of n random variables whose PDF has a known parametric

---

[1] Here we use the notation $f(x \mid \theta)$, more natural in the Bayesian context. The notation $f(x; \theta)$ is also used for parametric PDFs.

form which depends on m parameters, $\theta_1, \cdots, \theta_m$. The *likelihood function* is defined as the probability density, evaluated at the point $(x_1, \cdots, x_n)$ corresponding to the observed values, for fixed values of the parameters $\theta_1, \cdots, \theta_m$:

$$L(x_1, \cdots, x_n; \theta_1, \cdots, \theta_m) = \left. \frac{dP(x_1, \cdots, x_n)}{dx_1 \cdots dx_n} \right|_{\theta_1, \cdots, \theta_m}. \tag{5.29}$$

The notation $L(x_1, \cdots, x_n \,|\, \theta_1, \cdots, \theta_m)$ is sometimes used in place of $L(x_1, \cdots, x_n; \theta_1, \cdots, \theta_m)$, similarly to the notation used for conditional probability, which is more natural in the Bayesian context. The likelihood function is more extensively discussed in Sect. 6.6 in the context of frequentist inference.

The posterior Bayesian PDF for the parameters $\theta_1, \cdots, \theta_m$, given the observation of $(x_1, \cdots, x_n)$, can be defined using the likelihood function in Eq. (5.29):

$$p(\theta_1, \cdots, \theta_m \,|\, x_1, \cdots, x_n) = \frac{L(x_1, \cdots, x_n \,|\, \theta_1, \cdots, \theta_m)\, \pi(\theta_1, \cdots, \theta_m)}{\int L(x_1, \cdots, x_n \,|\, \theta_1', \cdots, \theta_m')\, \pi(\theta_1', \cdots, \theta_m')\, d^m\theta'}, \tag{5.30}$$

where the probability density function $\pi(\theta_1, \cdots, \theta_m)$ is the prior PDF of the parameters $\theta_1, \cdots, \theta_m$, i.e., our degree of belief about the unknown parameters before one knows about the observation of $(x_1, \cdots, x_n)$. The denominator in Eq. (5.30), derived from an extension of the law of total probability, is clearly interpreted now as a normalization of the posterior PDF.

Fred James et al. summarized the role of the posterior PDF given by Eq. (5.30) with following sentence:

The difference between $\pi(\theta)$ and $p(\theta \,|\, x)$ shows how one's knowledge (degree of belief) about θ has been modified by the observation x. The distribution $p(\theta \,|\, x)$ summarizes all one's knowledge of θ and can be used accordingly [3].

5.6 Bayesian Inference

As seen in Sect. 1.11, a common problem, both in statistics and in physics, is the *estimate* of one or more unknown parameters with their *uncertainties*. The estimate is interpreted, in physics, as the *measurement* of the parameter, and the uncertainty of the estimate is the measurement *error*. In this section and in the followings, the problem is addressed by means of the posterior Bayesian of the unknown parameters. For continuous cases, which are the most common, Eq. (5.30) is used. Using a vector notations for the n observations $\vec{x} = (x_1, \cdots, x_n)$ and the m parameters $\vec{\theta} = (\theta_1, \cdots, \theta_m)$, the posterior can be written as:

$$p(\vec{\theta} \,|\, \vec{x}) = \frac{L(\vec{x}; \vec{\theta})\, \pi(\vec{\theta})}{\int L(\vec{x}; \vec{\theta}')\, \pi(\vec{\theta}')\, d^m\theta'}. \tag{5.31}$$

As estimate of $\vec{\theta}$, usually the *most likely* set of parameters is taken. In practice, we compute the mode, i.e., the set of parameter values $\hat{\vec{\theta}}$ that maximize the posterior PDF. As alternative, the *average value* of the parameters vector, $\langle \vec{\theta} \rangle$, could also be determined from the posterior PDF, but it would clearly have a different meaning. The determination of uncertainties is discussed in Sect. 5.10.

If no information is available about the parameters $\vec{\theta}$ before we perform the inference, the prior density $\pi(\vec{\theta})$ should not privilege any particular parameter value. Such kinds of prior, in general, are called *uninformative priors*. A uniform distribution may appear the most natural choice, but it would not be invariant under reparameterization, as already discussed in Sect. 3.1. An invariant uninformative prior is defined in Sect. 5.13.

If a uniform prior distribution is assumed, the most likely parameter values are the ones that give the maximum likelihood, since the posterior PDF is equal, up to a normalization constant, to the likelihood function:

$$ p(\vec{\theta} \mid \vec{x}) \Big|_{\pi(\vec{\theta})=\text{const.}} = \frac{L(\vec{x}\,;\,\vec{\theta})}{\int L(\vec{x}\,;\,\vec{\theta}')\,\mathrm{d}^m\theta'} \,. \tag{5.32} $$

The case of a uniform prior gives identical results to the maximum likelihood estimator under the frequentist approach, as noted in Sect. 6.5. Anyway, it should be remarked that the interpretation of frequentist and Bayesian estimates are very different, as further discussed in the following. In case the prior $\pi(\vec{\theta})$ is not uniform, the most likely parameter values do not necessarily coincide with the maximum of the likelihood function, and the Bayesian estimate differs from the frequentist maximum likelihood estimate.

In Bayesian inference, the computation of posterior probabilities, and, more in general, the computation of many quantities of interest, requires integrations. In most of the realistic cases, such integrations can only be performed using computer algorithms. Markov chain Monte Carlo (MCMC, see Sect. 4.11) is one of the most performant numerical integrations method for Bayesian computations. MCMC do not require the normalization of the PDF, as seen in Sect. 4.11. This simplifies the problem and avoids to compute the denominator present in Eq. (5.31), which would require one further integration step. In practice, sampling a posterior PDF with MCMC results in a discrete sequence of random points whose distribution approximates the desired PDF.

Example 5.4 - Estimating Seller Reliability from Feedbacks
Most of online shops or auction sites provide a rating of sellers measured as the fraction of positive feedback received for the sold items, as reported by buyers. Imagine we want to select a seller and the only information we have is the reported fraction of positive feedbacks. Would we better pick a seller

---

(continued)

Example 5.4 (continued)

with 100% positive out of a total of 10 feedbacks, or one with 99.9% positive out of a total of 1000 feedbacks? Bayesian inference of the probability p for the seller to receive a positive feedback, given the present number of feedbacks, provides an answer.

We can assume for simplicity that each feedback is given with a Bernoulli process with a probability p to receive positive feedback. The total number n of positive feedbacks is distributed according to a binomial distribution, given the total number N of received feedbacks:

$$P(n; p, N) = \frac{N!}{n!(N-n)!} p^n (1-p)^{N-n} . \qquad (5.33)$$

Given the reported number of positive feedbacks, $n^\star$, the posterior probability density for p is:

$$p(p \mid n^\star; N) = \frac{P(n^\star; p, N) \, \pi(p)}{\int P(n^\star; p', N) \, \pi(p') \, dp'} . \qquad (5.34)$$

If we assume a uniform prior, $\pi(p) = 1$, the resulting posterior simplifies to:

$$p(p \mid n^\star; N) = C p^{n^\star} (1-p)^{N-n^\star} , \qquad (5.35)$$

where C is a normalization factor given by:

$$C^{-1} = \int_0^1 p^{n^\star} (1-p)^{N-n^\star} \, dp . \qquad (5.36)$$

Such a distribution is a special case of a *beta distribution* (Sect. 3.13). Including the normalization factor, a beta distribution is given by:

$$p(x; a, b) = \frac{\Gamma(a+b)}{\Gamma(a) \, \Gamma(b)} x^{a-1} (1-x)^{b-1} . \qquad (5.37)$$

In our case, $a = n^\star + 1$ and $b = N - n^\star + 1$. In particular, for $n^\star = N$, the posterior is:

$$p(p \mid N; N) = (N+1) p^N , \qquad (5.38)$$

which has a maximum at $p = 1$ but may have a broad left tail if N is small. This corresponds to a non-negligible probability that p may be much smaller than one. Vice versa, if N is large, even if p is somewhat lower than one,

(continued)

Example 5.4 (continued)

the distribution may be very narrow, and the chance that p is much lower than one could be much smaller than for the previous case. For this reason, it is important to also consider the total number of feedbacks N, not only the fraction of positive feedbacks.

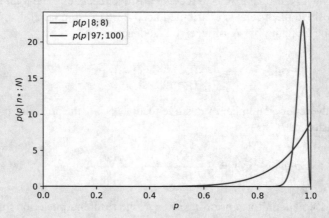

Fig. 5.4 Posterior for the binomial parameter p for two cases of positive feedbacks: $8/8 = 100\%$ (red) and $97/100 = 97\%$ (blue) of positive feedbacks. The most probable values of p are 1.0 and 0.97, respectively. The first case has a large probability that p is below 0.8 (0.13), while this probability is negligible (4.8×10^{-7}) for the second case

Figure 5.4 shows the posteriors for p in the cases of $8/8$ and $97/100$ positive feedbacks. While the most probable values of p are 1.0 and 0.97 for the two cases, respectively, the probability that p is smaller than, say, 0.8 is $P(p < 0.8) = 0.13$ for the first case and only $P(p < 0.8) = 4.8 \times 10^{-7}$ for the second case. The mean value of a beta distribution with parameters a and b, as in Eq. (5.37), is equal to:

$$\langle p \rangle = \frac{a}{a+b} = \frac{n^{\star}+1}{N+2} . \tag{5.39}$$

For the considered cases, $8/8$ positive feedbacks correspond to a mean value equal to $\langle p \rangle = 9/10 = 0.90$, while $97/100$ positive feedbacks correspond to $\langle p \rangle = 98/102 = 0.96$. Those results show that the mean value, as well as the mode, is also not sufficient to quantify the risk of a low value of p, which requires the evaluation of the area under the left tail of the posterior, below the considered threshold value of 0.8, in our example.

5.7 Repeated Observations of a Gaussian Variable

If we observe a number of repeated independent extractions $x_1, \cdots, x_N$ of a random variable x that follows the PDF $p(x; \theta)$, which depends on the unknown parameter θ, the joint PDF for the entire set of observations can be written as the following product:

$$p(x_1, \cdots, x_N; \theta) = \prod_{i=1}^{N} p(x_i; \theta) . \tag{5.40}$$

Consider the particular case where $p(x_i; \theta)$ is a Gaussian distribution g with unknown average μ and known standard deviation σ. We can write the joint probability distribution as:

$$p(x_1, \cdots, x_N; \mu, \sigma) = \prod_{i=1}^{N} g(x_i; \mu, \sigma) = \frac{1}{(2\pi\sigma^2)^{N/2}} \exp\left[-\sum_{i=1}^{N} \frac{(x_i - \mu)^2}{2\sigma^2} \right] . \tag{5.41}$$

If σ is known, we can write the posterior distribution for μ, assuming a prior $\pi(\mu)$, as:

$$p(\mu \mid x_1, \cdots, x_N) = \frac{p(x_1, \cdots, x_N; \mu, \sigma)\,\pi(\mu)}{\int p(x_1, \cdots, x_N; \mu', \sigma)\,\pi(\mu')\,d\mu'} . \tag{5.42}$$

Under the simplest assumption that $\pi(\mu)$ is constant, we can write:

$$p(\mu \mid x_1, \cdots, x_N) = C \exp\left[-\sum_{i=1}^{N} \frac{(x_i - \mu)^2}{2\sigma^2} \right] , \tag{5.43}$$

where the normalization factor C is given by:

$$\frac{1}{C} = \int_{-\infty}^{+\infty} \exp\left[-\sum_{i=1}^{N} \frac{(x_i - \mu')^2}{2\sigma^2} \right] d\mu' . \tag{5.44}$$

The argument of the exponential in Eq. (5.43) contains at the numerator the term:

$$\Sigma = \sum_{i=1}^{N} (x_i - \mu)^2 , \tag{5.45}$$

which can be expanded to write it as a second-order polynomial in μ:

$$\Sigma = \sum_{i=1}^{N} (\mu^2 - 2x_i\mu + x_i^2) = N\mu^2 - 2\mu \sum_{i=1}^{N} x_i + \sum_{i=1}^{N} x_i^2 \, . \tag{5.46}$$

If we introduce the arithmetic means $\bar{x}$ and $\overline{x^2}$:

$$\bar{x} = \frac{1}{N} \sum_{i=1}^{N} x_i \, , \tag{5.47}$$

$$\overline{x^2} = \frac{1}{N} \sum_{i=1}^{N} x_i^2 \, , \tag{5.48}$$

and we define:

$$\sigma_{\bar{x}}^2 = \frac{1}{N} \left(\overline{x^2} - \bar{x}^2 \right) = \frac{\sigma^2}{N} \, , \tag{5.49}$$

the polynomial Σ can be written as:

$$\Sigma = N \left(\mu^2 - 2\mu\bar{x} + \overline{x^2} \right) = N \left(\mu^2 - 2\mu\bar{x} + \bar{x}^2 - \bar{x}^2 + \overline{x^2} \right) =$$

$$= N \left[(\mu - \bar{x})^2 + (\overline{x^2} - \bar{x}^2) \right] = N \left[(\mu - \bar{x})^2 + \sigma_{\bar{x}}^2 \right] \, . \tag{5.50}$$

Plugging Σ into Eq. (5.43), we have:

$$p(\mu \mid x_1, \cdots, x_N) = C \exp\left[-\frac{\Sigma}{2\sigma^2} \right] = C \exp\left\{ -\frac{N \left[(\mu - \bar{x})^2 + \sigma_{\bar{x}}^2 \right]}{2\sigma^2} \right\} =$$

$$= C \exp\left[-\frac{N (\mu - \bar{x})^2}{2\sigma^2} \right] \exp\left(-\frac{N\sigma_{\bar{x}}^2}{2\sigma^2} \right) =$$

$$= \frac{C}{\sqrt{e}} \exp\left[-\frac{(\mu - \bar{x})^2}{2\sigma_{\bar{x}}^2} \right] \, . \tag{5.51}$$

In practice, the posterior PDF for μ is still a Gaussian distribution for μ with average value $\hat{\mu}$ and standard deviation $\sigma_{\hat{\mu}}$ equal to:

$$\hat{\mu} = \bar{x} \, , \tag{5.52}$$

$$\sigma_{\hat{\mu}} = \sigma_{\bar{x}} = \frac{\sigma}{\sqrt{N}} \, . \tag{5.53}$$

Writing explicitly the normalization factor, we have the complete expression for the posterior:

$$p(\mu \mid x_1, \cdots, x_N) = \frac{1}{\sqrt{2\pi\sigma_{\bar{x}}^2}} \exp\left[-\frac{(\mu - \bar{x})^2}{2\sigma_{\bar{x}}^2}\right]. \tag{5.54}$$

5.8   Bayesian Inference as Learning Process

If we have initially a prior PDF $\pi(\theta) = p_0(\theta)$ of an unknown parameter θ, Bayes' theorem can be applied after an observation x_1 to obtain the posterior probability:

$$p_1(\theta) \propto p_0(\theta)\, L(x_1 \mid \theta), \tag{5.55}$$

where the normalization factor $1/\int p_0(\theta')\, L(x_1 \mid \theta')\, d\theta'$ is the proportionality constant that is omitted. After a second observation x_2, independent on x_1, the combined likelihood function, corresponding to the two observations x_1 and x_2, is given by the product of the individual likelihood functions:

$$L(x_1, x_2 \mid \theta) = L(x_1 \mid \theta)\, L(x_2 \mid \theta). \tag{5.56}$$

Bayes' theorem can be applied again, giving:

$$p_2(\theta) \propto p_0(\theta)\, L(x_1, x_2 \mid \theta) = p_0(\theta)\, L(x_1 \mid \theta)\, L(x_2 \mid \theta), \tag{5.57}$$

where again a normalization factor is omitted.

Equation (5.57) can be interpreted as the application of Bayes' theorem to the observation x_2 having as prior probability $p_1(\theta)$, which is the posterior probability after the observation x_1 (Eq. (5.55)). Considering a third independent observation x_3, Bayes' theorem gives again:

$$p_3(\theta) \propto p_2(\theta)\, L(x_3 \mid \theta) = p_0(\theta)\, L(x_1 \mid \theta)\, L(x_2 \mid \theta)\, L(x_3 \mid \theta). \tag{5.58}$$

By adding more observations, Bayes' theorem can be applied repeatedly:

$$p_N(\theta) \propto p_{N-1}(\theta)\, L(x_N \mid \theta) = p_0(\theta)\, L(x_1 \mid \theta) \cdots L(x_N \mid \theta) =$$
$$= \pi(\theta)\, L(x_1 \cdots, x_N \mid \theta). \tag{5.59}$$

This allows to interpret Bayesian inference over repeated observations as *learning process*, where one's knowledge about an unknown parameter θ is influenced and improved by the subsequent observations x_1, x_2, x_3, and so on.

The more observations, $x_1, \cdots, x_N$, are added, the more the final posterior probability $p_N(\theta)$ is insensitive to the choice of the prior probability $\pi(\theta) = p_0(\theta)$, because the θ range in which $L(x_1, \cdots, x_N \mid \theta)$ is significantly different from zero gets smaller and smaller by adding more observations. Within a very small range of θ, a reasonably smooth prior $\pi(\theta)$ can be approximated by a constant value that cancels in the normalization of the posterior. In this sense, a sufficiently large number of observations may remove, asymptotically, any dependence on subjective choices of prior probability, if the prior is a sufficiently smooth and regular function. This was not the case with the extreme assumption considered in Exercise 5.3.

Figure 5.5 shows the result of the repeated application of Bayesian inference to the same sample of $N = 18$ randomly generated Bernoulli processes to estimate the Bernoulli parameter p. Each posterior $p_i(p)$ is determined using only the first i random extractions. The last posterior, $p_N(p)$, considers all $N = 18$ random extractions. Two different priors $\pi(p) = p_0(p)$ are used: a uniform distribution and an exponential distribution, $\pi(p) \propto e^{-4p}$. As the number of observations increases, the posteriors determined from the two different priors tend to become more and more similar. The posteriors $p_i(p)$ for a uniform prior are equivalent to the binomial inference described in Exercise 5.4 and are beta distributions.

5.9   Parameters of Interest and Nuisance Parameters

In many cases, our probability model depends on a number of parameters, but we are often interested, for our measurement, only in a subset of them, say $\vec{\theta} = (\theta_1, \cdots, \theta_h)$. The remaining parameters are required to provide an accurate description of the PDF but are not intended to be reported as part of our measurements. In this case, the parameters $\vec{\theta}$ are called *parameters of interest* or POI, while the remaining parameters, $\vec{\nu} = (\nu_1, \cdots, \nu_k)$, are called *nuisance parameters*, as already introduced in Sect. 1.11.

The posterior PDF for the sets of all parameters, including both POIs and nuisance parameters, can be written using Bayes' formula:

$$p(\vec{\theta}, \vec{\nu} \mid \vec{x}) = \frac{L(\vec{x}; \vec{\theta}, \vec{\nu}) \pi(\vec{\theta}, \vec{\nu})}{\int L(\vec{x}; \vec{\theta}', \vec{\nu}') \pi(\vec{\theta}', \vec{\nu}') \, \mathrm{d}^h \theta' \, \mathrm{d}^k \nu'} . \tag{5.60}$$

The posterior marginal PDF for the parameters $\vec{\theta}$ can be obtained by integrating Eq. (5.60) over all the remaining parameters $\vec{\nu}$:

$$p(\vec{\theta} \mid \vec{x}) = \int p(\vec{\theta}, \vec{\nu} \mid \vec{x}) \, \mathrm{d}^k \nu = \frac{\int L(\vec{x}; \vec{\theta}, \vec{\nu}) \pi(\vec{\theta}, \vec{\nu}) \, \mathrm{d}^k \nu}{\int L(\vec{x}; \vec{\theta}', \vec{\nu}) \pi(\vec{\theta}', \vec{\nu}) \, \mathrm{d}^h \theta' \mathrm{d}^k \nu} . \tag{5.61}$$

Using Eq. (5.61), nuisance parameters can be removed from the posterior PDF, under the Bayesian approach, with a simple integration. In jargon, we say that the nuisance parameters are *integrated away*. Section 12.20 discusses more in details the treatment of nuisance parameters, including the frequentist approach.

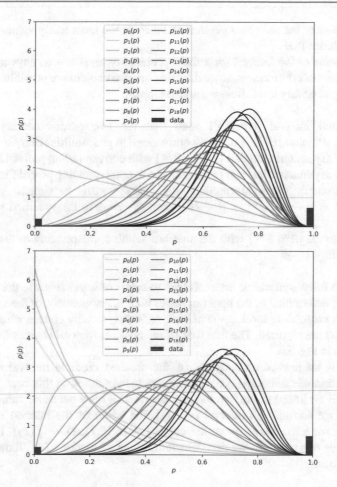

Fig. 5.5 Repeated application of Bayes' theorem to the same sample of randomly generated Bernoulli processes. Two different priors $\pi(p) = p_0(p)$ are considered: uniform (top) and exponential, $\pi(p) = \propto e^{-4p}$ (bottom). As i increases, the posteriors $p_i(p)$ become less sensitive on the assumed prior $\pi(p)$. The two red bars show on each plot the fractions of the 18 extracted values that are equal to 0 and 1. 13 out of 18 outcomes are equal to 1 in this example

5.10   Credible Intervals

Once we have computed the posterior PDF for an unknown parameter of interest θ, intervals $[\theta^{\,\mathrm{lo}}, \theta^{\,\mathrm{up}}]$ can be determined such that the integral of the posterior PDF from $\theta^{\,\mathrm{lo}}$ to $\theta^{\,\mathrm{up}}$ corresponds to a given probability value, usually indicated with $1 - \alpha$. The most frequent choice for $1 - \alpha$ is 68.27%, corresponding to a $\pm 1\sigma$ interval for a normal distribution. Probability intervals determined with the Bayesian approach from the posterior PDF are called *credible intervals* and reflect the *uncertainty* in the *measurement* of the unknown parameter, taken as the most likely value, according to the posterior PDF.

The choice of the interval for a fixed probability level $1 - \alpha$, anyway, has still some degrees of arbitrariness, since different interval choices are possible, all having the same probability level. Below some examples:

- A central interval $[\theta^{\text{lo}}, \theta^{\text{up}}]$ such that the two complementary intervals $] - \infty, \theta^{\text{lo}}[$ and $]\theta^{\text{up}}, +\infty[$ both correspond to probabilities of $\alpha/2$[2]
- A fully asymmetric interval $] - \infty, \theta^{\text{up}}]$ with corresponding probability $1 - \alpha$
- A fully asymmetric interval $]\theta_{\text{lo}}, +\infty]$ with corresponding probability $1 - \alpha$
- A symmetric interval around the value with maximum probability $\hat{\theta}$: $[\theta^{\text{lo}} = \hat{\theta} - \delta\hat{\theta}, \theta^{\text{up}} = \hat{\theta} + \delta\hat{\theta}]$, corresponding to the specified probability $1 - \alpha$[3]
- The interval $[\theta^{\text{lo}}, \theta^{\text{up}}]$ with the smallest width corresponding to the specified probability $1 - \alpha$

Cases with fully asymmetric intervals lead to *upper* or *lower limits* to the parameter of interest, determined as the upper or lower bound, respectively, of the asymmetric interval. A probability level $1 - \alpha$ of 90% or 95% is usually chosen when upper or lower limits are reported. The first four of the possible interval choices listed above are shown in Fig. 5.6.

There is an interesting property of the shortest credible interval of a PDF, assuming that it is unimodal, i.e., it has a single maximum. In this case, the values of the PDF evaluated at the two interval boundaries have the same value. In other words, if we consider an interval $[x_1, x_2]$, the width of the interval $x_2 - x_1$ is minimum, for a fixed area under the curve $1 - \alpha$, if $f(x_1) = f(x_2)$. In order to demonstrate this property, we first determine x_2 as a function of x_1, fixing the area α, by imposing:

$$\int_{x_1}^{x_2} f(x) \, dx = F(x_2) - F(x_1) = 1 - \alpha \,. \tag{5.62}$$

We have defined in this way the function:

$$x_2 : x_1 \longmapsto x_2(x_1) : \int_{x_1}^{x_2(x_1)} f(x) \, dx = 1 - \alpha \,. \tag{5.63}$$

---

[2] In some cases such credible intervals may not contain the point estimate. For instance, the red curve in Fig. 5.4 has maximum value corresponding to the upper extreme of the range that lies outside the credible interval.

[3] It is not always possible to find such interval. For instance, the red curve in Fig. 5.4 has maximum value corresponding to the upper extreme of the range, so any credible interval containing this value is fully asymmetric.

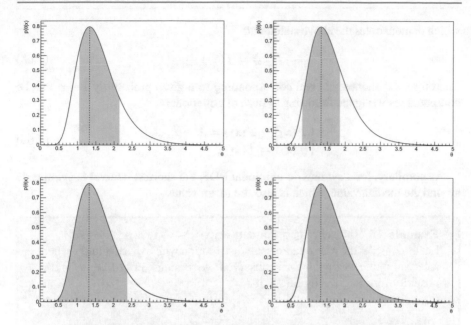

Fig. 5.6 Different interval choices, shown as shaded areas, corresponding to a 68.27% (top) or 90% (bottom) probability. Top left: central interval, the left and right tails have equal probability. Top right: symmetric interval; the most probable value lies at the center of the interval. Bottom left: fully asymmetric intervals for an upper limit. Bottom right: fully asymmetric interval for a lower limit

If the width of the interval $x_2(x_1) - x_1$ is minimum, the derivative of this width with respect to the independent variable x_1 must be zero:

$$\frac{d(x_2(x_1) - x_1)}{dx_1} = \frac{dx_2(x_1)}{dx_1} - 1 = 0 \, . \tag{5.64}$$

Therefore:

$$\frac{dx_2(x_1)}{dx_1} = 1 \, . \tag{5.65}$$

On the other hand, we have:

$$\frac{d}{dx_1} \int_{x_1}^{x_2(x_1)} f(x) \, dx = \frac{d(1 - \alpha)}{dx_1} = 0 \, , \tag{5.66}$$

hence:

$$0 = \frac{dF(x_2(x_1))}{dx_1} - \frac{dF(x_1)}{dx_1} = \frac{dF(x_2)}{dx_2} \frac{dx_2(x_1)}{dx_1} - f(x_1) = f(x_2) \cdot 1 - f(x_1) \, , \tag{5.67}$$

which demonstrates the above statement:

$$f(x_2) = f(x_1) .$$ (5.68)

Therefore, the shortest interval corresponding to a given probability $1 - \alpha$ can be computed resolving the following system of equations:

$$\begin{cases} F(x_2) - F(x_1) = 1 - \alpha , \\ f(x_2) = f(x_1) . \end{cases}$$ (5.69)

As corollary, for a symmetric unimodal PDF, the shortest interval is symmetric around the median value, which is also the mean value.

Example 5.5 - Inference of a Poisson Rate

Let us consider the case of a Poisson distribution with expected rate s where a certain number of counts n is observed. Assuming a prior PDF $\pi(s)$, the posterior for s is given by Eq. (5.30):

$$p(s \mid n) = \frac{\dfrac{s^n e^{-s}}{n!} \pi(s)}{\displaystyle\int_0^\infty \frac{s'^n e^{-s'}}{n!} \pi(s') \, ds'} .$$ (5.70)

If $\pi(s)$ is a constant, the normalization factor at the denominator becomes:

$$\frac{1}{n!} \int_0^\infty s^n e^{-s} \, ds = \left[-\frac{\Gamma(n+1, s)}{n!} \right]_{s=0}^{s=\infty} = 1 .$$ (5.71)

The posterior PDF is, therefore:

$$p(s \mid n) = \frac{s^n e^{-s}}{n!} .$$ (5.72)

Equation (5.72) has the same expression of the original Poisson distribution, but this time it is interpreted as posterior PDF of the unknown parameter s, given the observation n. As a function of s, the posterior is a gamma distribution (Sect. 3.12). Figure 5.7 shows the distributions of $p(s \mid n)$ for the cases $n = 5$ and $n = 0$ according to Eq. (5.72). For $n = 5$ the plot shows a 68.27% central probability; for $n = 0$ a 90% fully asymmetric interval is shown.

The most probable value of s can be determined as the maximum of the posterior in Eq. (5.72):

$$\hat{s} = n .$$ (5.73)

(continued)

Example 5.5 (continued)
The average value and variance of s can also be determined from the same posterior:

$$\langle s \rangle = n + 1 , \tag{5.74}$$

$$\text{Var}[s] = n + 1 . \tag{5.75}$$

Note that the most probable value $\hat{s}$ is different from the average value $\langle s \rangle$, since the distribution is not symmetric.

Those results depend of course on the choice of the prior $\pi(s)$. A constant prior is not necessarily the most natural choice. Jeffreys' prior, for instance, discussed in Sect. 5.13, is $\pi(s) \propto 1/\sqrt{s}$, instead of a constant.

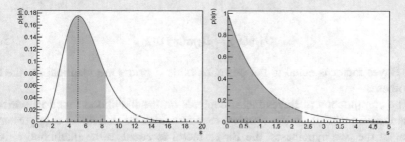

Fig. 5.7 Poisson posterior PDFs for $n = 5$ with a central 68.27% probability interval (left) and for $n = 0$ with a fully asymmetric 90% probability interval (right). Credible intervals are shown as shaded areas

5.11 Bayes Factors

As seen at the end of Exercise 5.1, there is a convenient way to compare the probability of two hypotheses using Bayes' theorem which does not require to consider all possible hypotheses. Using Bayes' theorem, one can write the ratio of posterior probabilities evaluated under two hypotheses H_0 and H_1, given our observation $\vec{x}$, called *posterior odds* of the hypothesis H_1 versus the hypothesis H_0:

$$O_{1/0}(\vec{x}) = \frac{P(H_1 \mid \vec{x})}{P(H_0 \mid \vec{x})} = \frac{p(\vec{x} \mid H_1)}{p(\vec{x} \mid H_0)} \times \frac{\pi(H_1)}{\pi(H_0)} , \tag{5.76}$$

where $\pi(H_1)$ and $\pi(H_0)$ are the priors for the two hypotheses. The ratio of the priors,

$$o_{1/0} = \pi(H_1)/\pi(H_0) \,, \tag{5.77}$$

is called *prior odds*. The ratio that multiplies the prior odds is:

$$B_{1/0}(\vec{x}) = \frac{p(\vec{x} \mid H_1)}{p(\vec{x} \mid H_0)} \tag{5.78}$$

and is called *Bayes factor* [4]. With this nomenclature, Eq. (5.76) reads as:

$$\text{posterior odds} = \text{Bayes factor} \times \text{prior odds}$$

or:

$$O_{1/0}(\vec{x}) = B_{1/0}(\vec{x}) \, o_{1/0} \,. \tag{5.79}$$

The Bayes factor is equal to the posterior odds if priors are identical for the two hypotheses.

The computation of Bayes factor depends on the likelihood function, which is present in Eq. (5.30). In the simplest case in which no parameter $\vec{\theta}$ is present in either of the two hypotheses, the Bayes factor is equal to the likelihood ratio of the two hypotheses. If parameters are present, the probability densities $p(\vec{x} \mid H_{0,\,1})$ should be computed by integrating the product of the likelihood function and the prior over the parameter space:

$$p(\vec{x} \mid H_0) = \int L(\vec{x} \mid H_0, \vec{\theta}_0) \, \pi_0(\vec{\theta}_0) \, d\theta_0 \,, \tag{5.80}$$

$$p(\vec{x} \mid H_1) = \int L(\vec{x} \mid H_1, \vec{\theta}_1) \, \pi_1(\vec{\theta}_1) \, d\theta_1 \,, \tag{5.81}$$

and the Bayes factor can be written as:

$$B_{1/0}(\vec{x}) = \frac{\int L(\vec{x} \mid H_1, \vec{\theta}_1) \, \pi_1(\vec{\theta}_1) \, d\vec{\theta}_1}{\int L(\vec{x} \mid H_0, \vec{\theta}_0) \, \pi_0(\vec{\theta}_0) \, d\vec{\theta}_0} \,. \tag{5.82}$$

Bayes factors have been proposed to assess the *evidence* of H_1 against H_0. An example is to assess the presence of a signal due to a new particle (H_1) against a null hypothesis (H_0) that no new signal is present in our data sample. Threshold values for Bayes factor have been proposed in [4] and are reported in Table 5.1.

Within the Bayesian approach to probability, Bayes factors are introduced as alternative to hypothesis testing, adopted under the frequentist approach and introduced in Chap. 10. In particular, Bayes factors have been proposed [4] as an alternative to p-values and significance levels (see Sects. 12.2 and 12.3).

Table 5.1 Threshold values for Bayes factors proposed to assess the evidence of an alternative hypothesis H_1 against a null hypothesis H_0 according to [4]

$B_{1/0}$	Evidence against H_0
1–3	Not worth more than a bare mention
3–20	Positive
20–150	Strong
>150	Very strong

5.12   Subjectiveness and Prior Choice

One main feature of Bayesian probability is its unavoidable and intrinsic dependence on a prior probability that could be chosen by different subjects in different ways. This feature makes Bayesian probability *subjective*, in the sense that it depends on one's choice of the prior probability. Example 5.3 demonstrated that, in extreme cases, drastic choices of prior PDFs may lead to insensitiveness of the posterior to the actual observations. It is also true, as remarked in Sect. 5.8, that, for reasonable choices of the prior, adding more and more observations increases one's knowledge about the unknown parameters, and the posterior probability becomes less and less sensitive to the choice of the prior. For this reason, when many observations are available, the application of Bayesian and frequentist calculations tend to give consistent results, despite the very different interpretations of their results. But it is also true that many interesting statistical problems arise in cases with a small number of observations.

The goal of an experiment is to extract the maximum possible information from the limited available data sample, which is in general precious because it is usually the outcome of a complex and labor-intensive experimental setup. In those cases, applying Bayesian or frequentist methods usually gives numerically different results, which should also be interpreted in very different ways. In those cases, when using the Bayesian approach, the choice of prior may play a crucial role and it may have relevant influence on the results.

One of the main difficulties arises when choosing a probability distribution to models one's complete ignorance about an unknown parameter θ, i.e., an uninformative prior. A frequently adopted prior distribution in physics is a uniform PDF in the interval of validity of θ. Imagine that we have to change parameterization, from the original parameter θ to a function of θ. The resulting transformed parameter no longer has a uniform prior. This is particularly evident in case of the measurement of a particle's lifetime τ: should one chose a prior uniform in τ, or in the particle's width, $\Gamma = 1/\tau$? There is no preferred choice provided by first principles.

This *subjectiveness* in the choice of the prior, intrinsic to the Bayesian approach, raises criticism by supporters of the frequentist approach, who object that results obtained under the Bayesian approach are to some extent arbitrary, while scientific results should not depend on any subjective assumptions. Supporters of the Bayesian approach reply that Bayesian result is not arbitrary, but *intersubjective* [5], in the sense that commonly agreed prior choices lead to common results, and a dependence on prior knowledge is unavoidable and intrinsic in the process of scientific progress.

The debate is in some cases still open, and literature still contains opposite positions about this issue.

5.13   Jeffreys' Prior

Harold Jeffreys [6] proposed an uninformative prior which is invariant under parameter transformation. Jeffreys' choice is, up to a normalization factor, given by:

$$p(\vec{\theta}) \propto \sqrt{\mathcal{J}(\vec{\theta})}, \tag{5.83}$$

where $\mathcal{J}(\vec{\theta})$ is the determinant of the *Fisher information* matrix defined below:

$$\mathcal{J}(\vec{\theta}) = \det \left[\left\langle \frac{\partial \log L(\vec{x} \mid \vec{\theta})}{\partial \theta_i} \frac{\partial \log L(\vec{x} \mid \vec{\theta})}{\partial \theta_j} \right\rangle \right]. \tag{5.84}$$

It is not difficult to demonstrate that Jeffreys' prior is invariant when changing parameterization, i.e., when transforming $\vec{\theta} \longmapsto \vec{\theta}' = \vec{\theta}'(\vec{\theta})$. The Jacobian determinant that appears in the PDF transformation, using Eq. (3.32), is:

$$p(\vec{\theta}') = \left| \det \left(\frac{\partial \theta_i}{\partial \theta_j'} \right) \right| p(\vec{\theta}) \tag{5.85}$$

and gets absorbed in the determinant that appears in Fisher information (Eq. (5.84)) once it is expressed in the transformed coordinates.

Jeffreys' priors corresponding to the parameters of some of the most frequently used PDFs are given in Table 5.2. Note that only the mean of a Gaussian corresponds to a uniform Jeffreys' prior. For instance, for a Poissonian counting experiment, like the one considered in Exercise 5.5, Jeffreys' prior is proportional to $1/\sqrt{s}$, not a uniform PDF, as it was assumed to determine Eq. (5.72).

Table 5.2 Jeffreys' priors corresponding to the parameters of some of the most frequently used PDFs

PDF parameter	Jeffreys' prior
Poissonian mean s	$p(s) \propto 1/\sqrt{s}$
Poissonian signal mean s with a background b	$p(s) \propto 1/\sqrt{s+b}$
Gaussian mean μ	$p(\mu) \propto 1$
Gaussian standard deviation σ	$p(\sigma) \propto 1/\sigma$
Binomial success probability p	$p(\varepsilon) \propto 1/\sqrt{p(1-p)}$
Exponential parameter λ	$p(\lambda) \propto 1/\lambda$

5.14   Reference Priors

Another approach, known as *reference analysis*, constructs priors that are invariant under reparameterization based on a procedure that minimizes the *informativeness* according to a proper mathematical definition. Such *reference priors* in some cases coincide with Jeffreys' priors. This method, which is beyond the purpose of the present book, has been described in [7] with examples of application to particle physics.

5.15   Improper Priors

In many cases, the priors we have encountered are not normalizable, since they have diverging integral over the entire parameter domain. This is the case with Jeffreys' priors in Table 5.2, but also with a uniform prior, in case of an unlimited parameter range. Such priors are called *improper prior distributions*. The integrals required in the evaluation of Bayesian posterior, which involve the product of the likelihood function and the prior, are anyway finite. An improper prior distribution can be regularized by setting the PDF to zero for values beyond an arbitrary large cutoff. The range beyond the cutoff would anyway be irrelevant if the likelihood function assumes negligible values at the extreme side, or sides, of the parameter range.

Example 5.6 - Posterior with an Exponential Distribution
A particle's mean lifetime τ can be determined from the measurement of a number N of decay times $t_1, \cdots, t_N$, which are expected to follow an exponential distribution:

$$f(t) = \frac{1}{\tau} e^{-t/\tau} = \lambda e^{-\lambda t}. \tag{5.86}$$

The two parameterizations given in Eq. (5.86) are equivalent if $\tau = 1/\lambda$. The likelihood function is given by the product of $f(t_1), \cdots, f(t_N)$. In the two parameterizations we can write it as:

$$L(\vec{t}\,;\,\lambda) = \prod_{i=1}^{N} \lambda e^{-\lambda t_i} = \lambda^N e^{-\lambda \sum_{i=1}^{N} t_i} = \frac{e^{-\sum_{i=1}^{N} t_i/\tau}}{\tau^N}. \tag{5.87}$$

The posterior distribution for the parameter τ, assuming a prior $\pi(\tau)$, is given by:

$$p(\tau|\vec{t}\,) = \frac{\pi(\tau) e^{-\sum_{i=1}^{N} t_i/\tau}/\tau^N}{\int \pi(\tau') e^{-\sum_{i=1}^{N} t_i/\tau'}/\tau'^N \, d\tau'}. \tag{5.88}$$

(continued)

Example 5.6 (continued)

The posterior as a function of the parameter λ is a gamma distribution (Sect. 3.12). A possible uninformative prior to model one's complete ignorance about τ is to assume a uniform distribution, i.e., $\pi(\tau) = $ const. But this is not the only possible choice. Another choice could be a uniform prior for $\lambda = 1/\tau$, i.e., $\pi(\lambda) = $ const. Using Eq. (3.33), we can determine $\pi(\tau)$ if a uniform prior on λ is assumed:

$$\pi(\tau) = \left| \frac{d\lambda}{d\tau} \right| \pi(\lambda) \propto \frac{1}{\tau^2} . \tag{5.89}$$

Alternatively, Jeffreys' prior could be used. From the likelihood function in Eq. (5.87), the Fisher information matrix, defined in Eq. (5.84), has a single element, that we can write as:

$$\mathcal{J}(\tau) = \left\langle \left(\frac{d \log L(\vec{t} \; ; \; \tau)}{d\tau} \right)^2 \right\rangle = \left\langle \left(\frac{d \left(-N \log \tau - \sum_{i=1}^{n} t_i/\tau \right)}{d\tau} \right)^2 \right\rangle =$$

$$= \left\langle \left(-\frac{N}{\tau} + \frac{\sum_{i=1}^{N} t_i}{\tau^2} \right)^2 \right\rangle = \left\langle \frac{N^2}{\tau^2} - 2 \frac{N \sum_{i=1}^{N} t_i}{\tau^3} + \frac{\left(\sum_{i=1}^{N} t_i \right)^2}{\tau^4} + \right\rangle =$$

$$= \frac{N^2}{\tau^2} - 2 \frac{N^2 \langle t \rangle}{\tau^3} + \frac{N^2 \langle t^2 \rangle}{\tau^4} . \tag{5.90}$$

For an exponential distribution, $\langle t \rangle = \tau$ and $\langle t^2 \rangle = 2\tau^2$, hence:

$$\mathcal{J}(\tau) \propto \frac{1}{\tau^2} , \tag{5.91}$$

and Jeffreys' prior is:

$$\pi(\tau) \propto \sqrt{\mathcal{J}(\tau)} \propto \frac{1}{\tau} . \tag{5.92}$$

Figure 5.8 shows the posterior distributions $p(\tau \mid \vec{t})$ for randomly extracted data sets with $\tau = 1$ using a uniform prior on τ, a uniform prior on $\lambda = 1/\tau$ and Jeffreys' prior, for $N = 5$, 10 and 50. The posteriors, evaluated assuming the three different priors, are more and more similar as the number of measurements N increases. The reduced sensitivity on the assumed prior as the number of measurements increases is a general property of Bayesian inference. The treatment of the same case with the frequentist approach is discussed in Exercise 6.4.

(continued)

Example 5.6 (continued)

Fig. 5.8 Posterior
distribution for τ using a
uniform prior (dashed line), a
uniform prior on $\lambda = 1/\tau$
(dotted line) and Jeffreys'
prior (solid line). Data values
$t_1, \cdots, t_N$ are shown as blue
histogram. Numbers of
measurements $N = 5$ (top),
10 (middle) and 50 (bottom)
have been considered.

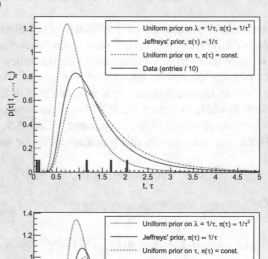

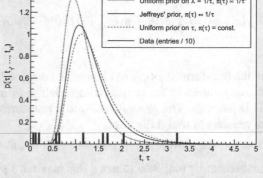

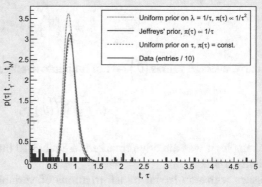

5.16   Transformations of Variables and Error Propagation

Measurement errors, related to credible intervals discussed in Sect. 5.10, need to be propagated when the original measured parameters $\vec{\theta}$ are transformed into a different set of parameters $\vec{\eta}$, and uncertainties on those new parameters must be quoted. Error propagation can be introduced in a natural way within Bayesian inference. The result of Bayesian inference is a posterior PDF for the unknown parameters of interest. In order to obtain the PDF for transformed parameters, it is sufficient to transform the posterior PDF as discussed in Sect. 3.5.

In the case of a two-variable transformation, for instance:

$$(\theta_1, \theta_1) \longmapsto (\eta_1, \eta_2) = (\eta_1(\theta_1, \theta_2), \ \eta_2(\theta_1, \theta_2)) \ , \tag{5.93}$$

a PDF $p(\theta_1, \theta_2)$ transforms according to:

$$p(\eta_1, \eta_2) = \int \delta(\theta_1 - \eta_1(\theta_1, \theta_2)) \, \delta(\theta_2 - \eta_2(\theta_1, \theta_2)) \, p(\theta_1, \theta_2) \, d\theta_1 \, d\theta_2 \ . \tag{5.94}$$

Once the transformed $p(\eta_1, \eta_2)$ has been computed, one can determine again the most likely values of the transformed variables η_1 and η_2, and the corresponding credible intervals. The generalization to a transformation $\vec{\theta} \longmapsto \vec{\eta}$ that involves more variables is straightforward.

Note that the most probable parameters $\hat{\vec{\theta}}$, i.e., the values that maximize $p\left(\vec{\theta}\right)$, do not necessarily map into values $\hat{\vec{\eta}}$ that maximize $p(\vec{\eta})$. That is,

$$\hat{\vec{\eta}} \neq \vec{\eta}\left(\hat{\vec{\theta}}\right) \ . \tag{5.95}$$

Similarly, average values $\left\langle \vec{\theta} \right\rangle$ do not map into average values $\langle \vec{\eta} \rangle$:

$$\langle \vec{\eta} \rangle \neq \vec{\eta}\left(\left\langle \vec{\theta} \right\rangle\right) \ . \tag{5.96}$$

As example, it was already remarked in Sect. 3.10 that $\langle e^y \rangle \neq e^{\langle y \rangle}$ if y is a normal random variable.

Issues with non-trivial transformations of variables and error propagation are also present in the frequentist approach. Sect. 6.23 briefly discusses the case of propagation of asymmetric uncertainties. Section 6.21 discusses how to propagate errors in the case of transformation of variables using a linear approximation. The results hold for Bayesian as well as for frequentist inference. Under the simplified assumption of a linear transformation, which is a sufficient approximation only in the presence of small uncertainties, the most probable parameters $\hat{\vec{\theta}}$ map into values $\hat{\vec{\eta}}$:

$$\hat{\vec{\eta}} = \vec{\eta}\left(\hat{\vec{\theta}}\right) \ . \tag{5.97}$$

Similarly, the average values $\left\langle \vec{\theta} \right\rangle$ map into the average values $\langle \vec{\eta} \rangle$.

References

1. G. Cowan, *Statistical Data Analysis* (Clarendon Press, Oxford, 1998)
2. G. D'Agostini, *Telling the Truth with Statistics* (CERN Academic Training, Meyrin, 2005)
3. W. Eadie, D. Drijard, F. James, M. Roos, B. Saudolet, *Statistical Methods in Experimental Physics* (North Holland, Amsterdam, 1971)
4. R. Kass, E. Raftery, Bayes factors. J. Am. Stat. Assoc. **90**, 773 (1995)
5. G. D'Agostini, *Bayesian Reasoning in Data Analysis: A Critical Introduction* (World Scientific, Hackensack, 2003)
6. H. Jeffreys, An invariant form for the prior probability in estimation problems. Proc. R. Soc. Lond. A Math. Phys. Sci. **186**, 453–461 (1946)
7. L. Demortier, S. Jain, H.B. Prosper, Reference priors for high energy physics. Phys. Rev. **D82**, 034002 (2010)

6.1   Frequentist Definition of Probability

The *frequentist probability* $P(E)$ of an event E is defined by the following limit:

$$P(E) = p \quad \text{if} \quad \forall \varepsilon \quad \lim_{N \to \infty} P\left(\left| \frac{N(E)}{N} - p \right| < \varepsilon \right) = 1 . \tag{6.1}$$

The limit is intended, in this case, as *convergence in probability*, and is justified by the law of large numbers (see Sect. 2.12). The limit only rigorously holds in the non-realizable case of an infinite number of experiments. F. James et al. report the following sentence:

> [$\cdots$] this definition is not very appealing to a mathematician, since it is based on experimentation, and, in fact, implies unrealizable experiments ($N \to \infty$) [3].

Rigorously speaking, the definition of frequentist probability in Eq. (6.1) is defined itself in terms of another probability, which introduces conceptual problems.

In real life, experiments are only reproducible for a finite range of time.[1] For the practical purposes of applications in physics, the frequentist definition of probability can be considered, beyond a possible exact mathematical meaning, as pragmatic

---

The original version of the chapter has been revised. Minor changes were made to equations 6.94, 6.99, 6.102, and 6.103. A correction to this chapter can be found at https://doi.org/10.1007/978-3-031-19934-9_13.

---

[1] On the planet Earth, for instance, experiments will be carried out until an intelligent life will exist. This will be very unlikely after the Sun will stop shining, or even before, if humans will succumb to their impact on the planet.

L. Lista, *Statistical Methods for Data Analysis*, Lecture Notes in Physics 1010, https://doi.org/10.1007/978-3-031-19934-9_6

definitions. It describes to a good level of approximation the concrete situations of a very large number of the cases we are interested in experimental physics.

Some Bayesian statisticians express very strong concerns about frequentist probability (see, for instance, [14]). This book does not enter such kind of debate. Nonetheless, the limitations of both frequentist and Bayesian approaches are remarked, whenever it is the case.

6.2   Estimators

An inference is a mathematical procedure to determine a *point estimate* (see Sect. 1.12) of an unknown parameter as a function of the observed data sample. In general, the function of the data sample that returns the point estimate of a parameter is called *estimator*. Estimators can be defined in practice as more or less complex mathematical procedures or numerical algorithms. Several statistical properties, introduced in Sect. 6.3, allow to compare the quality of adopted estimators.

Example 6.1 - A Very Simple Estimator of a Gaussian Mean

As a first and extremely simplified example, let us assume a Gaussian distribution whose standard deviation σ is known, e.g., the resolution of our apparatus, and whose average μ is the unknown parameter of interest. Consider the simplest possible data sample consisting of a single measurement $x = x^\star$ distributed according to the Gaussian distribution under consideration.

As estimator of μ, we take the function $\hat{\mu}$ that returns the single measured value $x^\star$:

$$\hat{\mu}(x^\star) = x^\star . \tag{6.2}$$

If the experiment is repeated many times, ideally an infinite number of times, different values of $\hat{\mu} = x^\star$ are obtained for every experiment, all distributed according to the original Gaussian. In 68.27% of the experiments, in the limit of an infinite number of experiments, the fixed and unknown true value μ lies in the *confidence interval* $[\hat{\mu} - \sigma, \ \hat{\mu} + \sigma]$, i.e., $\mu - \sigma < \hat{\mu} < \mu + \sigma$. In the remaining, 31.73% of the cases μ lies outside the same interval. This property expresses the *coverage* of the interval $[\hat{\mu} - \sigma, \ \hat{\mu} + \sigma]$ at the 68.27% *confidence level*.

The estimate

$$\mu = \hat{\mu} \pm \sigma \tag{6.3}$$

can be quoted in this sense. $\pm \sigma$ is the *error* or *uncertainty* assigned to the measurement $\hat{\mu}$, with the frequentist meaning defined in Sect. 1.14.

(continued)

Example 6.1 (continued)

In realistic cases, experimental data samples contain more information than a single measurement and more complex PDF models than a simple Gaussian are required. The definition of an estimator may require in general non-trivial mathematics and, in many cases, computer algorithms.

6.3 Properties of Estimators

Different estimators may have different statistical properties that make one or another estimator more suitable for a specific problem. In the following, some of the main properties of estimators are presented. Section 6.5 below introduces maximum likelihood estimators which have good properties in terms of most of the indicators described in the following. The quantitative study of the properties described in the following can be done analytically for the simplest cases but may require computer simulations in more complex ones. In general, it is not always possible to optimize simultaneously all the properties described in the following, and some trade-off may be necessary. The main properties of estimators are listed below:

- **Consistency**
  An estimator is said to be *consistent* if it converges, in probability, for number of measurements N that tends to infinity, to the true unknown parameter value:

$$\forall \varepsilon > 0 \quad \lim_{N \to \infty} P\left(\left|\hat{\theta}_n - \theta\right| < \varepsilon\right) = 1. \tag{6.4}$$

A good estimator must be consistent.
- **Bias**
  The *bias* of an estimator is the expected value of the deviation of the parameter estimate from the corresponding true value of that parameter:

$$\mathrm{Bias}\left[\hat{\theta}\right] = \left\langle\hat{\theta} - \theta\right\rangle = \left\langle\hat{\theta}\right\rangle - \theta. \tag{6.5}$$

A good estimator should have no bias, or at least the bias should be small.
- **Efficiency and minimum variance bound**
  The variance $\mathrm{Var}[\hat{\theta}]$ of any consistent estimator is subject to a lower bound due to Cramér [1] and Rao [2], which is given by

$$\mathrm{Var}[\hat{\theta}] \geq \mathrm{Var}_{\mathrm{CR}}(\hat{\theta}) = \frac{\left(1 + \dfrac{\partial\,\mathrm{Bias}\left[\hat{\theta}\right]}{\partial\theta}\right)^2}{\mathbb{E}\left[\left(\dfrac{\partial \log L(x_1, \cdots, x_n;\,\theta)}{\partial\theta}\right)^2\right]}, \tag{6.6}$$

where $\mathbb{Bias}(\hat{\theta})$ is the bias of the estimator (Eq. (6.5)) and the denominator is the *Fisher information*, already defined in Sect. 5.13. Under some regularity conditions, the Fisher information can also be written as

$$\mathbb{E}\left[\left(\frac{\partial \log L(x_1, \cdots, x_n; \theta)}{\partial \theta}\right)^2\right] = -\mathbb{E}\left[\frac{\partial^2 \log L(x_1, \cdots, x_n; \theta)}{\partial \theta^2}\right].$$

(6.7)

The ratio of the Cramér–Rao minimum variance bound to the estimator's variance is called estimator's *efficiency*:

$$\varepsilon(\hat{\theta}) = \frac{\mathbb{Var}_{CR}(\hat{\theta})}{\mathbb{Var}[\hat{\theta}]}.$$

(6.8)

Any consistent estimator $\hat{\theta}$ has efficiency $\varepsilon(\hat{\theta})$ lower than or equal to one, due to Cramér–Rao bound. The efficiency of a good estimator should be as large as possible, compatible with the Cramér–Rao bound upper bound.

Example 6.2 - Estimators with Variance Below the Cramér–Rao Bound Are Not Consistent

It is possible to find estimators that have variance lower than the Cramér–Rao bound, but this implies that they are not consistent. As example, consider the estimator of an unknown parameter that gives a constant value as estimate of the parameter. For instance, the estimated value could be always be zero or π, regardless of the data sample. This estimator has zero variance, but it is of course not consistent. An estimator of this kind is clearly not very useful in practice.

The *mean squared error* (MSE) of an estimator $\hat{\theta}$ is defined as

$$\mathbb{MSE}\left[\hat{\theta}\right] = \mathbb{E}\left[(\hat{\theta} - \theta)^2\right].$$

(6.9)

A more extensive discussion about the interplay of variance and bias is presented in Sect. 11.6. In particular, the following relation holds:

$$\mathbb{MSE}\left[\hat{\theta}\right] = \mathbb{E}\left[(\hat{\theta} - \theta)^2\right] = \mathbb{Var}\left[\hat{\theta} - \theta\right] + \left(\mathbb{E}\left[\hat{\theta} - \theta\right]\right) =$$
$$= \mathbb{Var}[\theta] + \mathbb{Bias}[\theta]^2.$$

(6.10)

The MSE combines variance and bias and is a measure of an estimator's overall quality.

6.4   Robust Estimators

The good properties of some estimators may be spoiled in case the real distribution of data deviates from the assumed PDF model. Entries in the data sample that introduce visible deviations from the theoretical PDF, such as data in extreme tails of the PDF where few entries are expected, are called *outliers*. If data exhibit deviations from the nominal PDF model, but we do not know an exact model of the real PDF that includes outliers, an important property of an estimator is to have a limited sensitivity to the presence of outliers. This property is in general defined as *robustness* and can be better quantified by some mathematical indicators.

An example of robust estimator of the central value of a symmetric distribution, given a sample $x_1, \cdots, x_N$, is the *median* $\tilde{x}$, defined in Eq. (2.42):

$$\tilde{x} = \begin{cases} x_{N+1/2} & \text{if } N \text{ is odd}, \\ \frac{1}{2}\left(x_{N/2} + x_{N/2+1}\right) & \text{if } N \text{ is even}. \end{cases} \tag{6.11}$$

Clearly, the presence of outliers at the left or right tails of the distribution does not significantly change the value of the median, if it is dominated by measurements in the core of the distribution. Conversely, the usual arithmetic mean (Eq. (2.41)) could be shifted from the true value by an amount that depends on how much the outliers' distribution is broader than the core part of the distribution. An average value computed by removing from the sample a given fraction f of data present in the rightmost and leftmost tails is called *trimmed average* and is also less sensitive to the presence of outliers.

It is convenient to define the *breakdown point* as the maximum fraction of incorrect measurements (i.e., outliers) above which the estimate may grow arbitrarily large in absolute value. A trimmed average that removes a fraction f of the events can be demonstrated to have a breakdown point equal to f, while the median has a breakdown point of 0.5. The mean of a distribution, instead, has a breakdown point of 0. A more detailed treatment of robust estimators is beyond the purpose of this text. Reference [3] contains a more extensive discussion about robust estimators.

6.5   Maximum Likelihood Method

The most frequently adopted estimation method is based on the construction of the joint probability distribution of all measurements in our data sample, called *likelihood function*, which was already introduced in Sect. 5.5. The estimate of the parameters we want to determine is obtained by finding the parameter set that corresponds to the maximum value of the likelihood function. This approach gives the name of *maximum likelihood method* to this estimator. More in general, a procedure that finds the optimal parameter set is also called *best fit*, or more simply *fit*, because it determines the parameters for which the theoretical PDF model best fits the experimental data sample.

Maximum likelihood fits are very frequently used because of very good statistical properties according to the indicators discussed in Sect. 6.3. The estimator discussed in Example 6.1 is a very simple application of the maximum likelihood method. A Gaussian PDF of a variable x with unknown average μ and known standard deviation σ was assumed. The estimate $\hat{\mu}$, equal to the single observed value of x, is indeed the value of μ that maximizes the likelihood function, which is just the Gaussian PDF in this simple case.

6.6   Likelihood Function

The *likelihood function* is the function that, for given values of the unknown parameters, returns the value of the PDF corresponding to the observed data. If the observed values of n random variables are $x_1, \cdots x_n$ and our PDF model depends on m unknown parameters $\theta_1, \cdots, \theta_m$, the likelihood function is the joint PDF of the random variables $x_1, \cdots, x_n$:

$$L(x_1, \cdots, x_n; \theta_1, \cdots, \theta_m) = \frac{dP(x_1, \cdots, x_n; \theta_1, \cdots, \theta_m)}{dx_1 \cdots dx_n}. \tag{6.12}$$

As already anticipated in Sect. 5.5, the notation $L(x_1, \cdots, x_n \mid \theta_1, \cdots, \theta_m)$ is also used, similarly to the notation adopted for conditional probability (see Sect. 1.8).

The *maximum likelihood estimator* of the unknown parameters $\theta_1, \cdots, \theta_m$ is the function that returns the values of the parameters $\hat{\theta}_1, \cdots, \hat{\theta}_m$ for which the likelihood function is maximum. If the maximum is not unique, the maximum likelihood estimate is ambiguous. If we have N repeated observations, each consisting of the n values of the random variables $\vec{x} = (x_1, \cdots, x_n)$, the likelihood function is the probability density corresponding to the entire sample $\{\vec{x}_1, \cdots, \vec{x}_N\}$ for given values of the parameters, $\vec{\theta} = (\theta_1, \cdots, \theta_m)$. If the observations are independent of each other and distributed according to the same PDF, the likelihood function of the sample consisting of the N independent observations[2] recorder by our experiment can be written as the product of the PDFs corresponding to individual observation:

$$L(\vec{x}_1, \cdots, \vec{x}_N; \vec{\theta}) = \prod_{i=1}^{N} p(\vec{x}_i; \vec{\theta}). \tag{6.13}$$

---

[2] In physics, often the word used to indicate an observation is *event*. This term has usually a different meaning with respect to what is intended in statistics. In physics, event refers to a collection of measurements of observable quantities $\vec{x} = (x_1, \cdots, x_n)$ corresponding to a physical phenomenon, like a collision of particles at an accelerator, or the interaction of a single particle from a beam, or a shower of particles from cosmic rays in a detector. Measurements performed for different events are usually uncorrelated, and each sequence of variables taken from N different events, $\vec{x}_1, \cdots, \vec{x}_N$, can be considered a sampling of independent and identically distributed, or IID, random variables, as defined in Sect. 4.2.

It is often convenient to compute the logarithm of the likelihood function, so that the product of terms that appears in the likelihood definition is transformed into the sum of logarithms. The logarithm of the likelihood function in Eq. (6.13) is

$$-\log L(\vec{x}_1, \cdots, \vec{x}_N; \vec{\theta}) = -\sum_{i=1}^{N} \log p(\vec{x}_i; \vec{\theta}). \tag{6.14}$$

6.7 Binned and Unbinned Fits

A fit using a data sample made of repeated independent observation of random variables all having the same distribution, like the one considered above, is also called *unbinned maximum likelihood fit*. Frequently, it may be convenient to subdivide the range of an observed variable into discrete interval, called *bins*. Instead of using all the individual observed values in the fit, the number of entries falling in each of the bins can be used as data sample to be fitted. This technique may allow to significantly reduce the number of variables used in a fit, from the number of observed values, down to the number of bins, improving in this way computing speed. If the binning is sufficiently fine and if the number of entries in each bin is sufficiently large, this reduction of information does not spoil the precision of the fit. Such fits are called *binned maximum likelihood fits*.

In practice, a *histogram* of the experimental distribution is built. This is usually done in one variable, but histograms in two variables are also sometimes used. If the data sample is made of independent extractions from a given random distribution, as usual, the number of entries in each bin follows a Poisson distribution whose expected value can be determined from the theoretical distribution and depends on the unknown parameters we want to estimate.

In order to write the likelihood function for a binned fit as the product of probabilities corresponding to each bin, the assumption that the content of each bin is independent on the contents of the other bins is implicit. This is usually true when independent entries are assigned to the bins. Anyway, in some cases, this assumption may not be true. For instance, if a histogram represents a cumulative distribution, all bin contents are correlated. For this reason, such situations require special care. For instance, fitting a histogram derived from a cumulative distribution with a χ^2 fit (see Sect. 6.15) where usually all bin contents are assumed to be uncorrelated is not correct.

6.8 Numerical Implementations

The maximization of the likelihood function L, or the equivalent minimization of $-\log L$, can be performed analytically only in the simplest cases. Most of the realistic cases require numerical methods implemented as computer algorithms. The minimization is, in many implementations, based on the so-called *gradient descent* method, which consists of following the steepest descent direction in the parameter space in steps, until the procedure converges to a point where the gradient is close to zero, which corresponds to a minimum value. The gradient of the negative log likelihood function is, in most of the computer algorithms, determined numerically from the likelihood function itself.

The negative log likelihood function in some cases may have local minima where the algorithm may converge, rather than reaching the absolute minimum. In those cases, the sensitivity of the result of the algorithm on initial conditions may be important, and finding the absolute minimum may require many attempts using different initial conditions. In the practice, finding a suitable set of initial parameters may require some experience.

The software package MINUIT [4] is one of the most widely used minimization tool in the field of particle physics since the years 1970s. MINUIT has been reimplemented from the original Fortran version in C++ and is available in the ROOT software toolkit [5]. Moreover, a python front-end for MINUIT is available with the package iminuit [6].

6.9 Likelihood Function for Gaussian Distribution

Consider N observations of a variable x distributed according to a Gaussian distribution with average μ and variance σ^2. The likelihood function can be written as the product of N Gaussian functions, as already written in Eq. (5.41):

$$L(x_1, \cdots, x_N; \mu, \sigma^2) = \frac{1}{(2\pi\sigma^2)^{N/2}} \exp\left[-\sum_{i=1}^{N} \frac{(x_i - \mu)^2}{2\sigma^2}\right]. \tag{6.15}$$

Twice the negative logarithm of the likelihood function is equal to

$$-2\log L(x_1, \cdots, x_N; \mu, \sigma^2) = \sum_{i=1}^{N} \frac{(x_i - \mu)^2}{\sigma^2} + N(\log 2\pi + 2\log\sigma). \tag{6.16}$$

The first term is χ^2 variable (see Sect. 3.9).

The minimization of $-2\log L$ can be performed analytically by finding the zeros of the derivatives with respect to μ and σ^2. The following maximum likelihood estimates for μ and σ^2 can be obtained:[3]

$$\hat{\mu} = \frac{1}{N} \sum_{i=1}^{N} x_i, \tag{6.17}$$

---

[3] Note that we use the notation $\widehat{\sigma^2}$ and not $\hat{\sigma}^2$, since we consider the variance σ^2 as a parameter of interest, rather than the standard deviation σ.

$$\widehat{\sigma^2} = \frac{1}{N} \sum_{i=1}^{N} (x_i - \hat{\mu})^2 .$$

(6.18)

The maximum likelihood estimate $\widehat{\sigma^2}$ is affected by bias, i.e., its expected value deviates from the true σ^2. The bias, anyway, decreases as $N \rightarrow \infty$. A way to correct the bias present in Eq. (6.18) is discussed in Example 6.5 and leads to the unbiased estimator:

$$s^2 = \frac{1}{N-1} \sum_{i=1}^{N} (x_i - \hat{\mu})^2 .$$

(6.19)

More in general, the estimates in Eqs. (6.17) and (6.19) are unbiased estimates of the mean and variance for any distribution, not only for a Gaussian distribution. The variance of $\hat{\mu}$ is easy to derive from Eq. (6.17):

$$\mathbb{V}\mathrm{ar}\left[\hat{\mu}\right] = \frac{\sigma^2}{N} .$$

(6.20)

This result coincides with the uncertainty determined with the Bayesian inference in Sect. 5.7.

For a Gaussian distribution, the variable

$$c^2 = \frac{(N-1)\,s^2}{\sigma^2}$$

(6.21)

is a χ^2 variable with $N-1$ degrees of freedom, because $\hat{\mu}$ is used in Eq. (6.19) in place of μ, and it has been determined from the N values x_i. Therefore, its variance is equal to $2(N-1)$, and the variance of s^2 is equal to

$$\mathbb{V}\mathrm{ar}\left[s^2\right] = \mathbb{V}\mathrm{ar}\left[\frac{\sigma^2 c^2}{N-1}\right] = \frac{\sigma^4}{(N-1)^2} \mathbb{V}\mathrm{ar}\left[c^2\right] = \frac{2\sigma^4}{N-1} .$$

(6.22)

More in general, without assuming that the x_i are normally distributed, the variance of s^2 can be demonstrated to be

$$\mathbb{V}\mathrm{ar}\left[s^2\right] = \frac{1}{N}\left(m_4 - \frac{N-3}{N-1}\sigma^2\right),$$

(6.23)

where m_4 is the 4th central moment of x. For a normal distribution, m_4 is equal to $3\sigma^4$, which gives again Eq. (6.22).

6.10   Errors of Maximum Likelihood Estimates

Once the point estimate $\hat{\theta}$ of a parameter θ is determined, a *confidence interval* must be assigned to the estimate, such that the probability that the interval contains the true value, also called *coverage* (see Sect. 1.14), is equal to a desired probability or *confidence level*. The confidence level is taken, in most of the cases, equal to 68.27%, corresponding to a Gaussian $1 \pm \sigma$ central interval.

Assuming a Gaussian distribution for $\hat{\theta}$, one can take the symmetric interval $[\hat{\theta} - \sigma_{\hat{\theta}}, \hat{\theta} + \sigma_{\hat{\theta}}]$, or $\hat{\theta} \pm \sigma_{\hat{\theta}}$, where $\sigma_{\hat{\theta}}$ is the square root of $\mathbb{V}\text{ar}\left[\hat{\theta}\right]$. For more parameters $\theta_1, \cdots, \theta_m$, the covariance matrix

$$C_{ij} = \mathbb{C}\text{ov}\left[\hat{\theta}_i, \hat{\theta}_j\right] \tag{6.24}$$

has to be determined.

If the distribution is not Gaussian, two approximate methods to determine parameter uncertainties for maximum likelihood estimates are presented in the following two sections. For both cases, the coverage is only approximate. Chapter 8 discusses a more rigorous treatment of uncertainty intervals that ensure the proper coverage.

6.11   Covariance Matrix Estimate

In the limit of very large number of observations N, it is possible to assume that the likelihood, as a function of the m unknown parameters, is close to a Gaussian function, as derived in Sect. 5.7. This approximation simplifies the estimate of uncertainty intervals. But in realistic cases with a finite number of measurements, the results obtained under a Gaussian assumption may only be an approximation, and deviation from the exact coverage may occur.

Taking as approximate likelihood function an m-dimensional Gaussian (Eq. (3.126)), an estimate of the m-dimensional covariance matrix C_{ij} of the parameter estimates may be obtained as the inverse of the second-order partial derivative matrix of the negative logarithm of the likelihood function,[4] which can be written as

$$C_{ij}^{-1} = -\left.\frac{\partial^2 \log L(x_1, \cdots, x_N; \theta_1, \cdots, \theta_m)}{\partial \theta_i \, \partial \theta_j}\right|_{\vec{\theta}=\hat{\theta}}. \tag{6.25}$$

This covariance matrix also gives an m-dimensional hyperellipsoidal confidence contour or simply ellipsoidal in $m = 2$ dimensions. Consider, for instance, the

---

[4] For the users of the program MINUIT, this estimate corresponds to call the method MIGRAD/HESSE.

Gaussian likelihood case seen in Sect. 6.9. The derivative of $-\log L$ with respect to the parameter μ is

$$\frac{1}{\sigma_{\hat{\mu}}^2} = \frac{\partial^2(-\log L)}{\partial \mu^2} = \frac{N}{\sigma^2}, \tag{6.26}$$

which gives the following error on the estimated average $\hat{\mu}$:

$$\sigma_{\hat{\mu}} = \frac{\sigma}{\sqrt{N}}. \tag{6.27}$$

This expression coincides with the standard deviation of the average from Eq. (6.17), which can be evaluated using the general formulae from Eqs. (2.19) and (2.20). It also coincides with the standard deviation of the Bayesian posterior for μ in Eq. (5.53).

Example 6.3 - Efficiency Estimate Using a Binomial Distribution

The *efficiency* ε of a device is defined as the probability that the device gives a positive signal when a process of interest occurs. Particle detectors are examples of such devices: a detector may produce a signal when a particle interacts with it with probability ε, but it may also fail, in a fraction of the cases, with probability $1 - \varepsilon$. The number of positive outcomes n out of N extractions, i.e., the number of detected particles out of the total number of particles that crossed the detector, follows a binomial distribution (Eq. (2.61)) with parameter $p = \varepsilon$.

The efficiency ε can be estimated using a large number N of particles crossing the detector by counting the number of times n the detector gives a signal, i.e., the number of cases it has been efficient. For a particle detector exposed to particles at a fixed rate, a sufficiently large number of particles is recorded if the data acquisition time is sufficiently long.

The likelihood function is given by binomial probability:

$$L(n; N, \varepsilon) = \frac{N!}{n!\,(N-n)!} \varepsilon^n (1 - \varepsilon)^{N-n}. \tag{6.28}$$

The negative logarithm of L diverges for $\varepsilon = 0$ or $\varepsilon = 1$, so it appears more convenient to find the maximum of the likelihood function directly by imposing a null derivative of L with respect to the parameter ε:

$$\frac{\partial L(n; N, \varepsilon)}{\partial \varepsilon} = \frac{N!}{n!\,(N-n)!} \varepsilon^{n-1} (1 - \varepsilon)^{N-n-1} (n - N\varepsilon) = 0. \tag{6.29}$$

(continued)

Example 6.3 (continued)

This derivative, anyway, also has divergences for $n = 0$ at $\varepsilon = 0$ and for $n = N$ at $\varepsilon = 1$. Solving Eq. (6.29) gives $\varepsilon = n/N$. The maximum likelihood estimate of the true efficiency ε is, therefore,

$$\hat{\varepsilon} = \frac{n}{N} . \tag{6.30}$$

By writing explicitly the likelihood function in Eq. (6.28) for $n = 0$ and $n = N$, one finds the maximum values at $\varepsilon = 0$ and $\varepsilon = 1$, respectively. This allows to extend Eq. (6.30) to all values of n.

The uncertainty on $\hat{\varepsilon}$ is obtained from the second-order derivative of $-\log L$. The negative log of L is

$$-\log L = -\log\left(\frac{N!}{n!(N-n)!}\right) - n\log\varepsilon - (N-n)\log(1-\varepsilon), \tag{6.31}$$

and its first and second derivatives are

$$\frac{\partial(-\log L)}{\partial\varepsilon} = \frac{N\varepsilon - n}{\varepsilon(1-\varepsilon)}, \tag{6.32}$$

$$\frac{\partial^2(-\log L)}{\partial\varepsilon^2} = \frac{N\varepsilon^2 - 2n\varepsilon + n}{\varepsilon^2(1-\varepsilon)^2}. \tag{6.33}$$

The second derivative, evaluated at $\varepsilon = \hat{\varepsilon} = n/N$, is

$$\left.\frac{\partial^2(-\log L)}{\partial\varepsilon^2}\right|_{\varepsilon=\hat{\varepsilon}} = \frac{\hat{\varepsilon}(1-\hat{\varepsilon})}{N}, \tag{6.34}$$

and its square root gives the uncertainty, equal to

$$\sigma_{\hat{\varepsilon}} = \sqrt{\frac{\hat{\varepsilon}(1-\hat{\varepsilon})}{N}}. \tag{6.35}$$

Note that the expression in Eq. (6.35) is equal to the standard deviation of $\hat{\varepsilon} = n/N$, given by the standard deviation of n, $\sigma_n = \sqrt{N\varepsilon(1-\varepsilon)}$ (Eq. (2.63)) divided by N, where ε is replaced by $\hat{\varepsilon}$:

$$\sigma_{\hat{\varepsilon}} = \left.\frac{\sqrt{\mathbb{V}\mathrm{ar}[n]}}{N}\right|_{\varepsilon=\hat{\varepsilon}} = \sqrt{\frac{\hat{\varepsilon}(1-\hat{\varepsilon})}{N}}. \tag{6.36}$$

The error estimate on $\hat{\varepsilon}$ is null for $n = 0$ and $n = N$ or equivalently for $\hat{\varepsilon} = 0$ and $\hat{\varepsilon} = 1$, respectively. This pathology is due to the approximate

(continued)

Example 6.3 (continued)
method adopted to determine the uncertainty. Section 8.5 discusses how to overcome this problem with a more rigorous treatment. A non-null uncertainty can be computed in this way also for cases with $n = 0$ and $n = N$. In those case, a fully asymmetric confidence interval is determined.

Example 6.4 - Maximum Likelihood Estimate for an Exponential Distribution
Assume an exponential PDF of a variable t with parameter λ:

$$f(t) = \lambda \, e^{-\lambda t} \,. \qquad (6.37)$$

λ is the inverse of the average lifetime $\tau = 1/\lambda$. Given N measurements of t: $t_1, \cdots, t_N$, the likelihood function can be written as the product of $f(t_1) \cdots f(t_N)$:

$$L(t_1, \cdots, t_N; \lambda) = \prod_{i=1}^{N} \lambda \, e^{-\lambda t_i} = \lambda^N e^{-\lambda \sum_{i=1}^{N} t_i} \,. \qquad (6.38)$$

The maximum of the likelihood function can be found by imposing that the derivative with respect to λ is null. The estimate of λ is therefore

$$\hat{\lambda} = \left(\frac{1}{N} \sum_{i=1}^{N} t_i \right)^{-1} , \qquad (6.39)$$

with uncertainty, from Eq. (6.25), uncorrected for a possible bias, equal to

$$\sigma_{\hat{\lambda}} = \hat{\lambda}/\sqrt{N} \,. \qquad (6.40)$$

The demonstration of this result is left as exercise to the reader. The same example using Bayesian inference was discussed in Example 5.6.

6.12 Likelihood Scan

A better approximation for uncertainty intervals is obtained considering the scan of $-2 \log L$ as a function of the unknown parameters around the minimum value, $-2 \log L_{\max} = -2 \log L(\hat{\vec{\theta}})$, that corresponds to the parameter set that maximizes L. The uncertainty interval, if we have one parameter, or, more in general, the

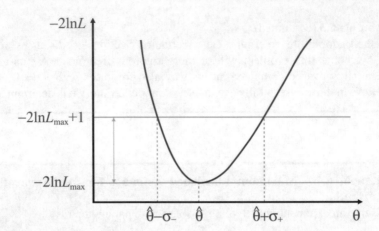

Fig. 6.1 Scan of $-2\log L$ as a function of a parameter θ. The bounds of the error interval are determined as the values for which $-2\log L$ increases by one unit with respect to its minimum, corresponding to $\theta = \hat{\theta}$

uncertainty region, for more parameters, corresponds to the set of parameter values for which $-2\log L$ increases at most by one unit with respect to its minimum value.[5] The coverage is usually improved using the $-2\log L$ scan with respect to the parabolic Gaussian approximation, but it may be still not perfect. This is illustrated in Fig. 6.1 for the case of a single parameter θ.

For a Gaussian PDF, this method is equivalent to the inverse of the second-order derivatives presented in the previous section. For a PDF model that deviates from the Gaussian approximation, the interval determined from the scan of $-2\log L$ may give *asymmetric errors*, as evident in Fig. 6.1, where the likelihood function is asymmetric around the minimum at $\hat{\theta}$. For more than one parameter, the 1σ contour of the uncertainty region corresponds to the set of parameter values $\vec{\theta}$ such that

$$-2\log L(\vec{\theta}) = -2\log L_{\max} - 1 \,. \tag{6.41}$$

Contours corresponding to Z standard deviations can be determined similarly by requiring

$$-2\log L(\vec{\theta}) = -2\log L_{\max} - Z^2 \,. \tag{6.42}$$

In the Gaussian case, where $-2\log L$ has a parabolic shape, this method gives (hyper)elliptic contours corresponding to the covariance matrix given by Eq. (6.25). But, in general, uncertainty contours given by Eq. (6.41) may not be elliptic.

---

[5] For MINUIT users, this procedure corresponds to the method MINOS.

6.13 Properties of Maximum Likelihood Estimators

Maximum likelihood estimators are the most widely used estimators because of their good properties, among the ones defined in Sect. 6.3. Below a number of properties of maximum likelihood estimators are listed:

- Maximum likelihood estimators are consistent.
- Maximum likelihood estimators may have a bias, but the bias tends to zero as the number of measurements N tends to infinity.
- Maximum likelihood estimators have efficiencies, compared to the Cramér–Rao bound (Eq. (6.8)) that tends to one for a large number of measurements; in other words, maximum likelihood estimators have, asymptotically, the lowest possible variance compared to any other consistent estimator.
- Maximum likelihood estimators are invariant under reparameterizations, that is, if a maximum of the likelihood is found in terms of some parameters $\vec{\theta} = \hat{\vec{\theta}}$, in a new parameterization, $\vec{\eta} = \vec{\eta}\,(\vec{\theta}\,)$, the transformed parameters $\hat{\vec{\eta}} = \vec{\eta}\,(\hat{\vec{\theta}}\,)$ also maximize the likelihood function. This property is not true for other quantities; for instance, the bias may change under reparameterization.

Example 6.5 - Bias of the Maximum Likelihood Estimate of a Gaussian Variance

The maximum likelihood estimate of the variance σ^2 of a Gaussian distribution, from Eq. (6.18), is given by

$$\widehat{\sigma^2} = \frac{1}{N} \sum_{i=1}^{N} (x_i - \hat{\mu})^2 \, . \tag{6.43}$$

The bias, defined in Eq. (6.5), is the difference between the expected value of $\widehat{\sigma^2}$ and the true value of σ^2. It is easy to show analytically that the expected value of $\widehat{\sigma^2}$ is

$$\left\langle \widehat{\sigma^2} \right\rangle = \frac{N-1}{N} \sigma^2 \, , \tag{6.44}$$

where σ^2 is the true variance. Hence, the maximum likelihood estimate $\widehat{\sigma^2}$ tends to underestimate the variance, and it has a bias given by

$$\mathbb{Bias}\left[\widehat{\sigma^2}\right] = \left\langle \widehat{\sigma^2} \right\rangle - \sigma^2 = \left(\frac{N-1}{N} - 1 \right) \sigma^2 = -\frac{\sigma^2}{N} \, . \tag{6.45}$$

(continued)

Example 6.5 (continued)

Bias $\left[\widehat{\sigma^2}\right]$ decreases with N, which is a general property of maximum likelihood estimates.

For this specific case, an unbiased estimate can be obtained by multiplying the maximum likelihood estimate by a correction factor $N/N-1$, which gives

$$s^2 = \frac{1}{N-1} \sum_{i=1}^{N} (x_i - \hat{\mu})^2 . \tag{6.46}$$

6.14   Extended Likelihood Function

If the number of recorded observations N is also a random variable, its distribution may be incorporated into the likelihood function to consider N as an additional random variable of our data sample. The distribution of N, say $P(N; \vec{\theta})$, may also depend on the unknown parameters $\vec{\theta}$, and including the information about the total number of observations N in the likelihood provides in general a reduction in the uncertainties of the estimate of $\vec{\theta}$. The *extended likelihood function* may be defined as the product of the usual likelihood function times the probability distribution for N:

$$L(\vec{x}_1, \cdots, \vec{x}_N; \vec{\theta}) = P(N; \vec{\theta}) \prod_{i=1}^{N} p(\vec{x}_i; \vec{\theta}) . \tag{6.47}$$

In almost all cases in physics, $P(N; \vec{\theta})$ is a Poisson distribution whose expected value ν may depend on the unknown parameters $\vec{\theta}$. Equation (6.47) becomes, in this case,

$$L(\vec{x}_1, \cdots, \vec{x}_N; \vec{\theta}) = \frac{e^{-\nu(\vec{\theta})} \nu(\vec{\theta})^N}{N!} \prod_{i=1}^{N} p(\vec{x}_i; \vec{\theta}) . \tag{6.48}$$

Consider the case where the PDF p is a mixture, with given proportions, of a PDF for signal, p_s, and a PDF for background, p_b. Expected signal and background yields are defined as s and b, respectively, and are two unknown parameters, together with a set of unknown parameters $\vec{\theta}$. The extended likelihood function, for the simplest case of a single dimension of $\vec{x}$, can be written as

$$L(x_1, \cdots, x_N; s, b, \vec{\theta}) = \frac{(s+b)^N e^{-(s+b)}}{N!} \prod_{i=1}^{N} (w_s p_s(x_i; \vec{\theta}) + w_b p_b(x_i; \vec{\theta})) . \tag{6.49}$$

The fractions of signal and background w_s and w_b are

$$w_s = \frac{s}{s+b}, \tag{6.50}$$

$$w_b = \frac{b}{s+b}. \tag{6.51}$$

Note that $w_s + w_b = 1$, and hence $p = w_s p_s + w_b p_b$ is normalized assuming that p_s and p_b are normalized. Replacing Eqs. (6.50) and (6.51) into Eq. (6.49) allows to simplify the expression for the likelihood function:

$$L(x_i; s, b, \vec{\theta}) = \frac{(s+b)^N e^{-(s+b)}}{N!} \prod_{i=1}^{N} \frac{(sp_s(x_i; \vec{\theta}) + bp_b(x_i; \vec{\theta}))}{s+b} = \tag{6.52}$$

$$= \frac{e^{-(s+b)}}{N!} \prod_{i=1}^{N} (sp_s(x_i; \vec{\theta}) + bp_b(x_i; \vec{\theta})). \tag{6.53}$$

The negative logarithm of the likelihood function gives a more convenient expression:

$$-\log L(x_i; s, b, \vec{\theta}) = s + b - \sum_{i=1}^{N} \log(sp_s(x_i; \vec{\theta}) + bp_b(x_i; \vec{\theta})) + \log N!. \tag{6.54}$$

The last term, $-\log N!$, is constant with respect to the unknown parameters and can be omitted when minimizing the function $-\log L(x_i; s, b, \vec{\theta})$.

An example of application of the extended maximum likelihood from Eq. (6.54) is the two-component fit shown in Fig. 6.2. The assumed PDF is the sum of a Gaussian component, modeling the signal, and an exponential component, modeling the background. The points with the error bars in the plot represent the data sample which is randomly extracted according to the assumed PDF. Data are shown as a binned histogram for convenience, but the fit is unbinned, and the individual randomly extracted values of the variable m, reported on the horizontal axis, are used in the likelihood function. The parameters determined simultaneously by the minimization of $-\log L$ are the mean μ and standard deviation σ of the Gaussian, the parameter λ of the exponential background component, in addition to the expected number of signal and background events, s and b.

Figure 6.3 shows the two-dimensional contour plot for the fit considered in Fig. 6.2. It corresponds to the points for which $-2\log L(x_1, \cdots, x_N; s, b, \vec{\theta})$ in Eq. (6.54) increases by one unit with respect to its minimum, shown as the central point. The shape is very close to an ellipse in this case, demonstrating that the Gaussian approximation is sufficiently precise, given the large number of observations.

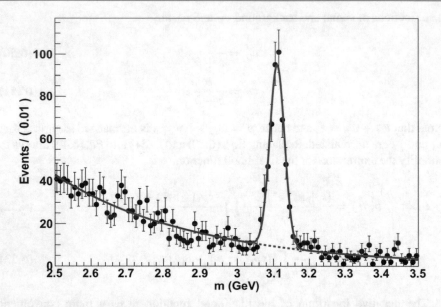

Fig. 6.2 Example of unbinned extended maximum likelihood fit of a simulated data set. The fit curve (blue solid line) is superimposed to the data points (black dots with error bars)

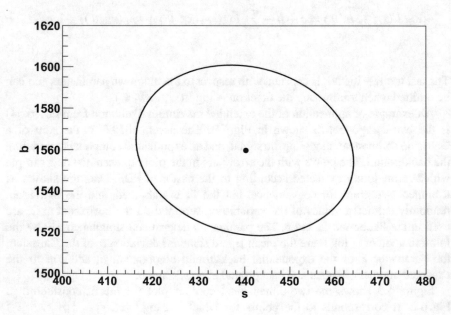

Fig. 6.3 Two-dimensional 1σ contour plot surrounding the uncertainty region for the parameters s and b, equal to the expected number of signal and background events, respectively. The plot corresponds to the fit in Fig. 6.2. The contour is determined from the equation $-2\log L(x_1, \cdots, x_N; s, b, \vec{\theta}) = -2\log L_{\max} - 1$

6.15   Minimum χ^2 and Least Squares Methods

Consider a number N of measurements affected by their uncertainties, $y_1 \pm \sigma_1, \cdots, y_N \pm \sigma_N$. Each measurement $y_i \pm \sigma_i$ corresponds to a value x_i of a variable x. Assume we have a model for the dependence of y on the variable x given by a function f:

$$y = f(x; \vec{\theta}), \tag{6.55}$$

where $\vec{\theta} = (\theta_1, \cdots, \theta_m)$ is a set of unknown parameters. If the measurements y_i are distributed around the value $f(x_i; \vec{\theta})$ according to a Gaussian distribution with standard deviation σ_i, the likelihood function can be written as the product of N Gaussian PDFs, which gives

$$L(\vec{y}; \vec{\theta}) = \prod_{i=1}^{N} \frac{1}{\sigma_i^2 \sqrt{2\pi}} \exp\left[-\frac{(y_i - f(x_i; \vec{\theta}))^2}{2\sigma_i^2} \right], \tag{6.56}$$

where the notation $\vec{y} = (y_1, \cdots, y_N)$ was introduced.

Maximizing $L(\vec{y}; \vec{\theta})$ is equivalent to minimizing $-2 \log L(\vec{y}; \vec{\theta})$, which is equal to

$$-2 \log L(\vec{y}; \vec{\theta}) = \sum_{i=1}^{N} \frac{(y_i - f(x_i; \vec{\theta}))^2}{\sigma_i^2} + \sum_{i=1}^{N} \log 2\pi \sigma_i^2. \tag{6.57}$$

The last term does not depend on the parameters $\vec{\theta}$ if the uncertainties σ_i are known and fixed, and hence it is a constant that can be dropped when performing the minimization. The first term to be minimized in Eq. (6.57) is a χ^2 variable (see Sect. 3.9):

$$\chi^2(\vec{\theta}) = \sum_{i=1}^{N} \frac{(y_i - f(x_i; \vec{\theta}))^2}{\sigma_i^2}. \tag{6.58}$$

The terms that appear squared in the sum,

$$\varepsilon_i = y_i - \hat{y}_i = y_i - f(x_i; \hat{\vec{\theta}}), \tag{6.59}$$

evaluated at the fit values $\hat{\vec{\theta}}$ of the parameters $\vec{\theta}$, are called *residuals*.

In case the uncertainties σ_i are all equal, it is possible to minimize the expression:

$$S = \sum_{i=1}^{N} (y_i - f(x_i; \vec{\theta}))^2. \tag{6.60}$$

This minimization is called *least squares method*.

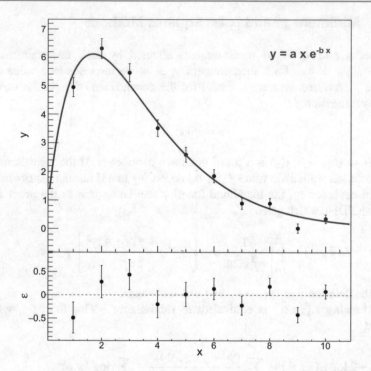

Fig. 6.4 Example of minimum χ^2 fit of a computer-generated data set. The points with the error bars are used to fit a function model of the type $y = f(x) = a\,x\,e^{-bx}$, where a and b are unknown parameters determined by the fit. The fit curve is superimposed as solid blue line. Residuals ε_i are shown in the bottom section of the plot

An example of fit performed with the minimum χ^2 method is shown in Fig. 6.4. Data are randomly extracted according to the assumed model and residuals are randomly distributed around zero, according to the data uncertainties.

6.16   Linear Regression

In the simplest case of a polynomial function, the minimum χ^2 problem can be solved analytically. Here we discuss the case of a linear function f that can be written as

$$y = f(x; a, b) = a + bx , \tag{6.61}$$

a and b being free parameters. The χ^2 becomes

$$\chi^2(a, b) = \sum_{i=1}^{N} \frac{(y_i - a - b\,x_i)^2}{\sigma_i^2} \,. \tag{6.62}$$

Let us introduce the weights w_i, conveniently defined as

$$w_i = \frac{1/\sigma_i^2}{1/\sigma^2} \,, \tag{6.63}$$

with

$$\frac{1}{\sigma^2} = \sum_{i=1}^{N} \frac{1}{\sigma_i^2} \,, \tag{6.64}$$

so that $\sum_{i=1}^{N} w_i = 1$. In case of identical errors, weights are all equal to $1/N$. The χ^2 can be written in terms of the weights as

$$\chi^2(a, b) = \frac{1}{\sigma^2} \sum_{i=1}^{N} w_i \,(y_i - a - b\,x_i)^2 \,. \tag{6.65}$$

The analytical minimization is achieved by imposing

$$\frac{\partial \chi^2(a, b)}{\partial a} = 0 \,, \tag{6.66}$$

$$\frac{\partial \chi^2(a, b)}{\partial b} = 0 \,. \tag{6.67}$$

The partial derivative can be written as

$$\begin{aligned}
\frac{\partial \chi^2(a, b)}{\partial a} &= \frac{1}{\sigma^2} \sum_{i=1}^{N} w_i \frac{\partial}{\partial a}(y_i - a - b\,x_i)^2 = \\[2mm]
&= -2\frac{1}{\sigma^2} \sum_{i=1}^{N} w_i\,(y_i - a - b\,x_i) = \\[2mm]
&= \frac{2}{\sigma^2} \left(-\sum_{i=1}^{N} w_i\,y_i + a + b \sum_{i=1}^{N} w_i x_i \right) ,
\end{aligned} \tag{6.68}$$

$$\frac{\partial \chi^2(a, b)}{\partial b} = \frac{1}{\sigma^2} \sum_{i=1}^{N} w_i \frac{\partial}{\partial b} (y_i - a - b x_i)^2 =$$

$$= -2 \frac{1}{\sigma^2} \sum_{i=1}^{N} w_i (y_i - a - b x_i) x_i =$$

$$= \frac{2}{\sigma^2} \left(-\sum_{i=1}^{N} w_i x_i y_i + a \sum_{i=1}^{N} w_i x_i + b \sum_{i=1}^{N} w_i x_i^2 \right) . \tag{6.69}$$

The notation can be simplified by introducing weighted means for the quantities, x_i, y_i, $x_i y_i$, x_i^2, etc., that are indicated below generically as α_i, as

$$\langle \alpha \rangle = \sum_{i=1}^{N} w_i \alpha_i . \tag{6.70}$$

In terms of the weighted means, partial derivatives in Eqs. (6.68) and (6.69) can be written in a more compact form as

$$\frac{\partial \chi^2(a, b)}{\partial a} = \frac{2}{\sigma^2} \left(-\langle y \rangle + a + b \langle x \rangle \right) , \tag{6.71}$$

$$\frac{\partial \chi^2(a, b)}{\partial b} = \frac{2}{\sigma^2} \left(-\langle xy \rangle + a \langle x \rangle + b \langle x^2 \rangle \right) . \tag{6.72}$$

The χ^2 minimum can be obtained by imposing the partial derivatives to be null, which gives

$$\langle y \rangle = a + b \langle x \rangle , \tag{6.73}$$

$$\langle xy \rangle = a \langle x \rangle + b \langle x^2 \rangle , \tag{6.74}$$

or, in matrix form:

$$\begin{pmatrix} \langle y \rangle \\ \langle xy \rangle \end{pmatrix} = \begin{pmatrix} 1 & \langle x \rangle \\ \langle x \rangle & \langle x^2 \rangle \end{pmatrix} \begin{pmatrix} a \\ b \end{pmatrix} = C \begin{pmatrix} a \\ b \end{pmatrix} . \tag{6.75}$$

The matrix inversion gives the solution for a and b:

$$\begin{pmatrix} a \\ b \end{pmatrix} = C^{-1} \begin{pmatrix} \langle y \rangle \\ \langle xy \rangle \end{pmatrix} . \tag{6.76}$$

The determinant of the inverse matrix C^{-1} is

$$\det\left[C^{-1}\right] = \left\langle x^2 \right\rangle - \langle x \rangle^2 = \left\langle (x - \langle x \rangle)^2 \right\rangle = \mathbb{V}\mathrm{ar}[x] , \tag{6.77}$$

and therefore, the inverse matrix is

$$C^{-1} = \begin{pmatrix} 1 & \langle x \rangle \\ \langle x \rangle & \langle x^2 \rangle \end{pmatrix}^{-1} = \frac{1}{\mathbb{V}\mathrm{ar}[x]} \begin{pmatrix} \langle x^2 \rangle & -\langle x \rangle \\ -\langle x \rangle & 1 \end{pmatrix} . \tag{6.78}$$

With a bit of math, one has explicit expressions for the parameter estimates:

$$b = \hat{b} = \frac{\langle xy \rangle - \langle x \rangle \langle y \rangle}{\mathbb{V}\mathrm{ar}[x]} = \frac{\mathbb{C}\mathrm{ov}(x, y)}{\mathbb{V}\mathrm{ar}[x]} , \tag{6.79}$$

$$a = \hat{a} = \langle y \rangle - \hat{b} \langle x \rangle . \tag{6.80}$$

Equivalent formulae are

$$\hat{b} = \frac{\langle (x - \langle x \rangle)(y - \langle y \rangle) \rangle}{\langle (x - \langle x \rangle)^2 \rangle} = \frac{\langle xy \rangle - \langle x \rangle \langle y \rangle}{\langle x^2 \rangle - \langle x \rangle^2} = \frac{\mathbb{C}\mathrm{ov}(x, y)}{\sigma_x^2} , \tag{6.81}$$

$$\hat{a} = \langle y \rangle - \hat{b} \langle x \rangle . \tag{6.82}$$

The uncertainties on the estimates $\hat{a}$ and $\hat{b}$ can be determined as described in Sect. 6.11 from the second derivative matrix of $-\log L = \frac{1}{2} \chi^2$:

$$\frac{1}{\sigma_{\hat{a}}^2} = -\frac{\partial^2 \log L}{\partial a^2} = \frac{1}{2} \frac{\partial^2 \chi^2}{\partial a^2} , \tag{6.83}$$

$$\frac{1}{\sigma_{\hat{b}}^2} = -\frac{\partial^2 \log L}{\partial b^2} = \frac{1}{2} \frac{\partial^2 \chi^2}{\partial b^2} . \tag{6.84}$$

The non-diagonal element of the inverse covariance matrix is

$$C_{ab}^{-1} = -\frac{\partial^2 \log L}{\partial a\, \partial b} = \frac{1}{2} \frac{\partial^2 \chi^2}{\partial a\, \partial b} . \tag{6.85}$$

Recalling partial derivatives written in Eqs. (6.71) and (6.72), the second derivatives are

$$\frac{\partial^2 \chi^2(a, b)}{\partial a^2} = \frac{2}{\sigma^2} , \tag{6.86}$$

$$\frac{\partial^2 \chi^2(a, b)}{\partial b^2} = \frac{2 \langle x^2 \rangle}{\sigma^2} , \tag{6.87}$$

$$\frac{\partial^2 \chi^2(a, b)}{\partial a \, \partial b} = \frac{2 \langle x \rangle}{\sigma^2} \,. \tag{6.88}$$

From those derivatives, the inverse covariance matrix is

$$C^{-1} = \frac{1}{\sigma^2} \begin{pmatrix} 1 & \langle x \rangle \\ \langle x \rangle & \langle x^2 \rangle \end{pmatrix} \,. \tag{6.89}$$

The inversion of the matrix is similar to what previously obtained and gives

$$C = \frac{\sigma^2}{\sigma_x^2} \begin{pmatrix} \langle x^2 \rangle & -\langle x \rangle \\ -\langle x \rangle & 1 \end{pmatrix} \,, \tag{6.90}$$

from which one obtains the uncertainties and the covariance term:

$$\sigma_{\hat{b}}^2 = \frac{\sigma^2}{\sigma_x^2} \,, \tag{6.91}$$

$$\sigma_{\hat{a}}^2 = \frac{\sigma^2}{\sigma_x^2} \langle x^2 \rangle = \sigma_{\hat{b}}^2 \langle x^2 \rangle \,, \tag{6.92}$$

$$\mathrm{Cov}(\hat{a}, \hat{b}) = -\frac{\sigma^2}{\sigma_x^2} \langle x \rangle = -\sigma_{\hat{b}}^2 \langle x \rangle \,. \tag{6.93}$$

In place of the uncertainties estimated above, it is often preferred to adopt a different estimate of σ^2. Namely, σ^2 is replaced by

$$\widehat{\sigma^2} = \frac{N}{N - D} \sum_{i=1}^{N} w_i \left(y_i - \hat{y}_i \right)^2 \,, \tag{6.94}$$

where $D = 2$ is the number of parameters determined from the fit, equal to one plus the degree of the polynomial, i.e., the estimates of the two parameters a and b. In Eq. (6.94), we have defined $\hat{y}_i = f(x_i; \hat{a}, \hat{b}) = \hat{a} + \hat{b} x$, i.e., the estimated values of y_i along the fit line. The denominator has $N - D$ in places of N, which is the number of degrees of freedom, considering that we have N measurements and D parameters determined with the fit. This denominator ensures that we have no bias, as discussed in Example 6.5. We can write $\widehat{\sigma^2}$ in terms of the residuals:

$$\varepsilon_i = y_i - \hat{y}_i, \tag{6.95}$$

and we obtain

$$\widehat{\sigma^2} = \frac{1}{N - D} \langle \varepsilon^2 \rangle \,. \tag{6.96}$$

Note that, if uncertainties on the measurement are determined using Eq. (6.94), it is no longer possible to use the minimum χ^2 for goodness-of-fit test (see Sect. 6.17).

A coefficient that is not very used in physics but appears rather frequently in linear regressions performed by commercial software is the *coefficient of determination*, or R^2, defined as

$$R^2 = \frac{\sum_{i=1}^{N} \left(\hat{y}_i - \langle y \rangle\right)^2}{\sum_{i=1}^{N} \left(y_i - \langle y \rangle\right)^2} \,. \tag{6.97}$$

R^2 may have values between 0 and 1 and is often expressed as percentage. $R^2 = 1$ corresponds to measurements perfectly aligned along the fitted regression line, indicating that the regression line accounts for all the measurement variations as a function of x, while $R^2 = 0$ corresponds to a perfectly horizontal regression line, indicating that the measurements are insensitive on the variable x.

6.17 Goodness of Fit and p-Value

One advantage of the minimum χ^2 method is that the minimum χ^2 value, which we indicate with $\hat{\chi}^2$, follows a χ^2 distribution (see Sect. 3.9) with a number of degrees of freedom equal to the number of measurements N minus the number of fit parameters D. We can define the p value as the probability that a χ^2 greater than or equal to $\hat{\chi}^2$ is obtained from a fit to a random data sample, extracted according to the assumed model:

$$p = P(\chi^2 \geq \hat{\chi}^2) \,. \tag{6.98}$$

See also Sect. 12.2 for a more general definition of p-values. If the data follow the assumed Gaussian distributions, the p-value is expected to be a random variable uniformly distributed from 0 to 1, from a general property of cumulative distributions discussed in Sect. 3.4.

If a small p-value is obtained from the fit, it could be a symptom of a poor description of the theoretical model $y = f(x; \vec{\theta})$. For this reason, the minimum χ^2 value can be used as a measurement of the goodness of the fit. Anyway, setting a threshold, say p-value more than 0.05, to determine whether a fit can be considered acceptable or not, implies the chance to discard on average 5% of the cases, even if the PDF model correctly describes the data, due to the possibility of statistical fluctuations.

Note also that the p-value cannot be considered as the *probability of the fit hypothesis to be true*. Such probability would only have a meaning in the Bayesian approach, as defined in Chap. 5. Under the Bayesian approach, such a probability would require a completely different type of evaluation.

In case the measurement uncertainties in a fit are estimated from the residuals, as in Eq. (6.94), the minimum χ^2 value no longer measures how data deviate from the fit model, within their uncertainties, and cannot be used to perform a goodness-of-fit test.

While the minimum χ^2 has a known distribution, more in general, for maximum likelihood fits, the value of $-2 \log L$ for which the likelihood function is maximum does not provide a measurement of the goodness of the fit. It is possible in some cases to obtain a goodness-of-fit measurement by finding a proper ratio of likelihood functions evaluated in two different hypotheses. Wilks' theorem (see Sect. 10.8) ensures that a properly defined likelihood ratio is asymptotically distributed as a χ^2 for a large number of observations. This approach is also discussed in Sect. 6.19 for what concerns fits of binned data that follow Poisson distribution.

6.18   Minimum χ^2 Method for Binned Histograms

In case of a sufficiently large number of entries in each bin, the Poisson distribution describing the number of entries in a bin can be approximated by a Gaussian with variance equal to the expected number of entries in that bin (see Sect. 2.11). In this case, the expression for $-2 \log L$ becomes

$$-2 \log L = \sum_{i=1}^{B} \frac{(n_i - \mu_i(\theta_1, \cdots, \theta_m))^2}{n_i} + B \log 2\pi + 2 \sum_{i=1}^{B} \log n_i , \qquad (6.99)$$

where

$$\mu_i(\theta_1, \cdots, \theta_m) = \int_{x_i^{\text{lo}}}^{x_i^{\text{up}}} f(x; \theta_1, \cdots, \theta_m) \, \mathrm{d}x \qquad (6.100)$$

and $[x_i^{\text{lo}}, x_i^{\text{up}}[$ is the interval corresponding to the i bin. If the binning is sufficiently fine, μ_i can be replaced by

$$\mu_i(\theta_1, \cdots, \theta_m) \simeq f(x_i; \theta_1, \cdots, \theta_m) \, \delta x_i , \qquad (6.101)$$

where $x_i = (x_i^{\text{up}} + x_i^{\text{lo}})/2$ is center of the i bin and $\delta x_i = x_i^{\text{up}} - x_i^{\text{lo}}$ is the bin's width. The quantity defined in Eq. (6.99), dropping the last two constant terms, is called *Neyman's* χ^2:

$$\chi_N^2 = \sum_{i=1}^{B} \frac{(n_i - \mu_i(\theta_1, \cdots, \theta_m))^2}{n_i} . \qquad (6.102)$$

It may be more convenient to replace at the denominator the *expected* number of entries with the *observed* number of entries. This gives the so-called *Pearson's* χ^2:

$$\chi_P^2 = \sum_{i=1}^{B} \frac{(n_i - \mu_i(\theta_1, \cdots, \theta_m))^2}{\mu_i(\theta_1, \cdots, \theta_m)} . \tag{6.103}$$

The value of χ^2 at the minimum can be used, as discussed in Sect. 6.17, as measurement of the goodness of the fit, where in this case the number of degrees of freedom is equal to the number of bins B minus the number of fit parameters m.

6.19 Binned Poissonian Fits

The Gaussian approximation assumed in Sect. 6.18 does not hold when the number of entries in some bin is small. A Poissonian model, which is also valid for a small number of entries, should be applied in those cases. The negative log likelihood function that replaces Eq. (6.99) is, for a Poissonian model,

$$-2 \log L = -2 \log \prod_{i=1}^{B} \text{Pois}(n_i; \mu_i(\theta_1, \cdots, \theta_m)) =$$

$$= -2 \log \prod_{i=1}^{B} \frac{e^{-\mu_i(\theta_1, \cdots, \theta_m)} \mu_i(\theta_1, \cdots, \theta_m)^{n_i}}{n_i!} . \tag{6.104}$$

Using the approach proposed in [7], in place of Eq. (6.104), one can use the ratio of the likelihood function divided by its maximum value, which does not depend on the unknown parameters and does not change, therefore, the fit result. This is equivalent to add a constant to $-2 \log L$. The denominator can be obtained by replacing μ_i with n_i, and the following negative log likelihood ratio is obtained, indicated with χ_λ^2:

$$\chi_\lambda^2 = -2 \log \prod_{i=1}^{B} \frac{L(n_i; \mu_i(\theta_1, \cdots, \theta_m))}{L(n_i; n_i)} = -2 \sum_{i=1}^{B} \log \frac{e^{-\mu_i} \mu_i^{n_i}}{n_i!} \frac{n_i!}{e^{-n_i} n_i^{n_i}}$$

$$= 2 \sum_{i=1}^{B} \left[\mu_i(\theta_1, \cdots, \theta_m) - n_i + n_i \log \left(\frac{n_i}{\mu_i(\theta_1, \cdots, \theta_m)} \right) \right] . \tag{6.105}$$

For bins with no entries, i.e., $n_i = 0$, the last term containing a logarithm should be set to zero.

There are cases where the total number of entries, $N = \sum_{i=1}^{B} n_i$, is fixed. In this case, the bin entries, $n_1, \cdots, n_B$, are distributed according to a multinomial distribution (see Sect. 2.10) where the parameters p_i depend on the unknown parameters $\vec{\theta}$. For fixed N, Eq. (6.105) becomes

$$\chi_\lambda^2 = 2 \sum_{i=1}^{B} n_i \log\left(\frac{n_i}{\mu_i(\theta_1, \cdots, \theta_m)}\right) , \tag{6.106}$$

where, as before, the logarithm term should be dropped in the sum for empty bins, i.e., for $n_i = 0$. If one of the unknown parameters is the normalization term, $\mu = \sum_{i=1}^{B} \mu_i$, we can write $\mu_i = \mu p_i(\vec{\theta})$, where the set of parameters $\vec{\theta}$ does not contain μ. In this case, where the m parameter set is $\mu, \theta_1, , \cdots, \theta_{m-1}$, the minimization of Eq. (6.105) provides the estimate of μ equal to

$$\hat{\mu} = N = \sum_{i=1}^{B} n_i . \tag{6.107}$$

In general, this property is not ensured for the least squares method.

From Wilks' theorem (see Sect. 10.8), if the model is correct, the distribution of the minimum value of χ_λ^2 can be asymptotically approximated with a χ^2 distribution (Eq. (3.53)) with a number of degrees of freedom equal to the number of bins B minus the number of fit parameters m or equal to $B - m - 1$ in the multinomial case with fixed N. χ_λ^2 can hence be used to determine a p-value (see Sect. 6.17) that provides a measure of the goodness of the fit.

If the number of observations is not sufficiently large, the distribution of χ_λ^2 for the specific problem may deviate from a χ^2 distribution but can still be determined by generating a sufficiently large number of Monte Carlo pseudo-experiments that reproduce the theoretical PDF, and the p-value can be computed accordingly.

In the limit where $B \to \infty$, the bin width tends to zero. Each number of entries n_i can have values equal to either zero or one, unless the PDF has singularities, and a number of entries greater than one have negligible probability, since $\mu_i \to 0$. The minimization of either Eq. (6.105) or Eq. (6.106) becomes equivalent, asymptotically, to the maximization of an unbinned extended likelihood function (Eq. (6.12)) or an unbinned ordinary likelihood function (Eq. (6.48)), respectively.

6.20 Constrained Fitting

In some cases, parameter estimates may be improved by taking in account constrains imposed by the specific problem. Examples of constrains are imposing energy and/or momentum conservation, or imposing that a set of tracks is generated at the same vertex, or fixing the invariant mass of a set of particles from the decay of a narrow resonance. Such constraints can be imposed as relations among the m parameters, expressed by a set of K equations:

$$c_k(\vec{\theta}) = 0, \tag{6.108}$$

with $k = 1, \cdots, K$.

The simplest approach to the minimization consists of expressing the parameters $\vec{\theta}$ as a function of a new set of $m - K$ parameters for which the constraints are satisfied. This is possible, in practice, only in a very limited number of cases. More in general, the minimization of a χ^2 or negative log likelihood function can be performed using the method of *Lagrange multipliers*, which consists of minimizing the following function instead of the usual χ^2 or negative twice the log likelihood function:

$$\mathcal{L}(\vec{y}; \vec{\theta}, \vec{\lambda}) = \chi^2(\vec{y}; \vec{\theta}) + \sum_{k=1}^{K} \lambda_k c_k(\vec{\theta}), \tag{6.109}$$

where $\vec{y}$ is the set of measurements and $\vec{\lambda} = (\lambda_1, \cdots, \lambda_K)$ is a set of additional parameters. The minimization should be performed simultaneously with respect to $\vec{\theta}$ and $\vec{\lambda}$ by solving the system of equations given by

$$\frac{\partial \mathcal{L}}{\partial \gamma_i} = 0, \tag{6.110}$$

where $(\gamma_1, \cdots, \gamma_{m+K}) = (\theta_1, \cdots, \theta_m, \lambda_1, \cdots, \lambda_K)$ is the set of all parameters, including the additional $\vec{\lambda}$. More details on constrained fit, in particular about the determination of the covariance matrix, can be found in [8] and [9].

6.21 Error Propagation

Given the measured values of the parameters $\theta_1, \cdots, \theta_m$ provided by an inference procedure, in some cases it may be necessary to evaluate a new set of parameters $\eta_1, \cdots, \eta_k$, determined as functions of the measured ones. The uncertainty on the original parameters propagates to the uncertainty on the new parameter set. The best option to determine the uncertainties on the new parameters would be to reparameterize the likelihood function using the new set of parameters, and then to perform a new maximum likelihood fit in terms of the new parameters. This procedure would directly provide estimates for $\eta_1, \cdots, \eta_k$ with their uncertainties. This is not possible when the details of the likelihood function and the original data set are not available. This is the case, for instance, when retrieving a measurement from a published paper.

In those cases, the simplest procedure may be to perform a local linear approximation of the function that transforms the measured parameters into the new ones.

Fig. 6.5 Transformation of variable $\eta = \eta(\theta)$ and visualization of the procedure of error propagation using local linear approximation

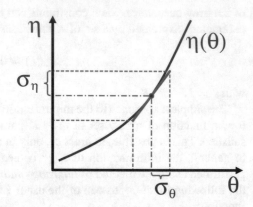

If the errors are sufficiently small, projecting them on the new variables, using the assumed linear approximation, leads to a sufficiently accurate result. This procedure is visualized in the simplest case of a one-dimensional transformation in Fig. 6.5.

In general, the covariance matrix H_{ij} of the transformed parameters can be obtained from the covariance matrix Θ_{kl} of the original parameters as follows:

$$H_{ij} = \sum_{k,l} \frac{\partial \eta_i}{\partial \theta_k} \frac{\partial \eta_j}{\partial \theta_l} \Theta_{kl} , \qquad (6.111)$$

or, in matrix form:

$$\boldsymbol{H} = \boldsymbol{A}^T \Theta \, \boldsymbol{A} , \qquad (6.112)$$

where

$$A_{ij} = \frac{\partial \eta_i}{\partial \theta_j}. \qquad (6.113)$$

6.22   Simple Cases of Error Propagation

Imagine multiplying a measured quantity x by a constant a:

$$y = ax . \qquad (6.114)$$

The corresponding uncertainty squared, applying Eq. (6.111), is

$$\sigma_y^2 = \left(\frac{dy}{dx}\right)^2 \sigma_x^2 = a^2 \sigma_x^2 . \qquad (6.115)$$

Hence the uncertainty on $y = ax$ is

$$\boxed{\sigma_{ax} = |a|\,\sigma_x\,.}$$

(6.116)

Consider a measured quantity z that is a function of two quantities x and y. Equation (6.111), also considering a possible correlation term, can be written as

$$\sigma_z^2 = \left(\frac{\partial z}{\partial x}\right)^2 \sigma_x^2 + \left(\frac{\partial z}{\partial y}\right)^2 \sigma_y^2 + 2\frac{\partial z}{\partial x}\frac{\partial z}{\partial y}\mathrm{cov}(x,\,y)\,.$$

(6.117)

For $z = x + y$, Eq. (6.117) gives

$$\boxed{\sigma_{x+y} = \sqrt{\sigma_x^2 + \sigma_y^2 + 2\,\rho\,\sigma_x\sigma_y}\,.}$$

(6.118)

and for $z = x - y$:

$$\boxed{\sigma_{x-y} = \sqrt{\sigma_x^2 + \sigma_y^2 - 2\,\rho\,\sigma_x\sigma_y}\,.}$$

(6.119)

Considering the product $z = x\,y$, the relative uncertainties should instead be added in quadrature, adding a possible correlation term:

$$\boxed{\left(\frac{\sigma_{x\,y}}{x\,y}\right) = \sqrt{\left(\frac{\sigma_x}{x}\right)^2 + \left(\frac{\sigma_y}{y}\right)^2 + \frac{2\,\rho\,\sigma_x\sigma_y}{x\,y}}\,.}$$

(6.120)

Similarly, for the ratio $z = x/y$, the correlation term should be subtracted from the sum in quadrature of relative uncertainties:

$$\boxed{\left(\frac{\sigma_{x/y}}{x/y}\right) = \sqrt{\left(\frac{\sigma_x}{x}\right)^2 + \left(\frac{\sigma_y}{y}\right)^2 - \frac{2\,\rho\,\sigma_x\sigma_y}{x\,y}}\,.}$$

(6.121)

If x and y are uncorrelated, Eq. (6.117) simplifies to

$$\sigma_z^2 = \left(\frac{\partial z}{\partial x}\right)^2 \sigma_x^2 + \left(\frac{\partial z}{\partial y}\right)^2 \sigma_y^2\,,$$

(6.122)

which, for the sum or difference of two uncorrelated variables, gives

$$\boxed{\sigma_{x+y} = \sigma_{x-y} = \sqrt{\sigma_x^2 + \sigma_y^2}}$$

(6.123)

and for the product and ratio gives

$$\boxed{\left(\frac{\sigma_{x\,y}}{x\,y}\right) = \left(\frac{\sigma_{x/y}}{x/y}\right) = \sqrt{\left(\frac{\sigma_x}{x}\right)^2 + \left(\frac{\sigma_y}{y}\right)^2}.}$$ (6.124)

For a power law $y = x^\alpha$, the error propagates as

$$\boxed{\left(\frac{\sigma_{x^\alpha}}{x^\alpha}\right) = |\alpha| \left(\frac{\sigma_x}{x}\right).}$$ (6.125)

The error of $\log x$ is equal to its relative error of x:

$$\boxed{\sigma_{\log x} = \frac{\sigma_x}{x}.}$$ (6.126)

6.23 Propagation of Asymmetric Errors

In Sect. 6.10, we observed that maximum likelihood fits may lead to asymmetric errors. The propagation of asymmetric errors and the combination of more measurements having asymmetric errors may require special care. If we have two measurements:

$$x = \hat{x}^{+\sigma_x^+}_{-\sigma_x^-} \quad \text{and} \quad y = \hat{y}^{+\sigma_y^+}_{-\sigma_y^-},$$ (6.127)

a naive extension of the sum in quadrature of errors, derived in Eq. (6.123), would lead to the (incorrect!) sum in quadrature of the positive and negative errors:

$$x + y = (\hat{x} + \hat{y})^{+\sqrt{(\sigma_x^+)^2+(\sigma_y^+)^2}}_{-\sqrt{(\sigma_x^-)^2+(\sigma_y^-)^2}}.$$ (6.128)

Though sometimes Eq. (6.128) has been used in real applications, it has no statistical motivation.

One reason why Eq. (6.128) is incorrect may be found in the central limit theorem. Uncertainties are related to the standard deviation of the distribution of a sample. In the case of an asymmetric (skew) distribution, asymmetric errors may be related to the skewness (Eq. (2.35)) of the distribution. Adding more random variables, each characterized by an asymmetric PDF should lead to a resulting PDF that approaches a Gaussian more than the original PDFs. Hence, a likelihood fit based on the combined PDF should lead to more symmetric errors compared to the individual fits. In Eq. (6.128), instead, the error asymmetry never decreases by adding more and more measurements all having the same error asymmetry.

One statistically correct way to propagate asymmetric errors on quantities, say $\vec{x}'$, that are expressed as functions of some original parameters, say $\vec{x}$, is to reformulate the fit problem in terms of the new parameters $\vec{x}'$ and to perform again the fit and error evaluation for the new quantities. This approach is sometimes not feasible when taking the result of a previous measurement, e.g.: from a publication, that does not specify the complete underlying likelihood model. In those cases, the treatment of asymmetric errors requires some assumptions on the underlying PDF model which is missing in the available documentation of the model's description.

Discussions about how to treat asymmetric errors are provided by Barlow in [10–12] using a frequentist approach. D'Agostini also discusses this subject in [13] using the Bayesian approach, reporting the method presented in Sect. 5.16, and demonstrating that potential problems, including bias, are present with naive error combination that consists in the sum in quadrature of positive and negative errors. In Sect. 6.24 below, the derivation from [12] is briefly presented as an example to demonstrate peculiar features of propagation and combination of asymmetric uncertainties.

6.24 Asymmetric Error Combination with a Linear Model

The following example presents a possible method to propagate asymmetric uncertainty that arises from a non-linear dependency on a nuisance parameter, e.g.: some source of systematic uncertainty. Consider that the uncertainty on a measured quantity x' arises from a non-linear dependence on another parameter x, i.e., $x' = f(x)$, which has a symmetric uncertainly σ. Figure 6.6 shows a simple case where x is a random variable distributed according to a Gaussian PDF, which is transformed into a variable x' through a piece-wise linear transformation, leading to an asymmetric PDF. The two straight-line sections, with different slopes, join with continuity at the central value of the original PDF. The resulting PDF of the transformed variable consists of two half-Gaussians, each corresponding to a 50% probability, having different standard deviation parameters, σ'_+ and σ'_-. Such a PDF is also called *bifurcated Gaussian* in some literature.

If the original measurement is

$$x = \hat{x} \pm \sigma ,$$

(6.129)

the transformation leads to a resulting measurement of the transformed variable:

$$x' = \hat{x}'{}^{+\sigma'_+}_{-\sigma'_-} ,$$

(6.130)

where σ'_+ and σ'_- depend on σ through factors equal to the two different slopes:

$$\begin{cases} \sigma'_+ = \sigma \cdot s_+ , \\ \sigma'_- = \sigma \cdot s_- , \end{cases}$$

(6.131)

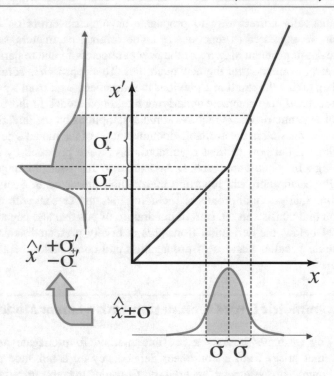

Fig. 6.6 Transformation of a variable x into a variable x' through a piece-wise linear transformation characterized by two different slopes. If x follows a Gaussian distribution with standard deviation σ, x' follows a bifurcated Gaussian, made of two Gaussian halves having different standard deviation parameters, σ'_+ and σ'_-

as evident in Fig. 6.6. One consequence of the shape of the transformed PDF is that the expected value of the transformed variable $\langle x' \rangle$ is different from the most likely value, $\hat{x}'$. The expected value of x' can be computed for a bifurcated Gaussian and is

$$\langle x' \rangle = \hat{x}' + \frac{1}{\sqrt{2\pi}} \left(\sigma'_+ - \sigma'_- \right) . \tag{6.132}$$

While the expected value of the sum of two variables is equal to the sum of the individual expected values (Eq. (2.24)), this addition rule does not hold for the most likely value of the sum of the two variables. Using the naive error combination from Eq. (6.128) could lead, for this reason, to a bias in the estimate of the sum of two quantities with asymmetric uncertainties, as evident from Eq. (6.132).

In the assumed case of a piece-wise linear transformation, in addition to Eq. (6.132), the expression for the variance of the transformed x' can also be considered:

$$\mathbb{V}\mathrm{ar}[x'] = \left(\frac{\sigma'_+ + \sigma'_-}{2}\right)^2 + \left(\frac{\sigma'_+ - \sigma'_-}{2}\right)^2 \left(1 - \frac{2}{\pi}\right), \qquad (6.133)$$

as well as the expression for the unnormalized skewness, defined in Eq. (2.36):

$$\gamma[x'] = \frac{1}{2\pi}\left[2(\sigma_+^3 - \sigma_-^3) - \frac{3}{2}(\sigma_+ - \sigma_-)(\sigma_+^3 + \sigma_-^3) + \frac{1}{\pi}(\sigma_+ - \sigma_-)^3\right]. \qquad (6.134)$$

The three equations, Eqs. (6.132), (6.133), and (6.134), allow to transform the three quantities given by the measured value $\hat{x}'$ and the two asymmetric error components σ'_+ and σ'_- into three other quantities $\langle x' \rangle$, $\mathbb{V}\mathrm{ar}[x']$, and $\gamma[x']$. The advantage of this transformation is that the expected value, the variance, and the unnormalized skewness add linearly when adding random variables, and this allows an easier combination of uncertainties.

Imagine we want to add two measurements affected by asymmetric errors, say:

$$x_1 = \hat{x}_1 {}^{+\sigma_1^+}_{-\sigma_1^-}, \qquad (6.135)$$

$$x_2 = \hat{x}_2 {}^{+\sigma_2^+}_{-\sigma_2^-}. \qquad (6.136)$$

The addition of the two measurements and the combination of their uncertainties can proceed through the following steps:

- The expected value, variance, and unnormalized skewness of x_1 and x_2 are computed, individually, from $\hat{x}_1$ and $\hat{x}_2$ and their corresponding asymmetric uncertainties using Eqs. (6.132), (6.133), and (6.134).
- The expected value, variance, and unnormalized skewness of the sum of x_1 and x_2 are computed as the sum of the corresponding individual quantities:

$$\langle x_1 + x_2 \rangle = \langle x_1 \rangle + \langle x_2 \rangle, \qquad (6.137)$$

$$\mathbb{V}\mathrm{ar}[x_1 + x_2] = \mathbb{V}\mathrm{ar}[x_1] + \mathbb{V}\mathrm{ar}[x_2], \qquad (6.138)$$

$$\gamma[x_1 + x_2] = \gamma[x_1] + \gamma[x_2]. \qquad (6.139)$$

- Using numerical techniques, the system of the three equations that express the relations between $\langle x_1 + x_2 \rangle$, $\mathbb{V}\mathrm{ar}[x_1 + x_2]$, and $\gamma[x_1 + x_2]$, and $\hat{x}_{1+2}, \sigma_{1+2}^+$, and σ_{1+2}^- from Eqs. (6.132), (6.133), and (6.134) is inverted to obtain the estimate for $x_{1+2} = x_1 + x_2$ and its corresponding asymmetric uncertainty components, providing the estimate with its asymmetric uncertainty:

$$\hat{x}_{1+2} \, {}^{+\sigma^+_{1+2}}_{-\sigma^-_{1+2}} \, . \tag{6.140}$$

The case of a parabolic dependence is also considered in [12], and a procedure to estimate $\hat{x}_{1+2} \, {}^{+\sigma^+_{1+2}}_{-\sigma^-_{1+2}}$ under this alternative assumption about the underlying PDF model is obtained. Any estimate of the sum of two measurements affected by asymmetric errors requires an assumption of an underlying PDF model. The result may be more or less sensitive to the assumed model, depending case by case. For this reason, it is not possible to define a unique model-independent prescription to add, or more in general combine, measurements having asymmetric uncertainties.

References

1. H. Cramér, *Mathematical Methods of Statistics* (Princeton University Press, Princeton, 1946)
2. C.R. Rao, Information and the accuracy attainable in the estimation of statistical parameters. Bull. Calcutta Math. Soc. **37**, 8189 (1945)
3. W. Eadie, D. Drijard, F. James, M. Roos, B. Saudolet, *Statistical Methods in Experimental Physics* (North Holland, London, 1971)
4. F. James, M. Roos, MINUIT: function minimization and error analysis. CERN Computer Centre Program Library, Geneve Long Write-up No. D506 (1989)
5. R. Brun, F. Rademakers, ROOT—an object oriented data analysis framework. Proceedings AIHENP96 Workshop, Lausanne (1996) Nucl. Inst. Meth. **A389** 81–86 (1997). http://root.cern.ch/
6. iminuit, Jupyter-friendly Python frontend for MINUIT2 in C++. https://pypi.org/project/iminuit/
7. S. Baker, R. Cousins, Clarification of the use of chi-square and likelihood functions in fit to histograms. Nucl. Instrum. Meth. **A221**, 437–442 (1984)
8. F. James, *Statistical Methods in Experimental Physics*, 2nd edn. (World Scientific, Singapore, 2006)
9. O. Behnke, et al. (ed.), *Data analysis in High Energy Physics* (Wiley-VCH, 2013)
10. R. Barlow, Asymmetric errors, in *Proceedings of PHYSTAT2003* (SLAC, Stanford, 2003). http://www.slac.stanford.edu/econf/C030908/
11. R. Barlow, Asymmetric statistical errors (2004). arXiv:physics/0406120v1
12. R. Barlow, Asymmetric systematic errors (2003). arXiv:physics/0306138
13. G. D'Agostini, Asymmetric uncertainties sources, treatment and potential dangers (2004). arXiv:physics/0403086
14. G. D'Agostini, *Bayesian Reasoning in Data Analysis: A Critical Introduction* (World Scientific, Hackensack, 2003)

Combining Measurements

7

7.1   Introduction

The problem of combining two or more measurements of the same unknown quantity θ can be addressed in general by building a likelihood function that combines two or more data samples used in the individual measurements. If the data samples are independent, the combined likelihood function is given by the product of the individual likelihood functions and depends on the unknown parameter present in each of them. At least some of which are in common among different measurements, in particular the parameter we want to combine. The minimization of the combined likelihood function provides an estimate of θ that takes into account all the individual data samples. This approach requires that the likelihood functions and the data sets are available for each individual measurement. This is the method usually adopted to extract information from multiple data samples within the same experiment, as discussed in Sect. 7.3.

Anyway, this strategy is not always pursuable, either because of the intrinsic complexity of the problem or because the original data samples and/or the probability models that are required to build the likelihood function are not available, and only the final individual results are known, with their uncertainties. This is very often the case when combining results taken from publications.

In case a Gaussian approximation is sufficiently accurate, as assumed in Sect. 6.18, the minimum χ^2 method can be used to perform a combination of the measurements considering their uncertainties and correlation. This is discussed in Sect. 7.4 and the followings.

L. Lista, *Statistical Methods for Data Analysis*, Lecture Notes in Physics 1010,
https://doi.org/10.1007/978-3-031-19934-9_7

7.2   Control Regions

Consider the model discussed in Sect. 6.14 with a signal peak around the value $m = 3.1\,\text{GeV}$, reported in Fig. 6.2. The two regions with $m < 3\,\text{GeV}$ and $m > 3.2\,\text{GeV}$ can be considered as *control regions*, where the amount of signal can be considered negligible, while the region with $3.04\,\text{GeV} < m < 3.18\,\text{GeV}$ can be taken as *signal region*.

The background yield in the signal region under the signal peak can be determined from the observed number of events in the control regions, which contain a negligible amount of signal, interpolated to the signal region by applying a scale factor given by the ratio of areas of the estimated background curve in the signal and control region areas. The predicted background distribution is an exponential distribution, in the considered case. If control regions are also contaminated by a significant signal fraction, this method may not lead to a correct signal subtraction in the signal region, and hence the estimate of the signal yield may not be correct.

When performing the fit described in Sect. 6.14, the amount of background is already effectively constraint by assuming an exponential shape for the background, whose slope parameter λ is determined, together with the other fit parameters, directly from the entire data sample.

7.3   Simultaneous Fits

A more general approach to the simultaneous fit of signal and background, or, more in general, to more components of the data sample, can also consider a possible signal contamination of the different data regions. The problem can be formalized in general as follows for the simplest case of two data samples, but the generalization to more samples is straightforward. Consider two data samples, $\vec{x} = (x_1, \cdots, x_h)$ and $\vec{y} = (y_1, \cdots, y_k)$. The likelihood functions $L_x(\vec{x}; \vec{\theta})$ and $L_y(\vec{y}; \vec{\theta})$ for the two individual data samples depend on a parameter set $\vec{\theta} = (\theta_1, \cdots, \theta_m)$. In particular, L_x and L_y may individually depend only on a subset of the comprehensive parameter set $\vec{\theta}$. For instance, some parameters may determine the yields of signal and background components, like in the example from the previous section. The combined likelihood function that comprises both data samples, assuming their independence, can be written as the product of the two likelihood functions:

$$L_{x,y}(\vec{x}, \vec{y}; \vec{\theta}) = L_x(\vec{x}; \vec{\theta}) L_y(\vec{y}; \vec{\theta}). \tag{7.1}$$

The likelihood function in Eq. (7.1) can be maximized to fit the parameters $\vec{\theta}$ simultaneously from the two data samples $\vec{x}$ and $\vec{y}$. Imagine now that an experiment wants to measure the production cross section of a rare signal affected by a large background, based on the different shapes of the distributions of an observable variable x in different physical processes. One of the easiest cases is when the distribution of x has a peaked shape for the signal and a smoother distribution for the backgrounds, like in Fig. 6.2. More complex situations may be present in realistic problems.

Imagine also that the background yield is not predicted with good precision. In order to measure the signal yield, one may select additional data samples, called *control samples*, enriched in background events with a negligible or anyway small contribution from the signal, and use those samples to determine the background yield from data, without relying on theory predictions.

A real case could be the measurement of the production cross section of events that contain a single top quark at the Large Hadron Collider. This signal is affected by background due to top-quark pair production. The selection of a *signal sample* containing single-top events relies on the presence of a single jet identified as produced by a b quark (the top quark decays into a b quark and a W boson). A control sample is defined by requiring the presence of two b quarks instead of one. With this requirement, the control sample contains a very small contribution from single-top events, which have a single b quark, unless a second one contaminates the event, either from the decay of another particle or from a spurious reconstruction and is dominated by the top-pair background. A small cross contamination of the signal and control sample is anyway expected, because the requirement on the number of b jets, one or two, does not perfectly select either the single-top signal or the top-pair background.

A simultaneous fit using both the signal and control samples allows to determine at the same time the yields of both the single-top signal and the top-pair background. The distribution of some observed quantities used in the fit can be taken from simulation, separately for signal and background. The signal and background yields are free parameters. The background yield is mainly determined from the control sample, where a small amount of signal is present. The signal yield is mainly determined from the signal sample, where the amount of background may be assumed to scale from the control region according to some constant ratio provided by simulation if the data sample is not sufficient to exploit the background distribution to separate signal and background in the signal region.

The modeling of signal and background distributions from simulation may be affected by uncertainties, whose propagation to the measured signal yield produces contributions to the total systematic uncertainty that needs to be estimated.

7.4 Weighted Average

Imagine we have two measurements of the same quantity x, which are

$$x = \hat{x}_1 \pm \sigma_1 , \quad x = \hat{x}_2 \pm \sigma_2 . \tag{7.2}$$

Assuming a Gaussian distribution for $\hat{x}_1$ and $\hat{x}_2$ and no correlation between the two measurements, the following χ^2 can be built:

$$\chi^2 = \frac{(x - \hat{x}_1)^2}{\sigma_1^2} + \frac{(x - \hat{x}_2)^2}{\sigma_2^2} . \tag{7.3}$$

The value $x = \hat{x}$ that minimizes the χ^2 can be found by imposing

$$\frac{\partial \chi^2}{\partial x} = 2\frac{(x - \hat{x}_1)}{\sigma_1^2} + 2\frac{(x - \hat{x}_2)}{\sigma_2^2} = 0 , \tag{7.4}$$

which gives

$$x = \hat{x} = \frac{\dfrac{\hat{x}_1}{\sigma_1^2} + \dfrac{\hat{x}_2}{\sigma_2^2}}{\dfrac{1}{\sigma_1^2} + \dfrac{1}{\sigma_2^2}} . \tag{7.5}$$

The uncertainty of the estimate $\hat{x}$ can be evaluated using Eq. (6.25):

$$\frac{1}{\sigma_{\hat{x}}^2} = -\frac{\partial^2 \log L}{\partial x^2} = \frac{1}{2}\frac{\partial^2 \chi^2}{\partial x^2} = \frac{1}{\sigma_1^2} + \frac{1}{\sigma_2^2} . \tag{7.6}$$

The estimate in Eq. (7.5) is a *weighted average* and can be generalized for N measurements as

$$\hat{x} = \sum_{i=1}^{N} w_i \hat{x}_i , \quad \text{where} \quad w_i = \frac{1/\sigma_i^2}{1/\sigma^2} , \tag{7.7}$$

with $1/\sigma^2 = \sum_{i=1}^{N} 1/\sigma_i^2$, to ensure that $\sum_{i=1}^{N} w_i = 1$. The variance of $\hat{x}$ is

$$\mathbb{V}\text{ar}[\hat{x}] = \sum_{i=1}^{N} w_i^2 \mathbb{V}\text{ar}[\hat{x}_i] = \sum_{i=1}^{N} w_i^2 \sigma_i^2 = \sum_{i=1}^{N} \left(\frac{1/\sigma_i^2}{1/\sigma^2}\right)^2 \sigma_i^2 = \sigma^2 . \tag{7.8}$$

Therefore, the error on $\hat{x}$ is given by

$$\sigma_{\hat{x}} = \sigma = \frac{1}{\sqrt{\sum_{i=1}^{N} 1/\sigma_i^2}} . \tag{7.9}$$

If all σ_i are identical and equal to σ_0, Eq. (7.9) gives the standard deviation of the arithmetic mean, already introduced in previous sections:

$$\sigma_{\hat{x}} = \frac{\sigma_0}{\sqrt{N}}. \tag{7.10}$$

7.5 Combinations with χ^2 in n Dimensions

The χ^2 defined in Eq. (7.3) can be generalized to the case of n measurements $\hat{\vec{x}} = (\hat{x}_1, \cdots, \hat{x}_n)$ of the n parameters $\vec{x} = (x_1, \cdots, x_n)$ which, on turn, depend on another parameter set $\vec{\theta} = (\theta_1, \cdots, \theta_m)$: $x_i = x_i(\theta_1, \cdots, \theta_m)$. If the covariance matrix of the measurements $(\hat{x}_1, \cdots, \hat{x}_n)$ is σ with elements σ_{ij}, the χ^2 can be written as follows:

$$\chi^2 = \sum_{i,\,j=1}^{n} (\hat{x}_i - x_i)\,\sigma_{ij}^{-1}(\hat{x}_j - x_j) = (\hat{\vec{x}} - \vec{x})^T \sigma^{-1}(\hat{\vec{x}} - \vec{x}) = \tag{7.11}$$

$$= (\hat{x}_1 - x_1, \cdots, \hat{x}_n - x_n) \begin{pmatrix} \sigma_{11} & \cdots & \sigma_{1n} \\ \vdots & \ddots & \vdots \\ \sigma_{n1} & \cdots & \sigma_{nn} \end{pmatrix}^{-1} \begin{pmatrix} \hat{x}_1 - x_1 \\ \cdots \\ \hat{x}_n - x_n \end{pmatrix}. \tag{7.12}$$

The χ^2 can be minimized to determine the estimates $\hat{\theta}_1, \cdots, \hat{\theta}_m$ of the unknown parameters $\theta_1, \cdots, \theta_m$, with the proper error matrix.

Examples of application of this method are the combinations of electroweak measurements performed with data taken at the Large Electron–Positron Colliderx (LEP) [3] at CERN and with the precision measurement performed at the Stanford Linear Collider (SLC) at SLAC [2, 4] in the context of the LEP Electroweak Working Group [5] and the GFitter Group [1]. The effect of radiative corrections that depend on the top-quark and Higgs-boson masses is considered in the predictions, above indicated in the relation $x_i = x_i(\theta_1, \cdots, \theta_m)$. The combination of precision electroweak measurements with this approach gave indirect estimates of the top-quark mass before its discovery at Tevatron and a less precise estimate of the Higgs-boson mass before the beginning of the LHC data taking, where eventually the Higgs boson was discovered. In both cases, the indirect estimates were in agreement with the measured masses. The global fit performed by the GFitter group is shown in Fig. 7.1 for the prediction of the mass of the W boson and the top quark.

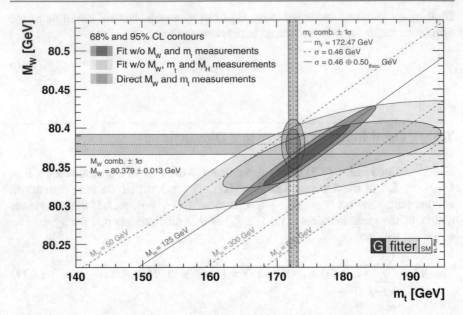

Fig. 7.1 Contours at the 68% and 95% confidence level in the plane mass of the W boson vs mass of the top quark, (m_W, m_t), from global electroweak fits including (blue) or excluding (gray) the measurement of the Higgs boson mass m_H. The direct measurements of m_W and m_t (green) are shown as comparison. The fit was performed by the GFitter group [1], and the figure Reprinted from [2], © The LEP Electroweak Working Group. Reproduced under CC-BY-4.0 license

7.6 The Best Linear Unbiased Estimator (BLUE)

Let us consider the case of two measurements of the same quantity x:

$$x = \hat{x}_1 \pm \sigma_1 , \quad x = \hat{x}_2 \pm \sigma_2 , \qquad (7.13)$$

which have a correlation coefficient ρ. A special case with $\rho = 0$ was discussed in Sect. 7.4. Taking into account the correlation term, the χ^2 can be written as

$$\chi^2 = (x - \hat{x}_1 \; x - \hat{x}_2) \begin{pmatrix} \sigma_1^2 & \rho \, \sigma_1 \sigma_2 \\ \rho \, \sigma_1 \sigma_2 & \sigma_2^2 \end{pmatrix}^{-1} \begin{pmatrix} x - \hat{x}_1 \\ x - \hat{x}_2 \end{pmatrix} . \qquad (7.14)$$

Applying the same minimization procedure used to obtain the weighted average in Eq. (7.5) gives, for a non-null correlation, the following combined estimate:

$$\hat{x} = \frac{\hat{x}_1(\sigma_2^2 - \rho \, \sigma_1 \sigma_2) + \hat{x}_2(\sigma_1^2 - \rho \, \sigma_1 \sigma_2)}{\sigma_1^2 - 2 \, \rho \, \sigma_1 \sigma_2 + \sigma_2^2} . \qquad (7.15)$$

The uncertainty on $\hat{x}$ is

$$\sigma_{\hat{x}}^2 = \frac{\sigma_1^2 \sigma_2^2 (1 - \rho^2)}{\sigma_1^2 - 2\rho\,\sigma_1\sigma_2 + \sigma_2^2} . \tag{7.16}$$

The general case of more than two measurements proceeds in a similar way.

It is possible to demonstrate that the χ^2 minimization is equivalent to find the *best linear unbiased estimate* (BLUE), i.e., the unbiased linear combination of the measurements $\hat{\vec{x}} = (\hat{x}_1, \cdots, \hat{x}_N)$ with known covariance matrix $\boldsymbol{\sigma} = (\sigma_{ij})$ that gives the lowest variance. Introducing the set of weights $\vec{w} = (w_1, \cdots, w_N)$, a linear combination of the $\hat{x}_i$ can be written as

$$\hat{x} = \sum_{i=1}^{N} w_i \hat{x}_i = \hat{\vec{x}} \cdot \vec{w} . \tag{7.17}$$

The condition of a null bias is $\langle \hat{x} \rangle = x$, where x is the unknown true value. If the individual measurements are unbiased, we have $x = \langle x_i \rangle$ for all i. Therefore, $\langle \hat{x} \rangle = x$ if

$$\sum_{i=1}^{N} w_i = 1 . \tag{7.18}$$

For $N = 2$, for instance, Eq. (7.15) defines a linear combination with weights

$$\begin{cases} w_1 = \dfrac{\sigma_2^2 - \rho\,\sigma_1\sigma_2}{\sigma_1^2 - 2\rho\,\sigma_1\sigma_2 + \sigma_2^2} , \\[2mm] w_2 = \dfrac{\sigma_1^2 - \rho\,\sigma_1\sigma_2}{\sigma_1^2 - 2\rho\,\sigma_1\sigma_2 + \sigma_2^2} . \end{cases} \tag{7.19}$$

The variance of $\hat{x}$ can be expressed as

$$\sigma_{\hat{x}}^2 = \vec{w}^{\mathrm{T}} \boldsymbol{\sigma}\, \vec{w} . \tag{7.20}$$

It can be shown [6] that the weights that minimize the variance $\sigma_{\hat{x}}^2$ in Eq. (7.20) are given by the following expression:

$$\vec{w} = \frac{\boldsymbol{\sigma}^{-1} \vec{u}}{\vec{u}^{\mathrm{T}} \boldsymbol{\sigma}^{-1} \vec{u}} , \tag{7.21}$$

where $\vec{u}$ is the vector having all elements equal to unity $\vec{u} = (1, \cdots, 1)$.

The interpretation of Eq. (7.15) becomes more intuitive [7] if we introduce the *common error*, defined as

$$\sigma_C = \rho \, \sigma_1 \sigma_2 \, . \tag{7.22}$$

Imagine, for instance, that the two measurements are affected by a common uncertainty, while the other uncertainties are uncorrelated. For instance, it could be the uncertainty on the integrated luminosity in case of a cross section measurement. In that case, the two measurements can also be written as

$$x = \hat{x}_1 \pm \sigma_1' \pm \sigma_C \, , \quad x = \hat{x}_2 \pm \sigma_2' \pm \sigma_C \, , \tag{7.23}$$

where $\sigma_1'^2 = \sigma_1^2 - \sigma_C^2$ and $\sigma_2'^2 = \sigma_2^2 - \sigma_C^2$. This is clearly possible only if $\sigma_C \leq \sigma_1$ and $\sigma_C \leq \sigma_2$, which is equivalent to require that the weights w_1 and w_2 in Eq. (7.19) are both positive. Equation (7.15), in that case, can also be written as

$$\hat{x} = \frac{\dfrac{\hat{x}_1}{\sigma_1^2 - \sigma_C^2} + \dfrac{\hat{x}_2}{\sigma_2^2 - \sigma_C^2}}{\dfrac{1}{\sigma_1^2 - \sigma_C^2} + \dfrac{1}{\sigma_2^2 - \sigma_C^2}} = \frac{\dfrac{\hat{x}_1}{\sigma_1'^2} + \dfrac{\hat{x}_2}{\sigma_2'^2}}{\dfrac{1}{\sigma_1'^2} + \dfrac{1}{\sigma_2'^2}} \tag{7.24}$$

with variance

$$\sigma_{\hat{x}}^2 = \frac{1}{\dfrac{1}{\sigma_1^2 - \sigma_C^2} + \dfrac{1}{\sigma_2^2 - \sigma_C^2}} + \sigma_C^2 = \frac{1}{\dfrac{1}{\sigma_1'^2} + \dfrac{1}{\sigma_2'^2}} + \sigma_C^2 \, . \tag{7.25}$$

Equation (7.24) is equivalent to the weighted average in Eq. (7.5), where the errors $\sigma_1'^2$ and $\sigma_2'^2$ are used to determine the weights instead of σ_1 and σ_2. Equation (7.25) shows that the uncertainty contribution term σ_C has to be added in quadrature to the expression that gives the error for the ordinary weighted average when correlation is present (Eqs. (7.6) and (7.9)).

More in general, as evident from Eq. (7.15), weights with the BLUE method can also become negative. This may lead to counter-intuitive results. In particular, a combination of two measurements may also lie outside the range delimited by the two individual measured values. Also, when combining two measurements which have a correlation coefficient $\rho = \sigma_1/\sigma_2$, the weight of the second measurement is zero, and hence the combined value is not influenced by $\hat{x}_2$. Conversely, if $\rho = \sigma_2/\sigma_1$, the combined value is not influenced by $\hat{x}_1$.

> **Example 7.1 - Reusing Multiple Times the Same Measurement Does Not Improve a Combination**
> Assume we have a single measurement, $\hat{x} \pm \sigma$, and we want to use it twice to determine again the same quantity. If one neglects the correlation coefficient $\rho = 1$, the uncertainty σ would be reduced, for free, by a factor of $\sqrt{2}$, which clearly would not make sense. The correct use of Eqs. (7.15) and (7.16) leads, when $\rho = 1$, to $\hat{x} \pm \sigma$, i.e., as expected, no precision is gained by using the same measurement twice in a combination.

7.7 Quantifying the Importance of Individual Measurements

The BLUE method, as well as the standard weighted average, provides weights of individual measurements that determine the combination of the measured values. BLUE weights, anyway, can become negative, and this does not allow to use weights to quantify the importance of each individual measurement used in a combination. One approach sometimes adopted in literature consists in quoting as *relative importance* (RI) of the measurement $\hat{x}_i$ a quantity proportional to the absolute value of the corresponding weight w_i, usually defined as

$$\mathrm{RI}_i = \frac{|w_i|}{\displaystyle\sum_{i=1}^{N} |w_i|} .$$

(7.26)

The choice of the denominator is justified by the normalization condition:

$$\sum_{i=1}^{N} \mathrm{RI}_i = 1 .$$

(7.27)

RIs were quoted, for instance, in combinations of top-quark mass measurements at the Tevatron and at the LHC [8, 9].

The questionability of the use of RIs was raised [10] because it violates the *combination principle*: in case of three measurements, say $\hat{x}$, $\hat{y}_1$, and $\hat{y}_2$, the RI of the measurement $\hat{x}$ changes whether the three measurements are combined all together, or if $\hat{y}_1$ and $\hat{y}_2$ are first combined into $\hat{y}$, and then the measurement $\hat{x}$ is combined with the partial combination $\hat{y}$.

Proposed alternatives to RI are based on the Fisher information, defined in Eq. (5.84), which is also used to determine a lower bound to the variance of an estimator (see Sect. 6.3). For a single parameter, the Fisher information is given by

$$J = J_{11} = \vec{u}^{\mathrm{T}} V \vec{u} = \frac{1}{\sigma_{\hat{x}}^2} .$$

(7.28)

Two quantities are proposed in [10] to replace RI, the intrinsic information weight (IIW), defined as

$$\text{IIW}_i = \frac{1/\sigma_i^2}{1/\sigma_{\hat{x}}^2} = \frac{1/\sigma_i^2}{J} \qquad (7.29)$$

and the marginal information weight (MIW), defined as follows:

$$\text{MIW}_i = \frac{\Delta J_i}{J} = \frac{J - J_{\{1, \cdots, n\}-\{i\}}}{J} , \qquad (7.30)$$

i.e., the relative difference of the Fisher information of the combination and the Fisher information of the combination obtained by excluding the i measurement. Both IIW and MIW do not obey a normalization condition.

For IIW, the quantity IIW_{corr} can be defined such that

$$\sum_{i=1}^{N} \text{IIW}_i + \text{IIW}_{\text{corr}} = 1 . \qquad (7.31)$$

IIW_{corr} represents the weight assigned to the correlation interplay, not assignable to any individual measurement and is given by

$$\text{IIW}_{\text{corr}} = \frac{1/\sigma_{\hat{x}}^2 - \sum_{i=1}^{N} 1/\sigma_i^2}{1/\sigma_{\hat{x}}^2} = \frac{J - \sum_{i=1}^{N} 1/\sigma_i^2}{J} . \qquad (7.32)$$

The quantities IIW and MIW are reported, instead of RI, in papers presenting the combination of LHC and Tevatron measurements related to top-quark physics [11–13]. The properties of the BLUE weights, RI, and the two proposed alternatives to RI are reported in Table 7.1.

Table 7.1 Properties of the different definitions of the importance of a measurement within a BLUE combination

Weight type		≥ 0	$\sum_{i=1}^{N} = 1$	Consistent with partial combination
BLUE coefficient	w_i	✗	✓	✓
Relative importance	$|w_i|/\sum_{i=1}^{N} |w_i|$	✓	✓	✗
Intrinsic information weight	IIW_i	✓	✗	✓
Marginal information weight	MIW_i	✓	✗	✓

7.8 Negative Weights

Negative weights may arise in BLUE combinations in the presence of large correlation. Figure 7.2 shows the dependence of the BLUE coefficient w_2 and the ratio of uncertainties $\sigma_{\hat{x}}^2/\sigma_1^2$ as from Eqs. (7.15) and (7.20) as a function of the correlation ρ for different values of σ_2/σ_1 in the combination of two measurements. The uncertainty ratio $\sigma_{\hat{x}2}/\sigma_1^2$ increases as a function of ρ for $\rho < \sigma_1/\sigma_2$; for $\rho = \sigma_1/\sigma_2$, it reaches a maximum, which also corresponds to $w_2 = 0$ and

Fig. 7.2 BLUE coefficient for the second measurement w_2 (top) and combined BLUE variance $\sigma_{\hat{x}}^2$ divided by σ_1^2 (bottom) as a function of the correlation ρ between the two measurements for various fixed values of the ratio σ_2/σ_1

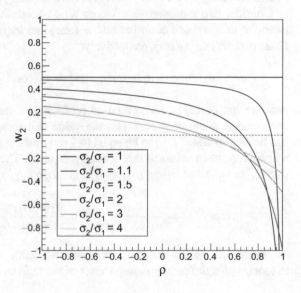

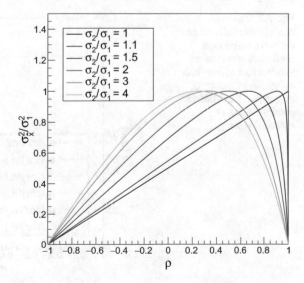

$\mathrm{MIW}_2 = 0$. For $\rho > \sigma_1/\sigma_2$, w_2 becomes negative and $\sigma_{\hat{x}^2}/\sigma_1^2$ decreases for increasing ρ. The dependence on ρ may also be very steep for some value range of σ_2/σ_1. The case for which $w_2 = 0$ should not be interpreted as the measurement $\hat{x}_2$ not being used in the combination.

In cases where the correlation coefficient ρ is not known with good precision, the assumption $\rho = 1$ is a conservative choice only if the uncorrelated contributions are dominant in the total uncertainty. As seen above, in case of negative weights, ρ should be accurately determined, because the uncertainty may strongly depend on ρ, and assuming $\rho = 1$ may result in underestimating the combination uncertainty.

Consider two measurements whose total uncertainties are given by the combination of uncorrelated contributions, $\sigma_1(\mathrm{unc})$ and $\sigma_2(\mathrm{unc})$, and correlated contributions, $\sigma_1(\mathrm{cor})$ and $\sigma_2(\mathrm{cor})$, respectively:

$$\sigma_1^2 = \sigma_1(\mathrm{cor})^2 + \sigma_1(\mathrm{unc})^2 \,, \quad \sigma_2^2 = \sigma_2(\mathrm{cor})^2 + \sigma_2(\mathrm{unc})^2 \,. \tag{7.33}$$

Assume that $\rho(\mathrm{cor})$ is the correlation coefficient of the correlated terms. The most conservative value of $\rho(\mathrm{cor})$, i.e., the value that maximizes the total uncertainty, can be demonstrated [10] to be equal to 1 only for $\sigma_2(\mathrm{cor})/\sigma_1(\mathrm{cor}) < (\sigma_1/\sigma_1(\mathrm{cor}))^2$, where it has been assumed that $\sigma_1(\mathrm{cor}) \leq \sigma_2(\mathrm{cor})$. The most conservative choice of $\rho(\mathrm{cor})$, for values of $\sigma_2(\mathrm{cor})/\sigma_1(\mathrm{cor})$ larger than $(\sigma_1/\sigma_1(\mathrm{cor}))^2$, is equal to

$$\rho(\mathrm{cor})^{\mathrm{cons.}} = \frac{\sigma_1^2}{\sigma_1(\mathrm{cor})\sigma_2(\mathrm{cor})} = \frac{\sigma_1(\mathrm{cor})/\sigma_2(\mathrm{cor})}{(\sigma_1(\mathrm{cor})/\sigma_1)^2} < 1 \,. \tag{7.34}$$

Figure 7.3 shows the most conservative value of $\rho(\mathrm{cor})$ as a function of $\sigma_1(\mathrm{cor})/\sigma_1$ for different possible values of the ratio $\sigma_1(\mathrm{cor})/\sigma_2(\mathrm{cor})$.

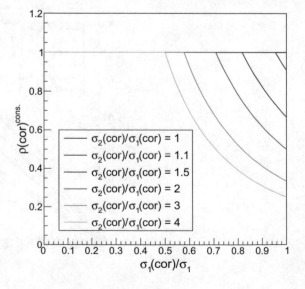

Fig. 7.3 The most conservative value of an unknown correlation coefficient $\rho(\mathrm{cor})$ of the uncertainties $\sigma_1(\mathrm{cor})$ and $\sigma_2(\mathrm{cor})$ as a function of $\sigma_1(\mathrm{cor})/\sigma_1$, for different possible values of $\sigma_2(\mathrm{cor})/\sigma_1(\mathrm{cor}) \geq 1$

7.9 Iterative Application of the BLUE Method

The BLUE method is unbiased by construction if the true uncertainties and their correlations are known. Anyway, it can be proven that BLUE combinations may exhibit a bias if uncertainty estimates are used in place of the true ones and in particular if the uncertainty estimates depend on measured values. For instance, when contributions to the total uncertainty are known as relative uncertainties, the actual uncertainty estimates are obtained as the product of the relative uncertainties times the measured values. An iterative application of the BLUE method can be implemented to mitigate such a bias.

L. Lyons et al. remarked in [14] the limitations of the BLUE method in the combination of lifetime measurements where uncertainty estimates $\hat{\sigma}_i$ of the true unknown uncertainties σ_i were used, and those estimates had a dependency on the measured lifetime. They also demonstrated that the application of the BLUE method violates, in that case, the combination principle: if the set of measurements is split into a number of subsets, then the combination is first performed in each subset, and finally all subset combinations are combined into a single grand combination, the obtained result differs from the single combination of all individual results of the entire set.

Reference [14] recommends applying iteratively the BLUE method, rescaling at each iteration the uncertainty estimates according to the combined value obtained with the BLUE method in the previous iteration, until the sequence converges to a stable result. It was also proven that the bias of the BLUE estimate is reduced compared to the standard application of the BLUE method. A more extensive study of the iterative application of the BLUE method is also available in [15].

References

1. A Generic Fitter Project for HEP Model Testing. http://project-gfitter.web.cern.ch/project-gfitter/
2. The Gfitter Group., J. Haller, A. Hoecker, et al., Update of the global electroweak fit and constraints on two-Higgs-doublet models. Eur. Phys. J. C **78**, 675 (2018)
3. The ALEPH, DELPHI, L3, OPAL Collaborations, the LEP Electroweak Working Group, Electroweak measurements in electron-positron collisions at W-Boson-pair energies at LEP. Phys. Rep. **532**, 119 (2013)
4. The ALEPH, DELPHI, L3, OPAL, SLD Collaborations, the LEP Electroweak Working Group, the SLD Electroweak and Heavy Flavour Groups, Precision electroweak measurements on the Z resonance. Phys. Rep. **427**, 257 (2006)
5. The LEP Electroweak Working Group. http://lepewwg.web.cern.ch/LEPEWWG/
6. L. Lyons, D. Gibaut, P. Clifford, How to combine correlated estimates of a single physical quantity. Nucl. Instrum. Meth. **A270**, 110–117 (1988)
7. H. Greenlee, *Combining CDF and D0 Physics Results*. Fermilab Workshop on Confidence Limits (2000)
8. The CDF and D0 Collaborations, *Combination of CDF and D0 Results on the Mass of the Top Quark Using up to 5.8 fb^{-1} of Data*. FERMILAB-TM-2504-E, CDF-NOTE-10549, D0-NOTE-6222 (2011)

9. The ATLAS and CMS Collaborations, *Combination of ATLAS and CMS Results on the Mass of the Top Quark Using up to* $4.9fb^{-1}$ *of Data*. ATLAS-CONF-2012-095, CMS-PAS-TOP-12-001 (2012)

10. A. Valassi, R. Chierici, Information and treatment of unknown correlations in the combination of measurements using the BLUE method. Eur. Phys. J. C **74** 2717 (2014)

11. ATLAS and CMS Collaborations, *Combination of ATLAS and CMS Results on the Mass of the Top-Quark Using up to 4.9 fb -1 of* $\sqrt{s}$ = 7 TeV LHC Data. ATLAS-CONF-2013-102, CMS-PAS-TOP-13-005 (2013)

12. ATLAS and CMS Collaborations, *Combination of ATLAS and CMS ttbar Charge Asymmetry Measurements Using LHC Proton-Proton Collisions at* $\sqrt{s}$ = 7 TeV. ATLAS-CONF-2014-012, CMS-PAS-TOP-14-006 (2014)

13. ATLAS, CMS, CDF and D0 Collaborations, *First Combination of Tevatron and LHC Measurements of the Top-Quark Mass*. ATLAS-CONF-2014-008, CDF-NOTE-11071, CMS-PAS-TOP-13-014, D0-NOTE-6416, FERMILAB-TM-2582, arXiv:1403.4427 (2014)

14. L. Lyons, A.J. Martin, D.H. Saxon, On the determination of the B lifetime by combining the results of different experiments. Phys. Rev. **D41**, 982–985 (1990)

15. L. Lista, The bias of the unbiased estimator: a study of the iterative application of the BLUE method. Nucl. Instrum. Meth. **A764**, 82–93 (2014) and corr. ibid. **A773**, 87–96 (2015)

Confidence Intervals

8

8.1 Introduction

Section 6.10 presented two approximate methods to determine uncertainties of maximum likelihood estimates. Basically, either the negative log likelihood function is approximated to a parabola at the minimum, corresponding to a Gaussian PDF approximation, or the excursion of the negative log likelihood around the minimum is considered to obtain the possibly asymmetric uncertainty. None of those methods guarantees an exact coverage of the uncertainty interval. In many cases, the provided level of approximation is sufficient, but for measurements with few observations and PDF models that exhibit large deviation from the Gaussian approximation, the uncertainty determined with those approximate methods may not be sufficiently accurate.

8.2 Neyman Confidence Intervals

A more rigorous and general treatment of *confidence intervals* under the frequentist approach is due to Neyman [1]. The method is discussed in the following in the simplest case of a single parameter. Let us consider a variable x distributed according to a PDF that depends on an unknown parameter θ. We have in mind that x could be the value of an estimator of the parameter θ. Neyman's procedure to determine confidence intervals proceeds in two steps, illustrated in Fig. 8.1:

1. The construction of a *confidence belt*
2. The inversion of the confidence belt to determine the confidence interval

© The Author(s), under exclusive license to Springer Nature Switzerland AG 2023
L. Lista, *Statistical Methods for Data Analysis*, Lecture Notes in Physics 1010,
https://doi.org/10.1007/978-3-031-19934-9_8

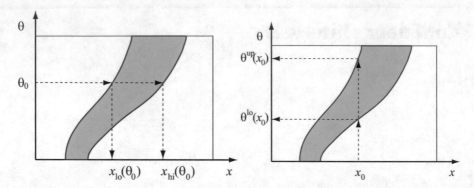

Fig. 8.1 Illustration of Neyman belt construction (left) and inversion (right)

8.3   Construction of the Confidence Belt

In the first step of the determination of Neyman confidence interval, the confidence belt is constructed by scanning the parameter space, varying θ within its entire allowed range. For each fixed value of the parameter $\theta = \theta_0$, the corresponding PDF $f(x \mid \theta_0)$, which describes the distribution of x, is known. According to the PDF $f(x \mid \theta_0)$, an interval $[x^{lo}(\theta_0),\, x^{up}(\theta_0)]$ is determined ensuring that the corresponding probability is equal to the specified *confidence level*, defined as $\mathrm{CL} = 1 - \alpha$, and usually equal to 68.27%, i.e., $\pm 1\sigma$, 90 or 95%:

$$1 - \alpha = \int_{x^{lo}(\theta_0)}^{x^{up}(\theta_0)} f(x \mid \theta_0)\, dx \,. \tag{8.1}$$

Neyman's construction of the confidence belt is illustrated in Fig. 8.1, left.

Equation (8.1) can be satisfied exactly for a continuous random variable x. In case of a discrete variable, instead, it is not possible, for most of the cases, to find an interval that corresponds exactly to the desired confidence level. The interval is therefore constructed ensuring that it corresponds to a probability at least equal to the desired confidence level. This introduces some degree of overcoverage of the confidence interval.

The choice of $x^{lo}(\theta_0)$ and $x^{up}(\theta_0)$ has still some arbitrariness, since there are different possible intervals having the same probability, according to the condition in Eq. (8.1). The choice of the interval is often called *ordering rule*. This arbitrariness was already encountered in Sect. 5.10 when discussing Bayesian credible intervals. For instance, one could choose a symmetric interval around a central value given by the mode or median of x corresponding to θ_0, i.e., an interval:

$$[x^{lo}(\theta_0), x^{up}(\theta_0)] = [\bar{x}(\theta_0) - \delta,\, \bar{x}(\theta_0) + \delta] \,, \tag{8.2}$$

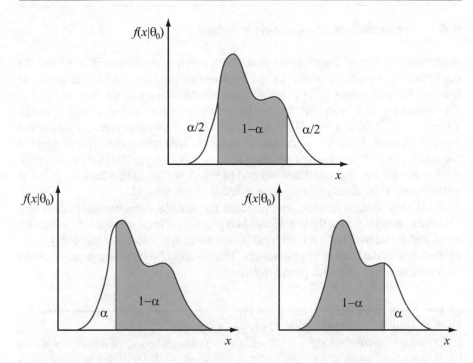

Fig. 8.2 Three possible choices of ordering rule: central interval (top) and fully asymmetric intervals (bottom left, right)

where δ is determined to ensure the coverage condition in Eq. (8.1). Alternatively, one could choose the interval with equal areas of the PDF tails at the two extreme sides, i.e.: such that:

$$\int_{-\infty}^{x^{\mathrm{lo}}(\theta_0)} f(x \mid \theta_0)\, \mathrm{d}x = \frac{\alpha}{2} \quad \text{and} \quad \int_{x^{\mathrm{up}}(\theta_0)}^{+\infty} f(x \mid \theta_0)\, \mathrm{d}x = \frac{\alpha}{2} . \tag{8.3}$$

Figure 8.2 shows three of the possible cases described above. Other possibilities are the interval having the smallest size, or fully asymmetric intervals on either side: $[x^{\mathrm{lo}}(\theta_0), +\infty[$ or $] -\infty, x^{\mathrm{up}}(\theta_0)]$. More options are also considered in the literature. A special ordering rule was introduced by Feldman and Cousins based on a likelihood ratio criterion and is discussed in Sect. 8.7.

Given a choice of the ordering rule, the intervals $[x^{\mathrm{lo}}(\theta), x^{\mathrm{up}}(\theta)]$, for all possible values of θ, define the Neyman belt in the plane (x, θ) as shown in Fig. 8.1, left.

8.4   Inversion of the Confidence Belt

In the second step of the Neyman procedure, given a measurement $x = x_0$, the confidence interval for θ is determined by inverting the Neyman belt, in the sense that two extreme values $\theta^{\text{lo}}(x_0)$ and $\theta^{\text{up}}(x_0)$ are determined as the intersections of the vertical line at $x = x_0$ with the two boundary curves of the belt (Fig. 8.1, right). The interval $[\theta^{\text{lo}}(x_0), \theta^{\text{up}}(x_0)]$ has, by construction, a coverage equal to the desired confidence level, $1 - \alpha$ for a continuous variable. This means that, if θ is equal to the true value θ^{true}, randomly extracting $x = x_0$ according to the PDF $f(x \mid \theta^{\text{true}})$, θ^{true} is included in the confidence interval $[\theta^{\text{lo}}(x_0), \theta^{\text{up}}(x_0)]$ in a fraction $\geq 1 - \alpha$ of the cases, in the limit of a very large number of extractions.

As already said, overcoverage is possible for discrete distributions. In case of a discrete random variable, the confidence belt provides at least the desired confidence level, and the interval $[\theta^{\text{lo}}(x_0), \theta^{\text{up}}(x_0)]$ contains the true value in a fraction $\geq 1-\alpha$ of the cases, in the limit of large numbers. The amount of overcoverage may depend on the value of the unknown parameter θ.

Example 8.1 - Neyman Belt: Gaussian Case
The Neyman belt for the parameter μ of a Gaussian distribution with $\sigma = 1$ is shown in Fig. 8.3 for a 68.27% confidence level with the choice of a central interval. The inversion of the belt is straightforward and gives the usual result $\mu = x \pm \sigma$, as in Exercise 6.1.

Fig. 8.3 Neyman belt for the parameter μ of a Gaussian with $\sigma = 1$ at the 68.27% confidence level

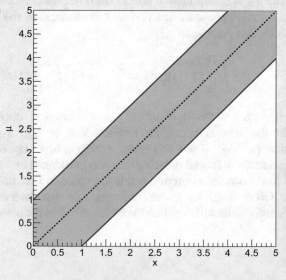

8.5 Binomial Intervals

The binomial distribution was introduced in Sect. 2.9, Eq. (2.61):

$$P(n; N, p) = \frac{n!}{N!(N-n)!} p^n (1-p)^{N-n} . \tag{8.4}$$

Given an extracted value of n, $n = n^\star$, the parameter p can be estimated as

$$\hat{p} = \frac{n^\star}{N} . \tag{8.5}$$

An approximate estimate of the uncertainty on $\hat{p}$ was obtained in Exercise 6.3 and has the same expression of the standard deviation of n divided by N, with $\hat{p}$ replacing p:

$$\delta \hat{p} \simeq \sqrt{\frac{\hat{p}(1-\hat{p})}{N}} . \tag{8.6}$$

This expression is consistent with the law of large numbers because, for $N \to \infty$, $\hat{p}$ and p coincide. Anyway, Eq. (8.6) gives a null error in case either $n^\star = 0$ or $n^\star = N$, i.e., for $\hat{p} = 0$ or $\hat{p} = 1$, respectively.

A solution to the problem of determining the correct confidence interval for a binomial distribution is due to Clopper and Pearson [2]. The interval $[p^{lo}, p^{up}]$ is determined to have at least the correct coverage $1 - \alpha$. The extremes p^{lo} and p^{up} of the interval are taken such that the following relations hold:

$$P(n \geq n^\star \mid N, p^{lo}) = \sum_{n=n^\star}^{N} \frac{n!}{N!(N-n)!} (p^{lo})^n (1-p^{lo})^{N-n} = \frac{\alpha}{2} , \tag{8.7}$$

$$P(n \leq n^\star \mid N, p^{up}) = \sum_{n=0}^{n^\star} \frac{n!}{N!(N-n)!} (p^{up})^n (1-p^{up})^{N-n} = \frac{\alpha}{2} . \tag{8.8}$$

The above conditions implement Neyman belt inversion in a discrete case. The corresponding Neyman belt is shown in Fig. 8.4 for the case $N = 10$.

Since n is a discrete variable, continuous variation of the discrete intervals $[n^{lo}(p_0), n^{up}(p_0)]$, for an assumed parameter value $p = p_0$, is not possible. In the construction of Neyman belt, one must choose an interval $[n^{lo}, n^{up}]$, consistently with the adopted ordering rule, that has at least the desired coverage. In this way, the confidence interval $[p^{lo}, p^{up}]$ determined by the inversion procedure for the parameter p could *overcover*, i.e., it may have a corresponding probability greater than the desired confidence level $1 - \alpha$. In this sense, the interval is said to be *conservative*.

Fig. 8.4 Neyman belt at the
68.27% confidence level for
the parameter p of a binomial
distribution with $N = 10$

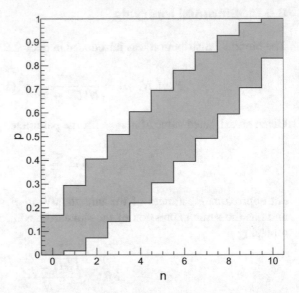

The coverage of Clopper–Pearson intervals as a function of p is shown in Fig. 8.5
for the cases with $N = 10$ and $N = 100$. The coverage is, by construction, always
greater than the nominal 68.27% and has a ripple structure, due to the discrete nature
of the problem, as also discussed in [3], whose amplitude is reduced for large N.
For $p = 0$ and $p = 1$, the coverage is necessarily 100%, by construction. See
Exercise 8.2 for more details.

Another case of a discrete application of the Neyman belt inversion in a
Poissonian problem can be found in Sect. 12.14.

Example 8.2 - Application of the Clopper–Pearson Method
As exercise, we compute with the Clopper–Pearson method the 90%
confidence level interval for a measurement $n^\star = N = 10$, i.e., equal to
the maximum possible outcome for n. The values p^{lo} and p^{up} have to be
determined such that Eqs. (8.7) and (8.8) hold. Considering $\alpha = 0.10$, we
have

$$P(n \geq N \mid N, p^{lo}) = \frac{N!}{N!\,0!}(p^{lo})^N(1 - p^{lo})^0 = (p^{lo})^N = 0.05\,,$$

$$P(n \leq N \mid N, p^{up}) = 1\,.$$

(continued)

Example 8.2 (continued)

For p^{up}, we should consider the largest allowed value $p^{\text{up}} = 1.0$, since the probability $P(n \leq N \mid N, p^{\text{up}})$ is equal to one. The first equation can be inverted and gives

$$p^{\text{lo}} = \exp\left[\log(0.05)/N\right].$$

For $N = 10$, we have $p^{\text{lo}} = 0.741$, and the confidence interval is [0.74, 1.00]. Symmetrically, the Clopper–Pearson confidence interval for $n^{\star} = 0$ is evaluated as [0.00, 0.26] for $N = 10$. The approximate expression in Eq. (8.6), instead, gives an interval of null size, which is a clear sign of a pathology.

Note that discrete intervals may overcover. In particular, if the true value is $p = 1$, $n^{\star}$ is always equal to N, i.e., the confidence interval [0.74, 1.00] contains p with 100% probability, while it was supposed to cover with 90% probability. Figure 8.5 shows that the coverage is 100% for $p = 0$ and $p = 1$.

8.6 The Flip-Flopping Problem

In order to determine confidence intervals, a consistent choice of the ordering rule has to be adopted. Feldman and Cousins remarked [4] that the ordering rule choice must not depend on the outcome of our measurement; otherwise, the quoted confidence interval or upper limit could not correspond to the correct confidence level, i.e., it does not respect the required coverage.

In some cases, experimenters searching for a rare signal decide to quote their result in different possible ways, switching from a central interval to an upper limit, depending on the outcome of the measurement. A typical choice is:

- Quote an upper limit if the measured signal yield is not greater than three times its uncertainty.
- Instead, quote the point estimate with its uncertainty if the measured signal exceeds three times its uncertainty.

This 3σ significance criterion is discussed in more detail in the next chapters. See Sect. 12.3 for the definition of *significance level*.

This problem is referred to in the literature as *flip-flopping* and can be illustrated in a simple example [4]. Consider a random variable x distributed according to a Gaussian with a fixed and known standard deviation σ and an unknown average μ, which is bound to be greater or equal to zero. This is the case of a signal yield or cross section measurement. For simplicity, we can take $\sigma = 1$, without loss of

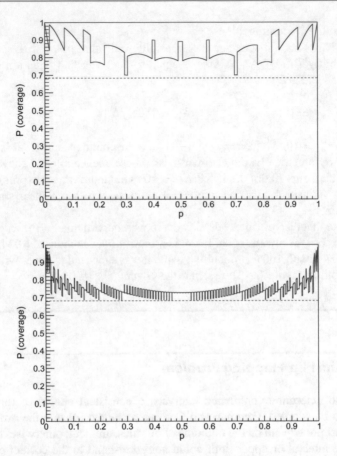

Fig. 8.5 Coverage of Clopper–Pearson intervals as a function of p for the cases with $N = 10$ (top) and $N = 100$ (bottom)

generality. The quoted measured value of μ must always be greater than or equal to zero, given the assumed constraint $\mu \geq 0$. Imagine one decides to quote, as measured value, zero if the significance level, defined as x/σ, is less than 3; the measured value is otherwise quoted as x. In a formula:

$$\hat{\mu}(x) = \begin{cases} x & \text{if} \quad x/\sigma \geq 3 \,, \\ 0 & \text{if} \quad x/\sigma < 3 \,. \end{cases} \tag{8.9}$$

If $x/\sigma \geq 3$, the confidence interval, given the measurement x, has a symmetric error equal to $\pm \sigma$ at the 68.27% confidence level (CL), or $\pm 1.645\sigma$ at 90% CL.

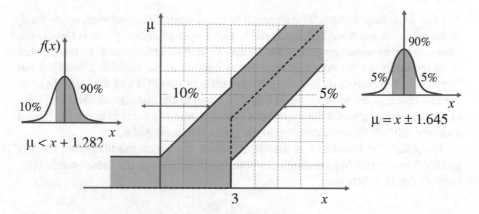

Fig. 8.6 Illustration of the flip-flopping problem. The plot shows the quoted point estimate of μ as a function of the measured x (dashed line), and the 90% confidence interval corresponding to the choice of quoting a central interval for $x/\sigma \geq 3$ and an upper limit for $x/\sigma < 3$. The coverage decreases from 90 to 85% for a value of μ corresponding to the horizontal lines with arrows

Instead, if $x/\sigma < 3$, the confidence interval is $[0, \mu^{\mathrm{up}}]$ with an upper limit $\mu^{\mathrm{up}} = x + 1.282\sigma$, because

$$\Phi(Z = 1.282) = \int_{-\infty}^{Z=1.282} \frac{1}{\sqrt{2\pi}} e^{-x^2/2} \, \mathrm{d}x = 0.9 \,, \tag{8.10}$$

considering the corresponding area under a Gaussian curve. In summary, the confidence interval at 90% CL is

$$[\mu^{\mathrm{lo}}, \ \mu^{\mathrm{up}}] = \begin{cases} [x - 1.645\sigma, \ x + 1.645\sigma] & \text{if} \quad x/\sigma \geq 3 \,, \\ [0, \ x + 1.282\sigma] & \text{if} \quad x/\sigma < 3 \,. \end{cases} \tag{8.11}$$

The situation is shown in Fig. 8.6.

The choice to switch from a central interval to a fully asymmetric interval, which gives an upper limit, based on the observation of x, produces an incorrect coverage. Looking at Fig. 8.6, depending on the value of μ, the coverage can be determined as the probability corresponding to the interval $[x^{\mathrm{lo}}, x^{\mathrm{up}}]$ obtained by crossing the confidence belt with a horizontal line. One may have cases where the coverage decreases from 90% to 85%, which is below the desired confidence level, indicated by the lines with arrows in the figure.

8.7   The Unified Feldman–Cousins Approach

In order to avoid the flip-flopping problem and to ensure the correct coverage, the ordering rule proposed by Feldman and Cousins [4] provides a Neyman confidence belt that smoothly changes from a central or quasi-central interval to an upper limit, in the case of a low observed signal yield. The proposed ordering rule is based on a likelihood ratio whose properties are further discussed in Sect. 10.5. Given a value θ_0 of the unknown parameter θ, the chosen interval of the variable x used for the Neyman belt construction is defined by the ratio of two PDFs of the variable x: at the numerator under the hypothesis that θ is equal to the considered fixed value θ_0, at the denominator under the hypothesis that θ is equal to the maximum likelihood estimate value $\hat{\theta}(x)$, corresponding to the given measurement x.

The x interval is defined as the set of values for which the likelihood ratio is greater than or equal to a constant k_α, whose value depends on the chosen confidence level $1 - \alpha$. In a formula:

$$\lambda(x \mid \theta_0) = \frac{f(x \mid \theta_0)}{f(x \mid \hat{\theta}(x))} \geq k_\alpha , \tag{8.12}$$

while the confidence interval R_α for a given value $\theta = \theta_0$ is defined by

$$R_\alpha(\theta_0) = \{x : \lambda(x \mid \theta_0) \geq k_\alpha\} . \tag{8.13}$$

The constant k_α should be taken such that the integral over the confidence interval R_α of the PDF $f(x \mid \theta_0)$ of x is equal to $1 - \alpha$:

$$\int_{R_\alpha} f(x \mid \theta_0) \, \mathrm{d}x = 1 - \alpha . \tag{8.14}$$

The case is illustrated in Fig. 8.7.

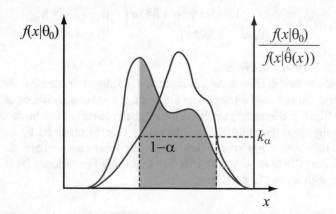

Fig. 8.7 Ordering rule in the Feldman–Cousins approach. The interval of the random variable x is determined on the basis of the likelihood ratio $\lambda(x \mid \theta_0) = f(x \mid \theta_0)/f(x \mid \hat{\theta}(x))$

Feldman and Cousins computed the confidence interval for the simple Gaussian case discussed in Sect. 8.6. The value $\mu = \hat{\mu}(x)$ that maximizes the likelihood function for a given x, under the constraint $\mu \geq 0$, is

$$\hat{\mu}(x) = \max(x, 0) . \tag{8.15}$$

The PDF for the value of the parameter $\mu = \hat{\mu}(x)$ that maximizes the likelihood function for the observed value x is, assuming f is the considered Gaussian function with $\sigma = 1$:

$$f(x \mid \hat{\mu}(x)) = \begin{cases} \dfrac{1}{\sqrt{2\pi}} & \text{if} \quad x \geq 0 , \\ \dfrac{1}{\sqrt{2\pi}} e^{-x^2/2} & \text{if} \quad x < 0 . \end{cases} \tag{8.16}$$

The likelihood ratio in Eq. (8.12) can be written in this case as

$$\lambda(x \mid \mu) = \frac{f(x \mid \mu)}{f(x \mid \hat{\mu}(x))} = \begin{cases} \exp(-(x - \mu)^2/2) & \text{if} \quad x \geq 0 , \\ \exp(x\mu - \mu^2/2) & \text{if} \quad x < 0 . \end{cases} \tag{8.17}$$

The interval $[x^{\text{lo}}(\mu_0), x^{\text{up}}(\mu_0)]$, for a given $\mu = \mu_0$, can be found numerically using the equation $\lambda(x \mid \mu) > k_\alpha$ and imposing the desired confidence level $1 - \alpha$, according to Eq. (8.14). The results are shown in Fig. 8.8 and can be compared with Fig. 8.6. Using the Feldman–Cousins approach, for large values of x, the usual symmetric confidence interval is obtained; as x moves toward lower values, the interval becomes more and more asymmetric, and, at some point, it becomes fully asymmetric (i.e. : $[0, \mu^{\text{up}}]$), determining an upper limit μ^{up}. For negative values of x, the result is always an upper limit, avoiding unphysical cases corresponding to negative values of μ. Negative values of μ would not be excluded using a Neyman belt construction with a central interval, like the one shown in Fig. 8.3. As seen, this approach smoothly changes from a central interval to an upper limit, yet correctly ensuring the required coverage (90% in this case). More applications of the Feldman–Cousins approach are presented in Chap. 12.

The application of the Feldman–Cousins method requires, in most of the cases, numerical evaluations, even for simple PDF models, such as the Gaussian case discussed above. The reason is that the non-trivial inversion of the integral in Eq. (8.14) is required. The inversion usually proceeds by scanning the parameter space, and in case of complex models, it may be very CPU intensive. For this practical reason, other methods are often preferred to the Feldman–Cousins.

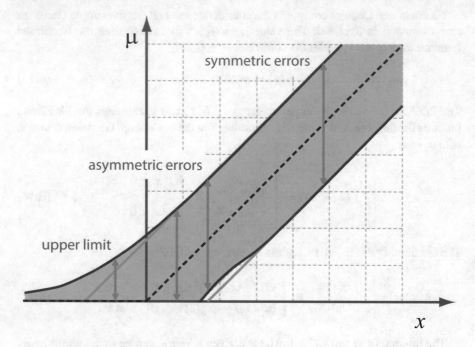

Fig. 8.8 Neyman confidence belt constructed using the Feldman–Cousins ordering rule for a Gaussian PDF

References

1. J. Neyman, Outline of a theory of statistical estimation based on the classical theory of probability. Philos. Trans. R. Soc. Lond. A Math. Phys. Sci. **236**, 333–380 (1937)
2. C.J. Clopper, E. Pearson, The use of confidence or fiducial limits illustrated in the case of the binomial. Biometrika **26**, 404–413 (1934)
3. R. Cousins, K.E. Hymes, J. Tucker, Frequentist evaluation of intervals estimated for a binomial parameter and for the ratio of Poisson means. Nucl. Instrum. Meth. **A612**, 388–398 (2010)
4. G. Feldman, R. Cousins, Unified approach to the classical statistical analysis of small signals. Phys. Rev. **D57**, 3873–3889 (1998)

Convolution and Unfolding

9

9.1 Introduction

This section presents two related problems:

- How to take into account realistic detector response, like resolution, efficiency, and background, into a probability model.
- How to remove those experimental effects from an observed distribution to recover the original distribution.

The first problem involves a combination of distributions called *convolution*, and the second problem is known as *unfolding*.

Unfolding is not always necessary to compare an experimental distribution with the expectation from theory since data can be compared with a realistic prediction that takes into account experimental effects. In some cases, anyway, it is desirable to produce a measured distribution that can be compared among different experiments, each introducing different experimental effects. For those cases, unfolding an observed distribution may be a necessary task.

9.2 Convolution

Detector effects distort the theoretical distribution of an observable quantity. Let us denote by y the true value of quantity we want to measure and x the corresponding measured quantity. The theoretical distribution of y is given by a PDF $g(y)$ and, for a given true value y, the PDF for x depends on y. We can write it as $r(x; y)$, and it is usually called *response function* or *kernel function*. The PDF that describes the distribution of the measured value x, considering both the original theoretical

L. Lista, *Statistical Methods for Data Analysis*, Lecture Notes in Physics 1010,
https://doi.org/10.1007/978-3-031-19934-9_9

distribution and the detector response, is given by the *convolution* of the two PDFs g and r, defined as follows:

$$f(x) = \int g(y)\,r(x;\,y)\,\mathrm{d}y\,. \tag{9.1}$$

Note that convolution may also be applied to a function $g(y)$ that is not normalized to unity. For instance, the integral of $g(y)$ may represent the overall number of random extractions of the variable y. The convolution of non-normalized distributions is useful for the discrete case, as discussed in Sect. 9.4. The convolution of a theoretical distribution with a finite resolution response is also called *smearing*. It typically broadens any peaking structure present in the original distribution. Examples of the effect of PDF convolution are shown in Fig. 9.1.

Fig. 9.1 Examples of convolution of two PDFs $f(x)$ (solid light blue lines) with Gaussian kernel functions $r(x, x')$ (dashed black lines). Top: $f(x)$ is the superposition of three Breit–Wigner functions (see Sect. 3.14); bottom: $f(x)$ is a piece-wise constant PDF. The result of the convolution is the solid dark blue line. The kernel function is scaled down by a factor of 2 and 20 in the top and bottom plots, respectively, for display convenience

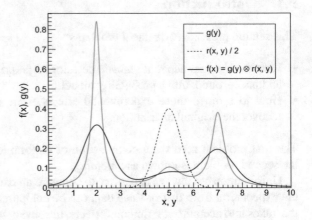

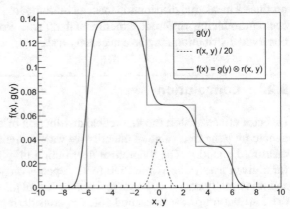

9.3 Convolution and Fourier Transform

In many cases, the response function r only depends on the difference $x - y$ and can be described by a function of one variable: $r(x; y) = r(x - y)$. Sometimes the notation $f = g \otimes r$ is also used in those cases to indicate the convolution of g and r.

Convolutions have interesting properties under the Fourier transform. Consider the Fourier transform of g:

$$\hat{g}(k) = \int_{-\infty}^{+\infty} g(y) \, e^{-iky} \, dy \qquad (9.2)$$

and conversely the inverse transform as

$$g(y) = \frac{1}{2\pi} \int_{-\infty}^{+\infty} \hat{g}(k) \, e^{iky} \, dk . \qquad (9.3)$$

It is possible to demonstrate that the Fourier transform of the convolution of two functions $g(y)$ and $r(x - y)$ is given by the product of the Fourier transforms of the two functions, i.e.,

$$\widehat{g \otimes r} = \hat{g} \cdot \hat{r} . \qquad (9.4)$$

Conversely, the Fourier transform of the product of two PDFs is equal to the convolution of the two Fourier transforms, i.e.,

$$\widehat{g \cdot r} = \hat{g} \otimes \hat{r} . \qquad (9.5)$$

This property allows to implement numerical convolutions based on the fast Fourier transform (FFT) algorithm [1]. The simplest and most common response function that models detector resolution is a Gaussian PDF:

$$r(x - y) = \frac{1}{\sigma\sqrt{2\pi}} e^{-(x-y)^2/2\sigma^2} . \qquad (9.6)$$

The Fourier transform of a Gaussian PDF can be computed analytically and is

$$\hat{r}(k) = e^{-ik\mu} e^{-\sigma^2 k^2/2} . \qquad (9.7)$$

This expression may be convenient to implement convolution algorithms based on FFT.

9.4   Discrete Convolution and Response Matrix

In many cases, data samples are available in the form of histograms. Consider a sample randomly extracted according to a continuous distribution $f(x)$. The range of x is split into N intervals, or bins, of given size, and the numbers of entries of each of the bins are $(n_1, \cdots, n_N)$. Each n_i is distributed according to Poissonian with expected value:

$$\nu_i = \langle n_i \rangle = \int_{x_i^{\text{lo}}}^{x_i^{\text{up}}} f(x) \, \mathrm{d}x \,, \tag{9.8}$$

where x_i^{lo} and x_i^{up} are the edges of the ith bin.

Consider that values of x are derived from original values of y, which are unaffected by the experimental effects, and are distributed according to a theoretical distribution $g(y)$. The range y can be split into M bins, with M not necessarily equal to N. The jth bin has an expected number of entries equal to

$$\mu_j = \int_{y_j^{\text{lo}}}^{y_j^{\text{up}}} g(y) \, \mathrm{d}y \,, \tag{9.9}$$

where y_j^{lo} and y_j^{up} are the edges of that bin.

For such cases, where continuous variables are discretized into bins, the effect of a realistic detector response is a discrete version of Eq. (9.1) and can be written as

$$\nu_i = \sum_{j=1}^{M} R_{ij} \, \mu_j \,. \tag{9.10}$$

The $N \times M$ matrix R_{ij} is called *response matrix* and is responsible for the *bin migration* from the original histogram of values of y to the observed histogram of values of x. The case of a kernel function that depends on the difference $x - y$ in a continuous case corresponds to a response matrix that only depends on the index difference $i - j$ in a discrete case. An example of the effect of bin migration is shown in Fig. 9.2, where, for simplicity of representation, $M = N$ has been chosen.

9.5   Effect of Efficiency and Background

The convolution presented in Eq. (9.1) represents the transformation of $g(y)$ into $f(x)$, where both g and f are normalized to unity. In Eq. (9.10), instead, the histograms of the expected values $(\nu_1, \cdots, \nu_N)$ and $(\mu_1, \cdots, \mu_M)$ do not obey a normalization condition, since their content does not represent a probability distribution but represent the expected yield in each bin.

Fig. 9.2 Effect of bin
migration. Top: response
matrix R_{ij}, and bottom: the
original distribution μ_j (light
blue line) with superimposed
the convoluted distribution
$v_i = \sum_{j=1}^{M} R_{ij}\,\mu_j$ (dark blue
line)

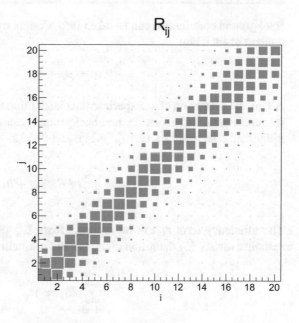

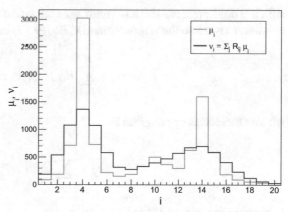

In those cases, in addition to the detector response, relevant experimental effects
may also be efficiency and background. A typical effect of histogram distortion
purely due to efficiency can be represented by a diagonal migration matrix, if $N =
M$ is assumed:

$$v_i = \sum_{j=1}^{M} \varepsilon_j\,\delta_{ij}\,\mu_j = \varepsilon_i\,\mu_i\,. \tag{9.11}$$

Background contribution can be taken into account by adding an offset b_i to the bin content of each bin:

$$v_i = \mu_i + b_i .$$ (9.12)

A realistic prediction of an experimental distribution that takes into account detector response, including efficiency and background, can be obtained by combining the individual effects in Eqs. (9.10), (9.11), and (9.12):

$$v_i = \sum_{j=1}^{M} \varepsilon_j R_{ij} \mu_j + b_i .$$ (9.13)

The efficiency term ε_i can be dropped from Eq. (9.13) and incorporated in the response matrix R_{ij} definition giving up the normalization condition:

$$\sum_{i=1}^{N} R_{ij} = 1 ,$$ (9.14)

which would preserve the histogram normalization in Eq. (9.10). Absorbing the efficiency term into the response matrix, Eq. (9.13) becomes

$$v_i = \sum_{j=1}^{M} R_{ij} \mu_j + b_i ,$$ (9.15)

where efficiencies ε_j are given by

$$\varepsilon_j = \sum_{i=1}^{N} R_{ij} .$$ (9.16)

In matrix form, Eq. (9.15) can be written as

$$\vec{v} = R \vec{\mu} + \vec{b} .$$ (9.17)

9.6   Unfolding by Inversion of the Response Matrix

Unfolding the response matrix from an experimental distribution $\vec{n} = (n_1, \cdots, n_N)$ consists in finding an estimate $\hat{\vec{\mu}}$ of the original distribution $\vec{\mu} = (\mu_1, \cdots, \mu_M)$. The most straightforward way is given by the inversion of Eq. (9.17), but this approach leads to results that are not usable in practice. The best estimate of $\vec{\mu}$ can be determined with the maximum likelihood method. The probability distribution for each n_i is a Poissonian with expected value ν_i. Hence, the likelihood function can be written as

$$L(\vec{n}\,;\,\vec{\mu}\,) = \prod_{i=1}^{N} \mathrm{Pois}(n_i;\,\nu_i(\vec{\mu}\,)) = \prod_{i=1}^{N} \mathrm{Pois}\left(n_i;\,\sum_{j=1}^{M} R_{ij}\,\mu_j + b_i\right). \qquad (9.18)$$

The maximum likelihood solution leads to the inversion of Eq. (9.17) [2]. Assuming for simplicity $N = M$, the estimate of $\hat{\vec{\mu}}$ that maximizes the likelihood function is

$$\hat{\vec{\mu}} = R^{-1}(\vec{n} - \vec{b})\,. \qquad (9.19)$$

The covariance matrix U_{ij} of the estimated expected number of bin entries $\hat{\mu}_j$ is given by

$$U = R^{-1}\,V\,\left(R^{-1}\right)^{T}\,, \qquad (9.20)$$

where the matrix V is the covariance matrix of the measurements n_i. Due to the Poisson distribution, sine the n_i are independent, V is a diagonal matrix with elements on the diagonal equal to ν_i:

$$V_{ij} = \delta_{ij}\,\nu_j\,. \qquad (9.21)$$

It is also possible to demonstrate that this estimate is unbiased and it has the smallest possible variance. Results are shown in Fig. 9.3 on a Monte Carlo example. The response matrix was taken from Fig. 9.2 (top), and a histogram containing the observations $\vec{n}$ was generated according to the convoluted distribution from Fig. 9.2 (bottom) with 10,000 entries. This histogram, generated by Monte Carlo, is shown in Fig. 9.3 (top) superimposed to the original convoluted distribution. The resulting estimated $\hat{\vec{\mu}}$ are shown in Fig. 9.3 (bottom), with uncertainty bars computed according to Eq. (9.20). The plot shows very large oscillations from one bin to the subsequent one, of the order of the uncertainty bars, which is very large as well.

The numerical values of the non-diagonal terms of the matrix U from Eq. (9.20) show that each bin $\hat{\mu}_j$ is very strongly anti-correlated with its adjacent bins, $\hat{\mu}_{j+1}$ and $\hat{\mu}_{j-1}$, the correlation coefficient being very close to -1. Statistical fluctuations in the generated number of entries n_i, which are bin-by-bin independent, turn into high-frequency bin-by-bin oscillations of the resulting estimates $\hat{\mu}_j$, due to the very large anti-correlation, rather than, as one may expect, reflecting into uncertainties of the estimates $\hat{\mu}_j$.

To overcome this problem present with the maximum likelihood solutions of unfolding, other techniques have been introduced to *regularize* the observed oscillation. The maximum likelihood solution has no bias, and it offers the smallest

Fig. 9.3 Top: Monte Carlo generated histogram with 10,000 entries randomly extracted according to the convoluted distribution $\vec{\nu}$ in Fig. 9.2 (bottom) superimposed to the original convoluted distribution, and bottom: the result of response matrix inversion, or the maximum likelihood estimate of the unconvoluted distribution $\hat{\vec{\mu}}$, with error bars. Note that the vertical scale on the bottom plot extends far beyond the range of the top plot

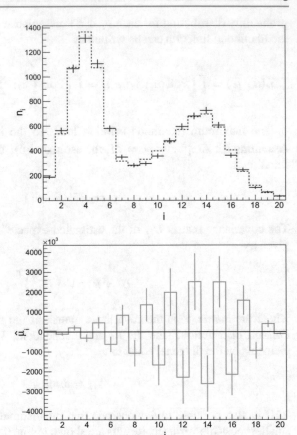

possible variance. For this reason, to reduce the large observed variance and the bin-by-bin correlations, regularization methods must unavoidably introduce some bias in the unfolded estimates. The problem of finding a trade-off between bias and variance is also typical in machine learning, see Sect. 11.6.

9.7   Bin-by-Bin Correction Factors

A solution to the unfolding problem, which also turns out to have disadvantages, consists in performing a correction to the yields n_i observed in each bin equal to the ratio of expected values before and after unfolding in each bin (assuming $N = M$), which are assumed to be known:

$$\hat{\mu}_i = \frac{\mu_i^{\text{est}}}{\nu_i^{\text{est}}} (n_i - b_i) \, . \tag{9.22}$$

μ_i^{est} and ν_i^{est} are usually estimates of the true expected yields determined from simulation. The estimate ν_i^{est} does not include the background contribution b_i.

This method produces results that, by construction, resemble the estimated expectation, and hence a simple visual inspection of unfolded distributions may not show any apparent problem. Anyway, this approach has the serious drawback that it introduces a bias that drives the estimates $\hat{\mu}_i$ toward the estimated expectations μ_i^{est} [2]. The bias, in fact, can be determined to be

$$\langle \hat{\mu}_i \rangle - \mu_i = \left(\frac{\mu_i^{\mathrm{est}}}{\nu_i^{\mathrm{est}}} - \frac{\mu_i}{(\nu_i - b_i)} \right) (\nu_i - b_i) . \tag{9.23}$$

Note that the expectation ν_i includes the background contribution b_i, and hence the term $\nu_i - b_i$ is the expected yield due to signal only. Due to Eq. (9.23), any comparison of unfolded data with simulation prediction, in this way, is biased toward good agreement, with the risk of hiding real discrepancies.

9.8 Regularized Unfolding

Acceptable unfolding methods should reduce the large variance which occurs in maximum likelihood estimates given by the matrix inversion discussed in Sect. 9.6. This has the unavoidable cost of introducing a bias in the estimate, since, as seen above, the only unbiased estimator with the smallest possible variance is the maximum likelihood one.

A trade-off between the smallest variance and the smallest bias is achieved by imposing an additional condition on the *smoothness* of the set of values to be estimated, $\vec{\mu}$, which is quantified by a function $S(\vec{\mu})$. The definition of S may vary according to the specific implementation. Such methods fall into the category of *regularized unfolding*. In practice, instead minimizing $\Lambda = -2 \log L$, which is a χ^2, for a Gaussian case, the function to be minimized is

$$\phi(\vec{\mu}) = \Lambda(\vec{\mu}) + \tau^2 S(\vec{\mu}) . \tag{9.24}$$

The parameter τ that appears in Eq. (9.24) is called *regularization strength*. The case $\tau = 0$ is equivalent to a maximum likelihood estimate, which has been already considered, and exhibits large bin-by-bin oscillations and large variance of the estimates. The other extreme, when τ is very large, provides an estimate $\hat{\vec{\mu}}$ that is extremely smooth, in the sense that it minimizes S, but it is insensitive to the observed data $\vec{n}$. The regularization procedure should find an optimal value of τ to achieve a solution $\hat{\vec{\mu}}$ that is sufficiently close to the minimum of $-2 \log L$ and, at the same time, ensures a sufficiently small S, i.e., a sufficient smoothness.

Another issue that should be considered is that, in general, the total observed yield $n = \sum_{i=1}^{N} n_i$ is not necessarily equal to the sum of the estimates for the expected values, $\hat{v} = \sum_{i=1}^{N} \hat{v}_i$. $\hat{\vec{v}}$ can be determined as

$$\hat{\vec{v}} = R\,\hat{\vec{\mu}} + \vec{b}\,, \tag{9.25}$$

where $\hat{\vec{\mu}}$ minimizes $\phi\,(\vec{\mu}\,)$ in Eq. (9.24). In order to impose the conservation of the overall normalization after unfolding, an extra term can be included in the function to be minimized, which contains a Lagrange multiplier λ:

$$\phi\,(\vec{\mu}\,) = \Lambda(\vec{\mu}\,) + \tau^2 S(\vec{\mu}\,) + \lambda\left(\sum_{i=1}^{N} n_i - \sum_{i=1}^{N} v_i\right)\,. \tag{9.26}$$

The choice of the regularization function $S(\vec{\mu}\,)$ defines the different regularization methods available in the literature, which are used for different problems.

9.9 Tikhonov Regularization

One of the most popular regularization techniques [3–5] takes the regularization function given by the following expression:

$$S(\vec{\mu}\,) = (L\,\vec{\mu}\,)^T L\,\vec{\mu}\,, \tag{9.27}$$

where L is a matrix with M columns and K rows. K could also be different from M. In the simplest implementation, L is a unit matrix, and Eq. (9.27) becomes

$$S(\vec{\mu}\,) = \sum_{i=1}^{M} \mu_j^2\,. \tag{9.28}$$

The term $\tau^2 S(\vec{\mu}\,)$ in the minimization just damps the cases with very large deviations of μ_j from zero.

Another commonly adopted L matrix is the following:

$$L = \begin{pmatrix} -1 & 1 & 0 & \cdots & 0 & 0 & 0 \\ 1 & -2 & 1 & \cdots & 0 & 0 & 0 \\ 0 & 1 & -2 & \cdots & 0 & 0 & 0 \\ \vdots & \vdots & \vdots & \ddots & \vdots & \vdots & \vdots \\ 0 & 0 & 0 & \cdots & -2 & 1 & 0 \\ 0 & 0 & 0 & \cdots & 1 & -2 & 1 \\ 0 & 0 & 0 & \cdots & 0 & 1 & -1 \end{pmatrix}\,. \tag{9.29}$$

The reason for this choice is that the derivative of a function, approximated by a discrete histogram h, can be computed from finite differences, divided by the bin size Δ. The first two derivatives can be approximated, in this way, as

$$h'_i = \frac{h_i - h_{i-1}}{\Delta}, \tag{9.30}$$

$$h''_i = \frac{h_{i-1} - 2h_i + h_{i+1}}{\Delta^2}. \tag{9.31}$$

The matrix L in Eq. (9.29) determines the approximate second derivative of the histogram $\vec{\mu}$ as approximation of the original function $g(y)$. Under this approximation, the function S in Eq. (9.27), with the choice of the matrix L in Eq. (9.29), is equal to the integral of the second derivative of g squared:

$$S \sim \int \left(\frac{d^2 g(y)}{dy^2} \right)^2 dy. \tag{9.32}$$

This regularization choice damps large second derivative terms which correspond to the large high-frequency oscillations of the unfolded distribution $\hat{\vec{\mu}}$, typical of the maximum likelihood solution. Note that, due to the matrix structure of the problem, a solution can be found only if $N > M$.

9.10   *L*-Curve Scan

The method usually adopted to determine the optimal value of the regularization strength parameter τ in a scan of the two-dimensional curve L defined by the following coordinates[6]:

$$\begin{cases} L_x = \log \Lambda(\hat{\vec{\mu}}(\tau)), \\ L_y = \log S(\hat{\vec{\mu}}(\tau)). \end{cases} \tag{9.33}$$

L_x measures how well $\hat{\vec{\mu}}(\tau)$ agrees with the model, while L_y measures how well $\hat{\vec{\mu}}(\tau)$ matches the regularization condition. Note that neither Λ nor S explicitly depend on τ, but their values at the minimum of $\phi(\vec{\mu})$ depend on the chosen value of τ. $\tau = 0$ corresponds to a minimum value of L_x and maximum of L_y, while, for increasing values of τ, L_x increases and L_y decreases. Usually, the L_y vs L_x curve exhibits a kink that makes the curve L-shaped. This motivates the name L *curve* for this method. The kink that corresponds to the maximum curvature of the L curve is taken as the optimal value.

An example of application of the Tikhonov regularization is shown in Figs. 9.4, 9.5, and 9.6. Figure 9.4 is similar to Fig. 9.2 and shows the assumed response matrix, the original unfolded, and convoluted distributions. In this case,

Fig. 9.4 Top: response matrix R_{ij}, and bottom: the original distribution μ_j is shown in light blue, while the convoluted distribution $v_i = \sum R_{ij}\,\mu_j$ is shown in dark blue

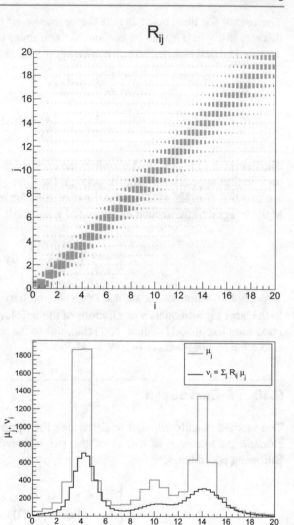

we assumed $M = 25$ bins in the original distribution and $N = 60$ bins in the convoluted distribution. Figure 9.5 shows a randomly extracted data sample and the corresponding L-curve scan, with the optimal point corresponding to the chosen value of τ. Figure 9.6 shows the unfolded distribution superimposed to the original one.

Fig. 9.5 Top: data sample n_i (dashed histogram with error bars) randomly generated according to the convoluted distribution ν_i (solid histogram, superimposed), and bottom: *L*-curve scan (black line). The optimal *L*-curve point is shown as blue cross

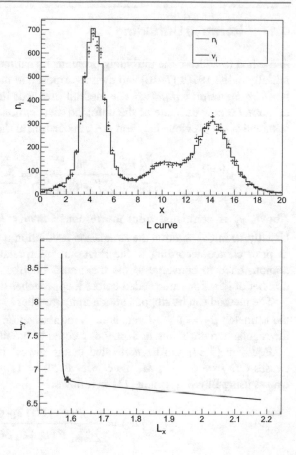

Fig. 9.6 Unfolded distribution using Tikhonov regularization, $\hat{\mu}_j$, shown as points with error bars, superimposed to the original distribution μ_i, shown as dashed line

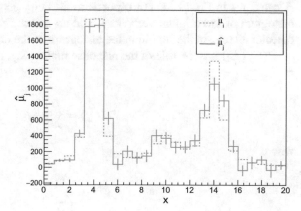

9.11   Iterative Unfolding

A method to address the unfolding problem by subsequent iteration was proposed
initially in the 1980s [7–10] and then reproposed in the 1990s [11] under the name
iterative Bayesian unfolding. The method proceeds iteratively computing at every
iteration l a new estimate of the unfolded distribution $\vec{\mu}^{\,(l+1)}$ from the distribution
obtained at the previous iteration $\vec{\mu}^{\,(l)}$, according to the following equation:

$$
\mu_j^{(l+1)} = \mu_j^{(l)} \sum_{i=1}^{N} \frac{R_{ij}}{\varepsilon_j} \frac{n_i}{\sum_{k=1}^{N} R_{ik}\, \mu_k^{(l)}} = \sum_{i=1}^{N} n_i\, M_{ij}^{(l)} . \tag{9.34}
$$

Above, ε_j is computed from the response matrix R_{ij} according to Eq. (9.16).
Usually, as initial solution the simulation prediction $\mu_j^{(0)} = \mu_j^{\text{est}}$ is taken, motivated
as prior choice according to the Bayesian interpretation. Equation (9.34) can be
demonstrated to converge to the maximum likelihood solution, but a very large
number of iterations are needed before it approaches the asymptotic limit.

The method can be stopped after a finite number $l = I$ of iterations and provides
the estimate: $\hat{\mu}_j = \mu_j^{(I)}$. I acts here as regularization parameter, similarly to τ for
the regularized unfolding in Sect. 9.8. I equal to infinity corresponds to $\tau = 0$.

Equation (9.34) can be motivated using Bayes' theorem. Given a number of
causes C_j, $j = 1, \cdots, M$, the effects E_i, $i = 1, \cdots, N$, can be related to the
causes using Bayes' formula [11] as follows:

$$
P(C_j \mid E_i) = \frac{P(E_i \mid C_j)\, P_0(C_j)}{\sum_{k=1}^{M} P(E_i \mid C_k)\, P_0(C_k)} . \tag{9.35}
$$

A cause C_j in Eq. (9.35) corresponds to an entry generated in the jth bin in the
histogram of the original variable y, and an effect E_i as an entry observed, after
detector effects, in the ith bin in the histogram of the observable variable x.

$P(C_j \mid E_i)$ has the role of the response matrix R_{ij}, and $P_0(C_j)$ can be rewritten
as

$$
P_0(C_j) = \frac{\mu_j^{(0)}}{n^{\text{obs}}} , \tag{9.36}
$$

where

$$
n^{\text{obs}} = \sum_{i=1}^{N} n_i . \tag{9.37}
$$

This allows to rewrite Eq. (9.35) as

$$
P(C_j \mid E_i) = \frac{R_{ij}\, \mu_j^{(0)}}{\sum_{k=1}^{M} R_{ik}\, \mu_k^{(0)}} . \tag{9.38}
$$

The observation of an experimental distribution $\vec{n} = (n_1, \cdots, n_N)$ is interpreted as the number of occurrences of each of the E_i effects: $n_i = n(E_i)$. The expected number of entries $\mu_j^{(1)}$ assigned to each cause C_j can be determined, taking also into account a term due to finite efficiency, as

$$\mu_j^{(1)} = \hat{n}(C_j) = \sum_{i=1}^{N} n(E_i) \frac{P(C_j \mid E_i)}{\varepsilon_j} = \mu_j^{(0)} \sum_{i=1}^{N} \frac{R_{ij}}{\varepsilon_j} \frac{n_i}{\sum_{k=1}^{M} R_{ik} \mu_k^{(0)}} . \qquad (9.39)$$

Equation (9.39) can be applied iteratively starting from an initial condition $\vec{\mu}^{(0)}$ due to the prior choice in Bayes' theorem, and the general iteration formula can be derived as anticipated in Eq. (9.34).

The evaluation of the covariance matrix may proceed using error propagation. The covariance matrix $U^{(l)}$ of the unfolded estimates $\vec{\mu}^{(l)}$ at a given iteration is given by

$$U_{pq}^{(l)} = \sum_{i,j} \left(\frac{\partial \mu_p}{\partial n_i} \right)^{(l)} V_{ij} \left(\frac{\partial \mu_q}{\partial n_j} \right)^{(l)} , \qquad (9.40)$$

where V is the covariance matrix of the measured distribution $\vec{n}$, usually due to independent Poissonian fluctuations, and can be estimated as

$$\hat{V}_{ij} = \delta_{ij} \, n_i . \qquad (9.41)$$

The derivative terms in Eq. (9.40) are, at the first iteration ($l = 1$), equal to [11]

$$\left(\frac{\partial \hat{\mu}_k}{\partial n_i} \right)^{(1)} = M_{ij}^{(0)} . \qquad (9.42)$$

At the following iteration, the derivative can be estimated again by iteration according to the following formula [12]:[1]

$$\left(\frac{\partial \hat{\mu}_k}{\partial n_i} \right)^{(l+1)} = M_{ij}^{(l)} + \frac{\mu_i^{(l+1)}}{\mu_i^{(l)}} \left(\frac{\partial \mu_i}{\partial n_j} \right)^{(l)} - \sum_{p,q} \frac{n_k \, \varepsilon_q}{\mu_q^{(l)}} M_{ip}^{(l)} M_{qp}^{(l)} \left(\frac{\partial \mu_q}{\partial n_j} \right)^{(l)} .$$

$$(9.43)$$

Figure 9.7 shows results of iterative unfolding with 100 iterations, which appears relatively smooth, for the same distribution considered in Fig. 9.6. Figure 9.8 shows unfolded distributions after 1000, 10,000, and 100,000 iterations. An oscillating

---

[1] The original derivation in [11] only reported the first term in Eq. (9.43), similarly to the case with $l = 0$, as in Eq. (9.42). The complete formula was reported in [12]. A revised and improved iterative method in [11] was also later proposed in [13].

Fig. 9.7 Unfolded
distribution using the iterative
unfolding, shown as points
with error bars, superimposed
to the original distribution,
shown as dashed line. 100
iterations have been used

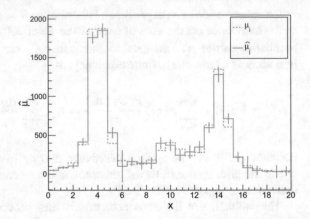

structure, as for the maximum likelihood solution, slowly begins to emerge as the
number of iterations increases.

Regularization is achieved using the iterative unfolding method by stopping the
procedure after a number $l = I$ of iterations. I should not be too large, otherwise,
the solution starts to exhibit oscillation, as seen in Fig. 9.8. On the other hand, if
I is small, the solution is biased toward the initial values, similarly to the bin-
by-bin corrections method (see Sect. 9.7). A prescription about the choice of the
regularization parameter I, for iterative unfolding, is not as obvious as it was for the
Tikhonhov regularization (see Sect. 9.9) with the L curve scan.

9.12   Adding Background in Iterative Unfolding

In case a background subtraction is needed ($b_i \neq 0$), two approaches could be used
to modify Eq. (9.34). The first is to subtract the background in the numerator:

$$\mu_j^{(l+1)} = \mu_j^{(l)} \sum_{i=1}^{N} \frac{R_{ij}}{\varepsilon_j} \frac{(y_i - b_i)}{\sum_{k=1}^{N} R_{ik} \mu_k^{(l)}} , \qquad (9.44)$$

the second is to add it in the folded term at the denominator:

$$\mu_j^{(l+1)} = \mu_j^{(l)} \sum_{i=1}^{N} \frac{R_{ij}}{\varepsilon_j} \frac{y_i}{\sum_{k=1}^{N} R_{ik} \mu_k^{(l)} + b_i} . \qquad (9.45)$$

The latter choice ensures non-negative yields: $\mu_j^{(l)} \geq 0$.

Fig. 9.8 Unfolded
distribution as in Fig. 9.7
using the iterative unfolding
with 1000 (top), 10,000
(middle), and 100,000
(bottom) iterations

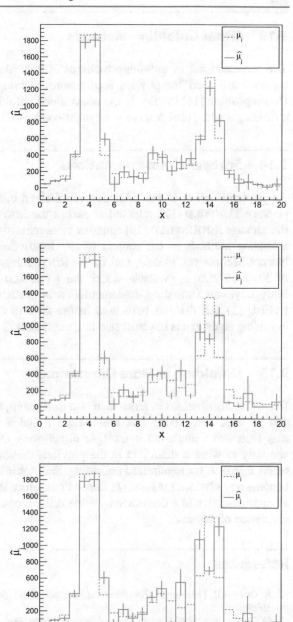

9.13 Other Unfolding Methods

The presented list of unfolding methods is not exhaustive, and other unfolding methods are used for physics applications. Among those, the Singular Value Decomposition [14] (SVD), Shape-constrained unfolding [15], and Fully Bayesian unfolding method [16] deserve to be mentioned.

9.14 Software Implementations

An implementation of the Tikhonov regularized unfolding is available with the package TUNFOLD [17], released as part of the ROOT framework [18]. Moreover, the package ROOUNFOLD [19] contains implementations and interfaces to multiple unfolding methods in the context of the ROOT framework [18]. This package has not become yet, to date, part of the ROOT release. A Python implementation of ROOUNFOLD is available under the PYNFOLD package [20]. Moreover, a Fully Bayesian Unfolding implementation is available with the Python package PYFBU [21, 22] that has been used in the ATLAS experiment [23]. An iterative unfolding algorithm is implemented in PYUNFOLD [24, 25].

9.15 Unfolding in More Dimensions

In case of distributions in more than one dimension, the methods described above can be used considering that the indices i and j in the response matrix R_{ij} may represent a single bin in multiple dimensions. Done that, unfolding proceeds similarly to what is described in the previous sections. One caveat is the choice of the matrix L for regularized unfolding. See, for instance, Eq. (9.29), which may become non-obvious for multiple dimensions, since the L matrix should represent an approximation of a derivative by finite differences. See [17] for a more detailed discussion of this case.

References

1. R. Bracewell, *The Fourier Transform and Its Applications*, 3rd. edn. (McGraw-Hill, New York, 1999)
2. G. Cowan, *Statistical Data Analysis* (Clarendon Press, Oxford, 1998)
3. D.L. Phillips, A technique for the numerical solution of certain integral equations of the first kind. J. ACM **9**, 84 (1962)
4. A.N. Tikhonov, On the solution of improperly posed problems and the method of regularization. Sov. Math. **5**, 1035 (1963)
5. A.N. Tikhonov, V.Y. Arsenin, *Solutions of Ill-Posed Problems* (John Wiley, New York, 1977)
6. P.C. Hansen, *The L-curve and Its Use in the Numerical Treatment of Inverse Problems, Computational Inverse Problems in Electrocardiology*. ed. by P. Johnston (2000)
7. L. Shepp, Y. Vardi, Maximum likelihood reconstruction for emission tomography. IEEE Trans. Med. Imaging **1**, 113–122 (1982)

8. A. Kondor, Method of converging weights – an iterative procedure for solving Fredholm's integral equations of the first kind. Nucl. Instrum. Meth. A **216**, 177 (1983)
9. H.N. Müulthei, B. Schorr, On an iterative method for a class of integral equations of the first kind. Mat. Meth. Appl. Sci. **9**, 137 (1987)
10. H.N. Müulthei, B. Schorr, On an iterative method for the unfolding of spectra. Nucl. Instr. Meth. A **257**, 371 (1983)
11. G. D'Agostini, A multidimensional unfolding method based on Bayes' theorem. Nucl. Instr. Meth. A **362**, 487–498 (1995)
12. T. Adye, Corrected error calculation for iterative Bayesian unfolding (2011). http://hepunx.rl. ac.uk/~adye/software/unfold/bayes_errors.pdf
13. G. D'Agostini, Improved iterative Bayesian unfolding. arXiv:1010.0632 (2010)
14. A. Hoecker, V. Kartvelishvili, SVD approach to data unfolding. Nucl. Instrum. Meth. A **372**, 469–481 (1996)
15. M. Kuusela, P.B. Stark., Shape-constrained uncertainty quantification in unfolding steeply falling elementary particle spectra (2015). arXiv:1512.00905
16. G. Choudalakis, Fully Bayesian unfolding (2012). arXiv:1201.4612
17. S. Schmitt, TUnfold: an algorithm for correcting migration effects in high energy physics JINST **7**, T10003 (2012)
18. R. Brun, F. Rademakers, ROOT—an object oriented data analysis framework, in *Proceedings ΛIHENP96 Workshop, Lausanne (1996)*. Nuclear Instruments and Methods, vol. A389 (1997), pp. 81–86. http://root.cern.ch/
19. T. Adye, R. Claridge, K. Tackmann, F. Wilson, RooUnfold: ROOT unfolding framework (2012). http://hepunx.rl.ac.uk/~adye/software/unfold/RooUnfold.html
20. PynFold – Unfolding With Python, https://pynfold.readthedocs.io/
21. PyFBU, https://pypi.org/project/fbu/
22. G. Choudalakis, Fully Bayesian unfolding (2012). arXiv:1201.4612. https://arxiv.org/abs/1201. 4612
23. S. Biondi On Behalf of the ATLAS Collaboration, Experience with using unfolding procedures in ATLAS. EPJ Web Conf. **137**, 11002 (2017). ATL-PHYS-PROC-2016-189. https://doi.org/ 10.1051/epjconf/201713711002
24. PyUnfold, a Python package for implementing iterative unfolding, https://jrbourbeau.github. io/pyunfold/
25. J. Bourbeau, Z. Hampel-Arias, PyUnfold: a python package for iterative unfolding. J. Open Source Softw. **3**(26), 741 (2018). https://doi.org/10.5281/zenodo.1258211, arXiv:1806.03350, https://arxiv.org/abs/1806.03350

Hypothesis Testing

10

10.1 Introduction

The goal of many important physics measurements consists of discriminating
between two or more different *hypotheses* based on observed experimental data.
This problem in statistics is known as *hypothesis testing*. Inputs to the test are the
followings:

- A set of observed values of some random variables, $x_1, \cdots, x_n$;
- Two or more alternative hypotheses that determine the joint probability distribution of $x_1, \cdots, x_n$.

We want to assign an observation to one of the hypotheses, considering the predicted
probability distributions of the observed quantities under either of the possible
hypotheses.

A typical example is the identification of a particle species in a detector with
capabilities to discriminate between different particle types. For instance, the
data used for the discrimination may be, depending on the detector technology,
the measured depth of penetration into an iron absorber, the energy released in
scintillator crystals, or the reconstructed direction of light emitted by the Cherenkov
radiation. Another example is to determine whether a recorded data sample is
composed of background only or it contains a particular signal, on top of the
typically unavoidable background. This procedure is used when assessing the
presence of a new signal which has never been observed before. In this case, if
data strongly favors the presence of the new signal, the measurement may lead to a
discovery, which is the most rewarding result of an experiment.

© The Author(s), under exclusive license to Springer Nature Switzerland AG 2023
L. Lista, *Statistical Methods for Data Analysis*, Lecture Notes in Physics 1010,
https://doi.org/10.1007/978-3-031-19934-9_10

10.2   Test Statistic

Consider the case of two different hypotheses. In the statistical literature, these are usually called *null hypothesis*, indicated with H_0, and *alternative hypothesis*, indicated with H_1. Assume that the observed data sample consists of the measurement of the variables $\vec{x} = (x_1, \cdots, x_n)$ randomly distributed according to some probability density function $f(\vec{x})$. The distribution $f(\vec{x})$ is, in general, different under the two hypotheses H_0 and H_1: $f(\vec{x}) = f(\vec{x} \mid H_0)$ or $f(\vec{x}) = f(\vec{x} \mid H_1)$, according to which of H_0 or H_1 is true. The goal of a test is to determine whether the observed data sample is more compatible with the distribution $f(\vec{x} \mid H_0)$ rather than with $f(\vec{x} \mid H_1)$.

Instead of using all the n available variables, $(x_1, \cdots, x_n)$, whose PDF $f(\vec{x})$ may be too complex to describe with accuracy, usually a test proceeds by determining the value of a function of the measured sample $\vec{x}$:

$$t = t(\vec{x}) , \tag{10.1}$$

which summarizes the information contained in the entire data sample. t is called *test statistic*, and its PDF, under either of the considered hypotheses, can be derived from the PDF $f(\vec{x})$.

Consider the simplest case of a single variable x. The simplest test statistic can be defined as the measured value itself:

$$t = t(x) = x . \tag{10.2}$$

In particular, if the observed value of x is $x^\star$, the corresponding value of the test statistic is $t^\star = x^\star$.

Imagine that t, or equivalently x in the simplest case of Eq. (10.2), has some discriminating power between two hypotheses, say *signal* versus *background*, as shown in Fig. 10.1. A good *separation* of the two cases can be achieved if the PDFs of t under the hypotheses $H_1 = signal$ and $H_0 = background$ are appreciably different. Note that which between signal and background is the null hypothesis is a conventional choice. Sometimes, the opposite choice is also adopted in the literature.

A *selection requirement* on t is sometimes called *cut* in physicists' jargon: we identify an observation $x^\star$ as signal if $\hat{t} \leq t_{cut}$ or as background if $t^\star > t_{cut}$, where the value t_{cut} is chosen a priori. The direction of the inequality sign is chosen because background tends to have values of t larger than signal, in the specific example (Fig. 10.1).

Consider that our signal is muons and background is pions in a particle identification problem. Not all real muons are correctly identified as muons according to the selection criterion, as well as not all real pions are correctly identified as pions. The expected fraction of selected signal particles (muons) is usually called signal *selection efficiency* and the expected fraction of selected background particles (pions) is called *misidentification probability*. Misidentified particles constitute a background to positively identified signal particles. Applying the required selection,

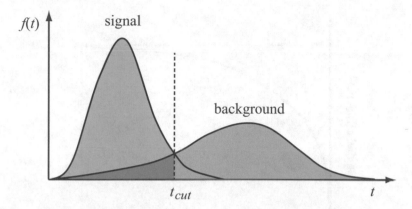

Fig. 10.1 Distribution of a test statistic t according to two possible PDFs for the signal (blue) and background (red) hypotheses under test

$t^\star \leq t_{\text{cut}}$, on a data sample containing different detected particles, we expect that the sample of selected particles is enriched of signal and has a reduced fraction of background, compared to the original unselected sample. This is true if the selection efficiency is larger than the misidentification probability, as achieved with the selection in Fig. 10.1, thanks to the separation of the two PDFs. This case was also discussed using Bayes' theorem in Exercise 5.1.

10.3 Type I and Type II Errors

Statistical literature defines the *significance level* α as the probability to reject the hypothesis H_0 if H_0 is true. Rejecting H_0 if it is true is called *error of the first kind* or *type-I error*. It corresponds to a *false positive*. In our example, this means selecting a particle as a muon in case it is a pion. The *background misidentification probability* corresponds to $1 - \alpha$.

The probability β to reject the hypothesis H_1 if it is true, called *error of the second kind* or *type-II error*, is equal to one minus the *signal efficiency*, i.e., the probability to incorrectly identify a muon as a pion, in our example. It corresponds to a *false negative*. $1 - \beta$ is also called *power of the test*.

By varying the value of the selection cut t_{cut}, different values of selection efficiency $1 - \beta$ and misidentification probability α are determined. A typical curve representing the signal efficiency versus the misidentification probability obtained by varying the selection requirement is shown in Fig. 10.2. This curve is called *receiver operating characteristic* or *ROC curve*. A good selection should have a low misidentification probability corresponding to a large selection efficiency. But clearly the background rejection cannot be perfect, i.e., the misidentification probability cannot drop to zero if the distributions $f(x \mid H_0)$ and $f(x \mid H_1)$ have some degree of overlap, as in Fig. 10.2.

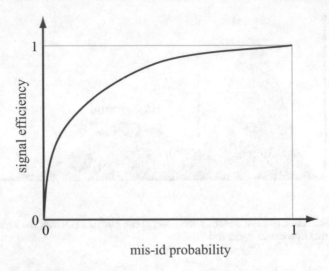

Fig. 10.2 Signal efficiency versus background misidentification probability (receiver operating characteristic or ROC curve)

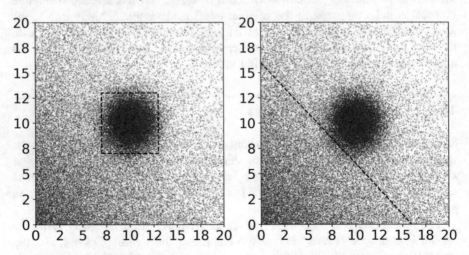

Fig. 10.3 Examples of two-dimensional selections of a signal (blue dots) against a background (red dots). A linear cut is chosen on the left plot, while a box cut is chosen on the right plot

In the presence of multiple variables, cut-based selections involve more complex requirements, which, in multiple dimensions, correspond to regions in the discriminating variable space. Observations are accepted as signal or background if they fall inside or outside the selection region. Finding an optimal selection in multiple dimensions is usually a non-trivial task. Two simple examples of selections with very different performances in terms of signal efficiency $1 - \beta$ and misidentification probability α are shown in Fig. 10.3.

The problem of classifying data according to the values of multiple variables is called *multivariate analysis* (MVA). The following sections discuss some basic techniques to approach this problem. Chapter 11 introduces machine learning techniques to address this problem.

10.4 Fisher's Linear Discriminant

A test statistic that allows discriminating two samples in more dimensions using a linear combination of n discriminating variables is due to Fisher [1]. A linear combination may be suboptimal, but in some cases it provides acceptable performances. The test consists of finding the optimal axis, corresponding to a direction vector $\vec{w}$ in the multidimensional space, along which the projections of the two PDFs, in the H_0 and H_1 hypotheses, achieve the best separation. The optimal separation between the one-dimensional projections of two PDFs along $\vec{w}$ is achieved by maximizing the squared difference of the means of the two projected PDFs divided by the sum in quadrature of the corresponding variances.

The Fisher discriminant, for a given axis direction $\vec{w}$, can be defined as

$$J(\vec{w}) = \frac{(\mu_1 - \mu_2)^2}{\sigma_1^2 + \sigma_2^2} , \qquad (10.3)$$

where μ_1 and μ_2 are the averages and σ_1 and σ_2 are the standard deviations of the two PDF projections, which depend on the projection direction $\vec{w}$. An illustration of Fisher projections along two different possible axes is shown in Fig. 10.4 for a two-dimensional case.

Different levels of separation of the two distributions can be achieved by changing the projection line. The goal is to find the projection direction $\vec{w}$ that achieves the optimal separation. The projected average difference can also be written as

$$\mu_1 - \mu_2 = \vec{w}^T (\vec{m}_1 - \vec{m}_2) , \qquad (10.4)$$

where $\vec{m}_1$ and $\vec{m}_2$ are the n-dimensional averages of the two samples. The square of Eq. (10.4) gives the numerator in Eq. (10.3):

$$(\mu_1 - \mu_2)^2 = \vec{w}^T (\vec{m}_1 - \vec{m}_2) (\vec{m}_1 - \vec{m}_2)^T \vec{w} = \vec{w}^T S_B \vec{w} , \qquad (10.5)$$

where the *between classes* scatter matrix S_B is defined as

$$S_B = (\vec{m}_1 - \vec{m}_2) (\vec{m}_1 - \vec{m}_2)^T . \qquad (10.6)$$

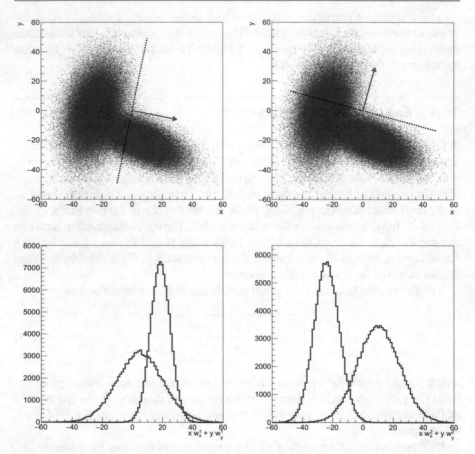

Fig. 10.4 Illustration of the projection of two-dimensional distributions along two different axes (top plots). The red and blue distributions are projected along the black dashed lines; the normal to that line is shown as a green line with an arrow. The bottom plots show the one-dimensional projections of the corresponding top plots

The projections along $\vec{w}$ of the two $n \times n$ covariance matrices S_1 and S_2 give the variances:

$$\sigma_1^2 = \vec{w}^{\,T} S_1 \, \vec{w} \,, \tag{10.7}$$

$$\sigma_2^2 = \vec{w}^{\,T} S_2 \, \vec{w} \,, \tag{10.8}$$

whose sum is the denominator in Eq. (10.3):

$$\sigma_1^2 + \sigma_2^2 = \vec{w}^{\,T} (S_1 + S_2) \, \vec{w} = \vec{w}^{\,T} S_W \, \vec{w} \,. \tag{10.9}$$

Above, the *within classes* scatter matrix S_W is defined as

$$S_W = S_1 + S_2 . \tag{10.10}$$

Fisher's discriminant in Eq. (10.3) can be written, in a compact form, as

$$J(\vec{w}) = \frac{\left[\vec{w}^{\,T}(\vec{m}_1 - \vec{m}_2)\right]^2}{\vec{w}^{\,T}(S_1 + S_2)\,\vec{w}} = \frac{\vec{w}^{\,T} S_B\,\vec{w}}{\vec{w}^{\,T} S_W\,\vec{w}} . \tag{10.11}$$

The problem of finding the vector $\vec{w}$ that maximizes $J(\vec{w})$ can be solved by performing the derivatives of $J(\vec{w})$ with respect to the components w_i of $\vec{w}$ and setting them to zero, or equivalently, by solving the following eigenvalues equation:

$$S_B\,\vec{w} = \lambda\,S_W\,\vec{w} , \tag{10.12}$$

i.e.,

$$S_W^{-1} S_B\,\vec{w} = \lambda\,\vec{w} . \tag{10.13}$$

This equation has solution:

$$\boxed{\vec{w} = S_W^{-1} S_B\,(\vec{m}_1 - \vec{m}_2) .} \tag{10.14}$$

A practical way to find Fisher's discriminant when the PDFs are not known exactly is to provide two *training samples* with a sufficiently large number of entries to represent approximately the two PDFs. The averages and covariance matrices determined from training samples can be used with Eq. (10.14) to find the direction $\vec{w}$ that maximizes Fisher's discriminant. Training samples can also be generated with Monte Carlo simulation.

10.5 The Neyman–Pearson Lemma

The performance of a selection criterion can be considered optimal if it achieves the selection that corresponds to the largest possible power or signal selection 142 efficiency $1 - \beta$ for fixed background misidentification probability α. According to the Neyman–Pearson lemma [2], the optimal test statistic, in this sense, is given by the ratio of the likelihood functions $L(\vec{x}\mid H_1)$ and $L(\vec{x}\mid H_0)$ evaluated for a given data sample $\vec{x}$ under the two hypotheses H_1 and H_0:

$$\lambda(\vec{x}) = \frac{L(\vec{x}\mid H_1)}{L(\vec{x}\mid H_0)} . \tag{10.15}$$

For a given observation $\vec{x} = \vec{x}^{\,\star}$, the test is

$$\lambda(\vec{x}^{\,\star}) = \frac{L(\vec{x}^{\,\star}|\,H_1)}{L(\vec{x}^{\,\star}|\,H_0)} \geq k_\alpha\,, \tag{10.16}$$

where, by varying the value of the selection cut k_α, the required value of α may be achieved. This corresponds to choose a point in the ROC curve (see Fig. 10.2) such that α corresponds to the required misidentification probability.

The Neyman–Pearson lemma provides the selection that achieves the optimal performances only if the joint multidimensional PDFs that characterize our problem under the two hypotheses are known. In many realistic cases, anyway, it is not easy to determine the multidimensional PDFs exactly, or at least with very good precision, and approximate solutions may necessary. Numerical methods and algorithms exist to find selections in the multivariate space that have performances in terms of selection efficiency and misidentification probability close to the optimal limit given by the Neyman–Pearson lemma. Among those approximate methods, machine-learning algorithms, such as *Artificial Neural Networks* and *Boosted Decision Trees*, are widely used in particle physics and are introduced in Chap. 11.

10.6   Projective Likelihood Ratio Discriminant

If the variables $x_1, \cdots, x_n$ that characterize our problem are independent, the likelihood function can be factorized into the product of one-dimensional marginal PDFs:

$$\lambda(\vec{x}) = \lambda(x_1, \cdots, x_n) = \frac{L(x_1, \cdots, x_n\,|\,H_1)}{L(x_1, \cdots, x_n\,|\,H_0)} = \frac{\prod_{i=1}^{n} f_i(x_i\,|\,H_1)}{\prod_{i=1}^{n} f_i(x_i\,|\,H_0)}\,. \tag{10.17}$$

If this factorization holds, optimal performances are achieved, according to the Neyman–Pearson lemma.

Even if it is not possible to factorize the PDFs into the product of one-dimensional marginal PDFs, i.e., if the variables are not independent, a test statistic inspired by Eq. (10.17) can be used as discriminant using the marginal PDFs f_i for the individual variables. This test statistic is called *projective likelihood ratio*:

$$\lambda(x_1, \cdots, x_n) = \frac{\prod_{i=1}^{n} f_i(x_i\,|\,H_1)}{\prod_{i=1}^{n} f_i(x_i\,|\,H_0)}\,. \tag{10.18}$$

If the PDFs cannot be exactly factorized, anyway, the test statistic defined in Eq. (10.18) differs from the exact likelihood ratio in Eq. (10.15) and has worse performances in terms of α and β compared with the best possible performances obtained using the Neyman–Pearson lemma.

In some cases, anyway, the simplicity of this method can justify its application, at the cost of the suboptimal performances. The marginal PDFs f_i can be obtained using Monte Carlo *training samples* with a large number of entries that allow to produce histograms corresponding to the distributions of the individual variables x_i that, to a good approximation, reproduce the marginal PDFs.

Some numerical implementations of projective likelihood ratio discriminant apply in a preliminary stage a suitable rotation in the variable space to eliminate the correlation among variable and to achieve the diagonalization of the covariance matrix. This usually improves the performance of the projective likelihood discriminant but does not necessarily allow to reach the optimal performance given by the Neyman–Pearson lemma. PDFs can be factorized only for independent variables (see Sect. 3.6), but uncorrelated variables are not necessarily independent. See, for instance, Exercise 3.4.

If one wants to determine the full multivariate distribution from a finite simulated sample, the approximation of a multidimensional PDF as a binned histogram in multiple dimensions may become intractable. The size of the training samples should increase proportionally to the total number of bins. If each dimension is subdivided into N bins, the total number of bins grows as N to the power of the number of dimensions.

10.7 Kolmogorov–Smirnov Test

A test due to Kolmogorov [3], Smirnov [4], and Chakravarti [5] can be used to assess the hypothesis that a data sample is compatible with being randomly extracted from a given distribution f. Consider a sample $(x_1, \cdots, x_N)$ of independent and identically distributed variables ordered by increasing value of x that has to be compared with a distribution $f(x)$, assumed to be a continuous function.

The discrete cumulative distribution of the sample can be defined as

$$F_N(x) = \frac{1}{N} \sum_{i=1}^{N} \theta\,(x - x_i)\,, \tag{10.19}$$

where θ is the step function defined as

$$\theta\,(x) = \begin{cases} 1 & \text{if } x \geq 0\,, \\ 0 & \text{if } x < 0\,. \end{cases} \tag{10.20}$$

The cumulative distribution $F_N(x)$ can be compared with the cumulative distribution of $f(x)$ defined, as usual, as

$$F(x) = \int_{-\infty}^{x} f(x')\,\mathrm{d}x'\,. \tag{10.21}$$

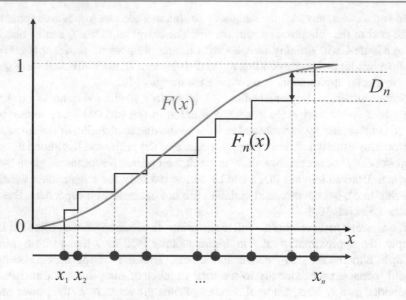

Fig. 10.5 Illustration of the Kolmogorov–Smirnov test

The maximum distance between the two cumulative distributions $F_N(x)$ and $F(x)$ is used to quantify the agreement of the data sample $(x_1, \cdots, x_N)$ with $f(x)$:

$$D_N = \sup_x |F_N(x) - F(x)| . \tag{10.22}$$

The procedure to define $F_N(x)$, $F(x)$, and D_N is illustrated in Fig. 10.5.

For large N, D_N converges to zero in probability. The distribution of the test statistic $K = \sqrt{N}\, D_N$, in the hypothesis that the values $x_1, \cdots, x_N$ are distributed according to $f(x)$, in the limit for $N \to \infty$, does not depend on $f(x)$. The probability that K is less than or equal to a given value k is known, due to Marsaglia et al. [6]:

$$P(K \le k) = 1 - 2 \sum_{i=1}^{\infty} (-1)^{i-1} e^{-i^2 k^2} = \frac{\sqrt{2\pi}}{k} \sum_{i=1}^{\infty} e^{-(2i-1)^2 \pi^2 / 8k^2} . \tag{10.23}$$

The distribution of K is usually called *Kolmogorov distribution*.

It is important to notice that Kolmogorov–Smirnov is a non-parametric test. If some parameters of the distribution $f(x)$ are determined, i.e., obtained from a fit, from the data sample $x_1, \cdots, x_N$, then the test cannot be applied. A pragmatic solution to this problem, in case parameters have been estimated from data, is to still use the test statistic K, but without relying on the Kolmogorov distribution given by Eq. (10.23). The distribution of K, in those cases, should be determined empirically with Monte Carlo for each specific problem. This approach is implemented, for instance, as optional method in the Kolmogorov–Smirnov test provided by the ROOT framework [7].

The Kolmogorov–Smirnov test can also be used to compare two ordered samples, say $(x_1, \cdots, x_N)$ and $(y_1, \cdots, y_M)$, and asses the hypothesis that both come from the same distribution. In this case, the maximum distance

$$D_{N, M} = \sup_x |F_N(x) - F_M(x)| \tag{10.24}$$

asymptotically converges to zero as N and M are sufficiently large, and the following test statistic asymptotically follows Kolmogorov distribution in Eq. (10.23):

$$\sqrt{\frac{N M}{N + M}} D_{N, M} \, . \tag{10.25}$$

Alternative tests to Kolmogorov–Smirnov are due to Stephens [8], Anderson and Darling [9], and Cramér [10] and von Mises [11].

10.8 Wilks' Theorem

When the number of available measurements is large, Wilks' theorem allows to find an approximate asymptotic distribution of a test statistic based on a likelihood ratio inspired by the Neyman–Pearson lemma (Eq. (10.15)). Assume that two hypotheses H_1 and H_0 can be defined in terms of a set of parameters $\vec{\theta} = (\theta_1, \cdots, \theta_m)$ on which the likelihood function depends, and the condition that H_1 is true can be expressed as $\vec{\theta} \in \Theta_1$, while the condition that H_0 is true can be expressed as $\vec{\theta} \in \Theta_0$. Let us also assume that H_0 and H_1 are *nested hypotheses*, i.e., $\Theta_0 \subseteq \Theta_1$.

Given a data sample made of independent observations $\vec{x}_1, \cdots, \vec{x}_N$, the following test statistic can be defined:

$$\chi_r^2 = -2 \log \frac{\displaystyle\sup_{\vec{\theta} \in \Theta_0} \prod_{i=1}^{N} L(\vec{x}_i ; \vec{\theta})}{\displaystyle\sup_{\vec{\theta} \in \Theta_1} \prod_{i=1}^{N} L(\vec{x}_i ; \vec{\theta})} \, . \tag{10.26}$$

Under certain regularity conditions of the likelihood function, Wilks' theorem [12] ensures that χ_r^2 has a distribution that can be approximated, for $N \to \infty$, and if H_0 is true, by a χ^2 distribution having a number of degrees of freedom equal to the difference between dimension of the set Θ_1 and the dimension of the set Θ_0.

As a more specific example, assume that μ is the only parameter of interest and the remaining parameters, $\vec{\theta} = (\theta_1, \cdots, \theta_m)$, are m nuisance parameters (see Sect. 1.11). For instance, μ could be the ratio of a signal cross section to its theoretical value. This ratio was introduced in Higgs search and in searches for physics beyond the Standard Model at the Large Hadron Collider, and is called *signal strength*, see also Sect. 10.9. Taking as H_0 the hypothesis that μ assumes a particular value $\mu = \mu_0$, while H_1 is the alternative hypothesis that μ may have any possible value greater or equal to zero, Wilks' theorem can be applied to the following test statistic:

$$-2 \log \lambda(\mu_0) = -2 \log \frac{\sup_{\vec{\theta}} \prod_{i=1}^{N} L(\vec{x}_i;\, \mu_0, \vec{\theta})}{\sup_{\mu, \vec{\theta}} \prod_{i=1}^{N} L(\vec{x}_i;\, \mu, \vec{\theta})}, \tag{10.27}$$

where the two sets of possible values of μ and $\vec{\theta}$ are $\Theta_0 = \{\mu_0\} \times \mathbb{R}^m$ and $\Theta_1 = [0, +\infty[\times \mathbb{R}^m$. The test statistic $-2 \log \lambda(\mu_0)$, if the hypothesis H_0 is true, i.e., if $\mu = \mu_0$, is asymptotically distributed as a χ^2 with one degree of freedom, since the dimensions of Θ_0 and Θ_1 differ by one unit. In particular, $-2 \log \lambda(0)$ can be used to test against the null hypothesis, i.e., to exclude that no signal is present.

The denominator in Eq. (10.27) is the likelihood function evaluated at the parameter values $\mu = \hat{\mu}$ and $\vec{\theta} = \hat{\vec{\theta}}$ that maximize it:

$$\prod_{i=1}^{N} L(\vec{x}_i;\, \hat{\mu}, \hat{\vec{\theta}}) = \sup_{\mu, \vec{\theta}} \prod_{i=1}^{N} L(\vec{x}_i;\, \mu, \vec{\theta}). \tag{10.28}$$

In the numerator of Eq. (10.27), only the nuisance parameters $\vec{\theta}$ are fit, keeping μ fixed to the constant value $\mu = \mu_0$. Taking as $\vec{\theta} = \hat{\hat{\vec{\theta}}}(\mu_0)$ the set of values of $\vec{\theta}$ that maximize the likelihood function for a fixed $\mu = \mu_0$, Eq. (10.27) can be written as

$$\lambda(\mu_0) = \frac{L(\vec{x} \mid \mu_0, \hat{\hat{\vec{\theta}}}(\mu_0))}{L(\vec{x} \mid \hat{\mu}, \hat{\vec{\theta}})}. \tag{10.29}$$

This test statistic is known as *profile likelihood*, and its applications to upper limits determination are discussed in Sect. 12.24.

10.9   Likelihood Ratio in the Search for New Signals

In the previous section, the likelihood function was defined for a set of independent observations $\vec{x}_1, \cdots, \vec{x}_N$, characterized by the parameter values $\vec{\theta}$:

$$L(\vec{x}_1, \cdots, \vec{x}_N;\, \vec{\theta}) = \prod_{i=1}^{N} f(\vec{x}_i;\, \vec{\theta}). \tag{10.30}$$

Two hypotheses H_1 and H_0 correspond to two possible sets Θ_1 and Θ_0 of the parameter values $\vec{\theta}$. Usually, the number of observations N can also be taken into account by introducing the extended likelihood function (see Sect. 6.14):

$$L(\vec{x}_1, \cdots, \vec{x}_N; \vec{\theta}) = \frac{e^{-\nu(\vec{\theta})} \nu(\vec{\theta})^N}{N!} \prod_{i=1}^{N} f(\vec{x}_i; \vec{\theta}), \qquad (10.31)$$

where in the Poissonian term the expected number of event ν may also depend on the parameters $\vec{\theta}$.

Typically, we want to discriminate between two hypotheses: H_1 represents the presence of both signal and background, i.e., $\nu = \mu s + b$, while H_0 represents the presence of only background events in our sample, i.e., $\nu = b$, or equivalently $\mu = 0$. Above, the multiplier μ is the signal strength introduced in the previous section. All possible values of the expected signal yield are obtained by varying μ, $\mu = 1$ corresponding to the theory prediction of the signal yield s.

The PDF $f(\vec{x}_i; \vec{\theta})$ can be written as superposition of two components, one PDF for the signal and another for the background, weighted by the expected signal and background fractions, respectively:

$$f(\vec{x}; \vec{\theta}) = \frac{\mu s}{\mu s + b} f_s(\vec{x}; \vec{\theta}) + \frac{b}{\mu s + b} f_b(\vec{x}; \vec{\theta}). \qquad (10.32)$$

In this case, the extended likelihood function in Eq. (10.31) becomes similar to Eq. (6.53):

$$L_{s+b}(\vec{x}_1, \cdots, \vec{x}_N; \mu, \vec{\theta}) = \frac{e^{-(\mu s(\vec{\theta}) + b(\vec{\theta}))}}{N!} \prod_{i=1}^{N} (\mu s f_s(\vec{x}_i; \vec{\theta}) + b f_b(\vec{x}_i; \vec{\theta})).$$

$$(10.33)$$

Note that s and b may also depend on the unknown parameters $\vec{\theta}$:

$$s = s(\vec{\theta}), \quad b = b(\vec{\theta}). \qquad (10.34)$$

For instance, in a search for the Higgs boson, the theoretical cross section may depend on Higgs boson's mass, and the PDF for the signal f_s represents a resonance peak, which also depends on Higgs boson's mass.

Under the hypothesis H_0, i.e., $\mu = 0$, or background only, the likelihood function becomes

$$L_b(\vec{x}_1, \cdots, \vec{x}_N; \vec{\theta}) = \frac{e^{-b(\vec{\theta})}}{N!} \prod_{i=1}^{N} b f_b(\vec{x}_i; \vec{\theta}). \qquad (10.35)$$

The term $1/N!$ cancels when performing the likelihood ratio in Eq. (10.15), which becomes

$$
\begin{aligned}
\lambda(\mu, \vec{\theta}) &= \frac{L_{s+b}(\vec{x}_1, \cdots, \vec{x}_N; \mu, \vec{\theta})}{L_b(\vec{x}_1, \cdots, \vec{x}_N; \vec{\theta})} = \\
&= \frac{e^{-(\mu s(\vec{\theta}) + b(\vec{\theta}))}}{e^{-b(\vec{\theta})}} \prod_{i=1}^{N} \frac{\mu s f_s(\vec{x}_i; \vec{\theta}) + b f_b(\vec{x}_i; \vec{\theta})}{b f_b(\vec{x}_i; \vec{\theta})} = \\
&= e^{-\mu s(\vec{\theta})} \prod_{i=1}^{N} \left(\frac{\mu s f_s(\vec{x}_i; \vec{\theta})}{b f_b(\vec{x}_i; \vec{\theta})} + 1 \right).
\end{aligned}
\tag{10.36}
$$

The negative logarithm of the likelihood ratio is

$$
-\log \lambda(\mu, \vec{\theta}) = \mu s(\vec{\theta}) - \sum_{i=1}^{N} \log \left(\frac{\mu s f_s(\vec{x}_i; \vec{\theta})}{b f_b(\vec{x}_i; \vec{\theta})} + 1 \right).
\tag{10.37}
$$

In the case of a simple *event counting experiment*, the likelihood function only accounts for the Poissonian probability term. This term, in turn, only depends on the number of observed events N, and the dependence on the parameters $\vec{\theta}$ only appears in the expected signal and background yields. In that simplified case, the likelihood ratio that defines the test statistic is

$$
\lambda(\vec{\theta}) = \frac{L_{s+b}(N; \vec{\theta})}{L_b(N; \vec{\theta})},
\tag{10.38}
$$

where L_{s+b} and L_b are Poissonian probabilities for N, corresponding to expected yields of $\mu s + b$ and b, respectively. $\lambda(\vec{\theta})$ can be written as

$$
\begin{aligned}
\lambda(\vec{\theta}) &= \frac{e^{-(\mu s(\vec{\theta}) + b(\vec{\theta}))}(\mu s(\vec{\theta}) + b(\vec{\theta}))^N}{N!} \frac{N!}{e^{-b(\vec{\theta})} b(\vec{\theta})^N} = \\
&= e^{-\mu s(\vec{\theta})} \left(\frac{\mu s(\vec{\theta})}{b(\vec{\theta})} + 1 \right)^N.
\end{aligned}
\tag{10.39}
$$

The negative logarithm of the above expression is

$$
-\log \lambda(\vec{\theta}) = \mu s(\vec{\theta}) - N \log \left(\frac{\mu s(\vec{\theta})}{b(\vec{\theta})} + 1 \right),
\tag{10.40}
$$

which is a simplified version of Eq. (10.37), where the terms f_s and f_b have been dropped. Equations (10.40) and (10.37) have been used to determine upper limits in searches for new signals at LEP and Tevatron, as discussed in Chap. 12.

If no nuisance parameter is present, note that the hypotheses of expected yields $\nu = b\,(H_0)$ and $\nu = \mu s + b\,(H_1)$ are nested, since b is a particular case of $\mu s + b$ with $\mu = 0$, but the likelihood ratio defined to determine the test statistics in Eqs. (10.37), as well as in Eq. (10.40), assumes $\mu s + b$ in the numerator and b in the denominator, which is the inverse convention of the likelihood ratio introduced in Eq. (10.26). Wilks' theorem hypotheses apply to this case as well, with an additional minus sign in the definition of test statistics, which is now $2 \log \lambda$ instead of the more usual $-2 \log \lambda$. The two sets of possible values of μ are $\Theta_0 = \{0\}$ and $\Theta_1 = [0, +\infty[$. Therefore, the observed test statistic $2 \log \lambda(\hat\mu)$ can be used to test against the null hypothesis, i.e., to exclude that no signal is present.

In case nuisance parameters are present and our parameter set is $\mu, \theta_1, \cdots, \theta_m$, Wilks' theorem can also be applied if the test statistic is modified as follows:

$$\lambda(\mu) = \frac{\sup_{\vec\theta} L_{s+b}(\mu, \vec\theta)}{\sup_{\vec\theta} L_b(\vec\theta)} = \frac{L_{s+b}(\mu, \hat{\hat\theta}(\mu))}{L_{s+b}(0, \hat{\hat\theta}(0))} , \qquad (10.41)$$

where $\hat{\hat\theta}(\mu)$ is the set of parameters that maximize L_{s+b} for a fixed μ. The negative logarithm of the test statistic defined in Eq. (10.41) is equal to the profile likelihood test statistic, defined in Eq. (10.29), up to a constant:

$$-2 \log \lambda_{s+b}(\mu) = -2 \log \lambda_{\mathrm{prof}}(\mu) + 2 \log \lambda_{\mathrm{prof}}(0) , \qquad (10.42)$$

where, to avoid ambiguity, two likelihood ratios are indicated as λ_{s+b} and λ_{prof} for L_{s+b}/L_b and for the profile likelihood test statistic, respectively. While $-\log \lambda_{s+b}(\mu)$ is null for $\mu = 0$, $-2 \log \lambda_{\mathrm{prof}}(\mu)$ is null at its minimum, $\hat\mu$. Equation (10.42) demonstrates that the test statistics used to exclude the null hypothesis $\mu = 0$ are the same for the L_{s+b}/L_b ratio and for the profile likelihood, since

$$2 \log \lambda_{s+b}(\hat\mu) = -2 \log \lambda_{\mathrm{prof}}(0) . \qquad (10.43)$$

An application of the test statistic L_{s+b}/L_b in a simplified case with a single nuisance parameter is presented in Exercise 12.4.

References

1. R.A. Fisher, The use of multiple measurements in taxonomic problems. Ann. Eugen. **7**, 179–188 (1936)
2. J. Neyman, E. Pearson, On the problem of the most efficient tests of statistical hypotheses. Philos. Trans. R. Soc. Lond. Ser. A **231**, 289–337 (1933)
3. A. Kolmogorov, Sulla determinazione empirica di una legge di distribuzione. Giorn. Ist. Ital. Attuar. **4**, 83–91 (1933)

4. N. Smirnov, Table for estimating the goodness of fit of empirical distributions. Ann. Math. Stat. **19**, 279–281 (1948)
5. I.M. Chakravarti, R.G. Laha, J. Roy, *Handbook of Methods of Applied Statistics*, vol. I (Wiley, New York, 1967)
6. G. Marsaglia, W.W. Tsang, J. Wang, Evaluating Kolmogorov's distribution. J. Stat. Softw. **8**, 1–4 (2003)
7. R. Brun, F. Rademakers, ROOT—an object oriented data analysis framework, in *Proceedings AIHENP96 Workshop, Lausanne (1996)*. Nuclear Instruments and Methods, vol. A389 (1997), pp. 81–86. http://root.cern.ch/
8. M.A. Stephens, EDF statistics for goodness of fit and some comparisons. J. Am. Stat. Assoc. **69**, 730–737 (1974)
9. T.W. Anderson, D.A. Darling, Asymptotic theory of certain "goodness-of-fit" criteria based on stochastic processes. Ann. Math. Stat. **23**, 193–212 (1952)
10. H. Cramér, On the composition of elementary errors. Scand. Actuar. J. **1928**(1), 13–74 (1928)
11. R.E. von Mises, *Wahrscheinlichkeit, Statistik und Wahrheit* (Julius Springer, Vienna, 1928)
12. S. Wilks, The large-sample distribution of the likelihood ratio for testing composite hypotheses. Ann. Math. Stat. **9**, 60–62 (1938)

Machine Learning 11

Machine learning is a rapidly developing field of computer science. It provides classes of algorithms that automatically learn from data to make predictions that are useful in many applications, in both science and society at large. Progress in computing technologies allows to implement complex and advanced machine learning algorithms that are currently used in a vast number of fields, such as image recognition, written text and speech recognition, human language translation, e-mail spam detection, self-driving car, social media, customized advertisements, and more, including of course applications for particle physics. In several problems, such as classifying images or playing board games, machine learning algorithms already outperform humans.

Unlike more traditional computing algorithms, machine learning methods define data models that are not hard-coded into an explicit software program, and the computer learns the structure of the model from data itself. In general, machine-learning algorithms implement very generic and possibly complex parametric models. After a learning phase, machine-learning algorithms provide outputs that addresses the desired problem.

The model's parameters, which can also be very numerous, are determined and optimized by *training* the algorithm from data samples where the desired output is already known. Either real or simulated data samples may be used in the training procedure. In practice, the algorithm learns from examples. Once the algorithm has been trained, its performances can be tested on data samples, usually independent of the data utilized for training. This approach is called *supervised learning*.

Other types of machine-learning algorithms identify features and structures in data with no need to provide independent training data samples. Those methods, which fall under the category of *unsupervised learning*, can be used to automatically cluster data or to identify anomalies in a data sample, for instance to spot frauds.

Since machine learning is an area in continuous development, any text risks to become quickly obsolete, and popular methods can be overcome by new developments. Some of the most popular software frameworks, to date, usable

© The Author(s), under exclusive license to Springer Nature Switzerland AG 2023
L. Lista, *Statistical Methods for Data Analysis*, Lecture Notes in Physics 1010,
https://doi.org/10.1007/978-3-031-19934-9_11

within a Python [1] environment, are TensorFlow [2], Keras [3], PyTorch [4], scikit-learn [5], and XGBoost [6]. This list is largely incomplete and may be superseded in the next years.

For users of the ROOT [7] framework, TMVA [8] implements several machine-learning algorithms with both a C++ and Python interface, including wrapped versions of third-party algorithms, such as the already mentioned Keras and PyTorch. TMVA implements the most frequently adopted multivariate analysis methods in particle physics, including Fisher's linear discriminant, (see Sect. 10.4), projective likelihood analysis (Sect. 10.6), as well as artificial neural network (Sect. 11.10) and boosted decision trees (Sect. 11.16).

In all implementations, each method is configurable, and all parameters of the configuration can be tuned. In general, default values, reasonably valid in most applications, are also provided.

For an overview of machine learning seen from the perspective of a physicist, see [9].

11.1   Supervised and Unsupervised Learning

In many applications, machine-learning strategy consists of tuning the parameters of an algorithm based on inputs provided in the form of large data sets. For instance, to train an algorithm that identifies the content of an image, large sets of images of various subjects, such as animals or faces, can be provided. In particle physics, one can provide large simulated or real data samples of particles of given types (e.g., muons, pions, protons, or jets originating from the decay of various particles, such as top quark, W, Z, etc.), or collision events produced via a specific interaction process (e.g., Drell–Yan, top–antitop pair, Higgs boson production, etc.).

Individual data entries from the training data set are labeled according to their original and known category. Provided labels are the output we would desire an ideal algorithm to produce for the corresponding inputs. By comparing the algorithm's output to the output corresponding to the true category of data, i.e., the label, the model's parameters are modified to improve the matching between the response of the algorithm and the expected output. We would like that, for new inputs, never provided in the training phase, the algorithms are able to produce the correct labels. In general, an algorithm is never able to perfectly classify any possible input, and it has unavoidably a possibly small misclassification probability.

Algorithms trained on provided specific data samples fall under the category of *supervised learning*. The most frequent applications of supervised learning are:

- **Classification**: Machine-learning algorithms are used to assign data of unknown origin to categories based on the training phase. Classification is related to hypothesis testing, as discussed in Sect. 11.2.
- **Regression**: The model determines a relationship between dependent variables, which represent the output of the algorithm, and independent variables, which are used as input. In this case, labels may assume continuous values, representing

the target output, while, for classification problems, labels assume discrete values that represent categories.

As opposite to supervised learning, *unsupervised learning* algorithms group data or find structures within a data sample that is provided unlabeled to the algorithm. Such algorithms may recognize patterns, similarities, or features of the data and may automatically organize data into clusters that are determined from the data sample itself.

In particle physics, supervised learning is the most frequently used mainly for physics data analysis, and there the interest in unsupervised learning as a strategy to unveil new physics or to improve reconstruction algorithms is growing.

In the next sections, supervised learning is mainly discussed, followed by a short overview of the main unsupervised learning algorithms.

11.2   Machine-Learning Classification from a Statistics Perspective

In general, while statistical methods are more focused on how to determine unknown quantities with their uncertainty from observations, assuming some underlying model, machine-learning algorithms tend to be more focused on the capability to perform a prediction, and, rather than assuming a specific model, optimize very generic models whose parameters are determined from data itself.

It is possible to use a common formal approach for both machine-learning and more classic statistical problems. In both cases, we have a model that can be specified as a function that takes a vector of variables $\vec{x}$ as input and determines an output $\vec{y} = f(\vec{x}; \vec{\theta})$ given a number of parameters $\vec{\theta}$. For machine-learning algorithms, the number of parameters specified in the vector $\vec{\theta}$ can become extremely large compared to usual fit problems.

The output $\vec{y}$ may be one or more continuous real values for a regression problem, or, for a classification problem, a set of one or more integer values that encode the identifier of a category. For a binary classification, the true value of $\vec{y}$ could be 0 or 1, but the output value of the algorithm, $f(\vec{x}; \vec{\theta})$, also called *classifier*, could be a continuous value ranging from 0 to 1. A discrete value, either 0 or 1, may be obtained by applying a cut on the value of $f(\vec{x}; \vec{\theta})$. An optimal classifier should have distributions peaked near 0 or 1 for inputs corresponding to either of the two categories.

Learning from examples, in supervised learning, corresponds to provide a number of observations, $\vec{x}_1 \cdots, \vec{x}_N$, whose true outputs $\vec{y}_1, \cdots \vec{y}_N$, or labels, are known. This constitutes our *training sample*. The parameters $\vec{\theta}$ are optimized to have the outputs of the algorithm, $f(\vec{x}_1; \vec{\theta}), \cdots, f(\vec{x}_N; \vec{\theta})$, as close as possible to the true labels, $\vec{y}_1, \cdots, \vec{y}_N$.

The parameters $\vec{\theta}$ are optimized by minimizing a *loss function* that compares each output $f(\vec{x}_i; \vec{\theta})$ to the corresponding true labels y_i:

$$C(\vec{\theta}) : \vec{x}_1 \cdots , \vec{x}_N ; \vec{y}_1 \cdots , \vec{y}_N \longmapsto C(\vec{y}_1, \cdots , \vec{y}_N; f(\vec{x}_1; \vec{\theta}), \cdots , f(\vec{x}_N; \vec{\theta})) . \tag{11.1}$$

For a best fit problem, the loss function C may be a χ^2, but other metrics exist for specific problems. The optimized model is the function $f(\vec{x}; \hat{\vec{\theta}})$, where $\hat{\vec{\theta}}$ is the set of parameters that minimize the loss function C for a given training sample.

In some cases, an extra term may be added to the loss function C to reduce a possible bias of the function $f(\vec{x}; \hat{\vec{\theta}})$. This procedure is called *regularization*, and the approach has similarities with the regularization discussed in Chap. 9 for unfolding problems. In this case, we talk about *cost function* rather than loss function. More in general, when a function is minimized in an optimization procedure, we talk about *objective function*. Anyway, in the literature, these terms tend to be used somewhat interchangeably, depending on the context.

As discussed in Sect. 11.6, if the expected cost function is estimated on the training set as

$$E = C(\vec{y}_1, \cdots , \vec{y}_N; f(\vec{x}_1; \hat{\vec{\theta}}), \cdots , f(\vec{x}_N; \hat{\vec{\theta}})) , \tag{11.2}$$

a bias may be introduced because the parameters $\hat{\vec{\theta}}$ are optimized on the same training set. An unbiased estimate of the expected cost function can be obtained by using an independent sample, $\vec{x}'_1 \cdots , \vec{x}'_{N'}$ with the corresponding true labels $\vec{y}'_1 \cdots \vec{y}'_{N'}$. This new sample is called *test sample* and gives an estimated loss equal to

$$E' = C(\vec{y}'_1, \cdots , \vec{y}'_{N'}; f(\vec{x}'_1; \hat{\vec{\theta}}), \cdots , f(\vec{x}'_{N'}; \hat{\vec{\theta}})) . \tag{11.3}$$

The cross validation of an algorithm performance on a test sample independent of the training sample is in practice a requirement to have an unbiased performance estimate. Note here that, since $\hat{\vec{\theta}}$ has been optimized on the training sample, not on the test sample, in general, we have $E' \geq E$.

We have seen that many fit problems assume specific functional form f that come from theory, whose parameters are unknown. In machine learning, instead, very generic functions are assumed, with many parameters, that can easily adapt to the most, if not all, possible models we want to fit. This is similar to fit problems where we do not have a theory model, and we try to determine the functional model from data. For instance, when modeling a background of unknown shape, like in many cases of combinatorial background, we often chose empiric functional models, such as a polynomial of a certain degree, an exponential, an exponential times a polynomial, etc. If the model is sufficiently flexible and if it has a sufficiently large number of free parameters, the fit tends to closely follow the data.

Classification problems are closely related to hypothesis testing. The Neyman–Pearson lemma, introduced in Sect. 10.5, sets an upper limit to the performance of any possible binary multivariate classifier, in the sense that a classifier based on the likelihood ratio test statistic achieves the largest possible power $1 - \beta$, or selection efficiency, for a fixed value of the misidentification probability α.

The exact evaluation of the likelihood function in the two hypotheses H_0 and H_1 is not always possible, and approximate methods are developed to approach as close as possible the performance of an ideal selection based on the likelihood ratio, which is the limit set by the Neyman–Pearson lemma. Some methods have already been discussed in Chap. 10.

The most powerful approximate methods are usually implemented by means of computer algorithms that receive as input a set of discriminating variables, or features, in the machine-learning terminology. Each feature, if used individually, does not allow to reach optimal statistical performances in terms of α and β. The classifier algorithm combines the input features to compute an output value that is taken as test statistic. A selection cut on the value of the output test statistic determines the classification for a given input, as shown in Fig. 10.1.

In machine-learning algorithms, the optimization of the model is performed with a training from large input data sets distributed according to either the H_0 or H_1 hypotheses. The known true hypothesis, either H_0 or H_1, is provided as label for the training data, typically encoded as 0 or 1, and defines the two possible categories.

Training samples are often produced using computer simulations that follow, at least approximately, the expected distributions in either of the two hypotheses. In the most favorable cases, instead of simulations, control samples from real data are available with very good purity. Using real data samples for training avoids uncertainties that are present in simulation due to modeling imperfections but may have uncertainty due to contamination in the data sample due to spurious sources.

11.3 Terminology

Machine-learning nomenclature sometimes differs from what physicists use for statistical applications. Some terms have already been introduced in the previous sections. The following is a non-exhaustive list of commonly used terms that define the key ingredients common to many machine-learning algorithms and their relations:

- **Feature**: One of the variables that describe our data. The ordered set of all features is called *feature vector* and is the input to machine-learning algorithms. For a particle identification algorithm, features may be specific responses from detectors; for an algorithm that identifies a particular physics interaction channel, features may be the components of four momenta of identified particles in the detectors; for an image recognition algorithm, features may be the values of luminosity for the three RBG color channels in each pixel that composes a digital image.

- **Example** or **observation**: The set of all relevant features, i.e., the feature vector, corresponding to one individual entry in our data set. It is provided with a label when the example is part of a training sample for supervised learning. In particle physics, an observation could be a collision event; it could be one event that simulates with Monte Carlo a desired interaction process used as part of a training sample; or it could be the digital portrait for a face recognition algorithm.
- **Label**: Specifies the desired output in the training phase of supervised algorithms. It may represent a category for classification, usually encoded as one or more integer values, or it may have a continuous value for regression. For an algorithm that recognizes images of animals, labels may represent the strings "dog," "cat," "fish," etc. For a particle identification algorithm, labels may encode the categories "muon," "pion," "kaon," etc.
- **Target**: The true value of a variable that is dependent on a set of other independent variables that inputs to our algorithm. A regression procedure aims at predicting the target value as close as possible to the real value, given the values of the independent input variables. In general, the word label is more frequently used for classification, and the word target is more used for regression. But the two terms are sometimes interchanged in the literature.
- **Predictor**: an independent variable that is used as input to predict a target value, possibly together with other input variables. The term predictor is more frequently used for regression than for classification. Sometimes it is used as synonymous of feature.
- **Model**: The implementation of a parametric algorithm that performs a prediction. Parameters are optimized to make the prediction as close as possible to true values, for classification or regression problems.
- **Loss**: A measure of the deviation of the predictions of an algorithm from the real output values for an entire data set. The loss, as a function of the parameters to be optimized, is called *loss function*. The curve that describes the loss function as a function of one of the parameters is called *loss curve*. For instance, it can be the sum over all examples provided in a training sample of the differences squared between the prediction and true values. For a least squares fit, intended as a machine-learning regression algorithm, the loss may be the mean square error, equal to the average of the residuals squared:

$$\varepsilon^2 = \frac{1}{N} \sum_{i=1}^{N} \left(y_i^\star - \hat{f}(x_i) \right)^2 . \tag{11.4}$$

- **Cost function**: In some application, a penalty term is added to the loss function. The sum of loss and penalty terms is called *cost function* and is minimized to achieve a regularization of the algorithm that results in a lower bias compared to the unregularized loss. This approach is adopted in the XGBoost [6] algorithm (see Sect. 11.16). Sometimes, the word cost is used in place of loss even if no regularization is applied.
- **Objective function**: In general, a function that is minimized in the optimization phase of a machine-learning algorithm is called *objective function*. It can be a

loss function, or a cost function, in case a penalty term is added to achieve a regularization. This term may be used more in general to indicate either a loss or a cost function.

- **Training**: In supervised learning, the training implements the optimization process of a model's parameters via the minimization of the objective function. During training, many labeled examples are provided as input to the algorithm, and the model's parameters are subsequently updated according to the specific training strategy. For a parametric fit, intended as a machine-learning regression algorithm, the minimization of the χ^2 function can be considered as a form of training.

- **Batch**: Training algorithms proceed iteratively. At each iteration, the objective function is evaluated on several examples that are provided as input. Subsequently, parameters are varied accordingly. The set of examples provided altogether in each iteration is called *batch*. The *batch size*, i.e., the number of examples provided for each batch, is an important parameter. It is in general more efficient to provide more examples at each iteration, with a batch size usually of order of 10 to 1000, rather than one at a time.

- **Gradient descent**: The minimization of the objective function is achieved by iteratively varying the model's parameter, following the gradient of the objective function, o progressively approaching the minimum. Many algorithms are conceptually similar to what is discussed in Sect. 6.8 for numerical implementations of maximum likelihood fits, adding several optimal implementations specialized for specific machine-learning algorithms. This process is called *gradient descent*. The algorithm is stopped when the objective function is estimated to be sufficiently close to the minimum. In case the objective function does not have a single minimum, the algorithm may fall into one of the possible secondary local minima, and this may provide suboptimal performances.

- **Stochastic gradient descent**: A more efficient algorithm to minimize the objective function consists of providing subsets of the training sample, called *minibatches*, and computing the gradient on this subset only. The batches are usually extracted randomly from the training sample, and this randomness improves the convergence, in particular reducing the chance that the algorithm remains trapped into a secondary minimum. This approach is called *stochastic gradient descent*. Details may depend on specific implementations. For instance, a parameter may be added to control the inertia with whom it can keep some memory of the direction along which parameters were changed in the previous iteration.

- **Learning rate**: The amount of variation applied to the model's parameters to follow the gradient of the objective function may be controlled in the configuration of the training algorithm by tuning a parameter called *learning rate*. If the variations applied to the parameters are too small, the process converges slowly. If the variations are too large, the algorithm may overshoot the minimum and could even diverge. The learning rate can also be varied during the training process and can be reduced as the algorithm approaches convergence. This method may provide a smoother approach to the minimum.

- **Epoch**: The same training sample can be passed to the training algorithm more than once to improve the convergence. A training pass, where the entire data set is submitted to the training algorithm, is called epoch, which corresponds to a number of iterations equal to the size of the training sample divided by the batch size. The complete training foresees a certain number of epochs.

11.4   Curve Fit as Machine-Learning Problem

Chapter 6 presented examples of parameter fits where the function assumed as fit model was identical to the one that was used to generate simulated samples. In some cases, a fit model may be known from theory. This is the case, for instance, when a fundamental parameter is determined from experimental data that are assumed to follow that theory model. In many cases, instead, there is no predefined theoretical model, and one must be inspired by the data sample itself to find a function that describes the data. The model should be tuned to describe data as closely as possible, and this is done by finding the parameters that minimize the loss function. In a fit procedure, we usually minimize the χ^2 or, assuming equal uncertainties on all points, we minimize the sum of squared residuals.

A best fit can be seen as a simple example of training a machine-learning regression algorithm. Once the parameters have been determined from the fit, the model predicts the value of a dependent variable y (target) as a function of an independent variable x (predictor). In machine-learning terminology, the set of points used as input to the fit, $(x_1, y_1), \cdots, (x_N, y_N)$, is the training set having *features* $x_1, \cdots, x_N$ corresponding to the *labels* $y_1, \cdots, y_N$.

The fit example provided in the following and the issues that are discussed have general validity also for more complex machine-learning models that are presented in the next sections. A rather flexible function model, for many cases, is a polynomial of given degree p. The training procedure consists of determining the polynomial coefficients that are the model's parameters. The more the degree of the polynomial p increases, the more closely the fit curve follows the data.

This model adopts polynomials as a family of fit functions, defining the *representational capacity* or *expressivity* of the model. An algorithm that uses a different family of functions, for instance a linear combination of sine and cosine terms with different frequencies, may have different performances, depending on the input data set.

Figure 11.1 shows a sample randomly extracted from an original function. The model used for the random generation is not a polynomial, and a polynomial with a finite degree can only approximate the original function. For instance, Fig. 11.2 shows a third-order polynomial fit applied to the previously generated data sample. The fit exhibits evident discrepancies from the original function. More in general, Fig. 11.3 shows fits using polynomials with degrees from zero (i.e., a constant function) up to $p = 11$.

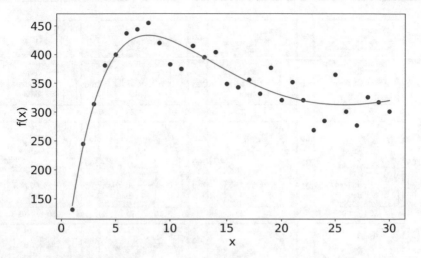

Fig. 11.1 Random data sample (red) extracted from a non-linear function model (blue) used in the example described in the text

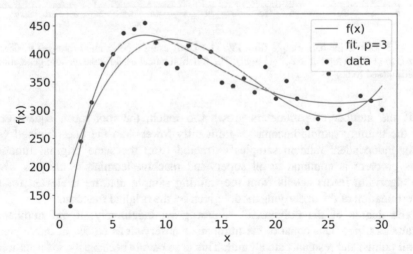

Fig. 11.2 Fit of the data sample from Fig. 11.1 (red) using a third-order polynomial (orange). The function model used for the random generation is superimposed (blue)

11.5 Undertraining and Overtraining

For a polynomial fit of degree p equal to zero, one, or two, the fit curve grossly deviates from the original curve. This behavior is called *underfitting*. The more parameters are used in the fit model, i.e., the larger the p, the more the fit function follows the statistical fluctuations in data, getting closer to the measured points even more than the original function. This behavior is called *overfitting*.

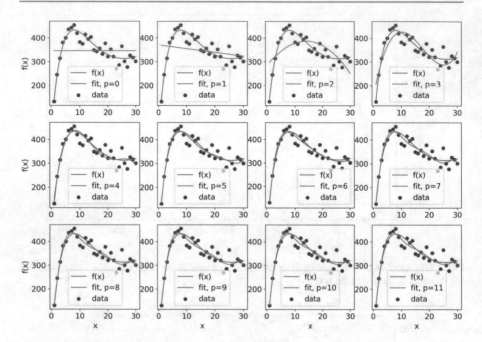

Fig. 11.3 Fits of the data sample from Fig. 11.1 (red) using a polynomial (orange) of degrees from 0 to 11 (from top left to bottom right). The function model used for the random generation is superimposed (blue)

If the number of parameters grows too much, the root mean square error for the training sample becomes significantly lower than the one obtained with other independent random samples extracted from the same original function. This problem is common to all supervised machine-learning techniques where the algorithm learns details from the training sample that are irrelevant for the determination of the underlying model given by the original function.

The degree of the polynomial cannot grow indefinitely: if the number of parameters, $p + 1$, is equal to the number of observations N, the fit curve passes by all points, and residuals are all null. This case occurs because the information in the training sample is particularly limited, being made of only N (x, y) value pairs. For more complex machine-learning problems, the training sample may consist of a richer amount of data. But this opportunity is often not guaranteed.

The example shown with this simple fit illustrates two general potential problems with the training of machine-learning algorithms:

- When a model's parameters are determined in a suboptimal way, we talk about *undertraining*. This issue may be addressed by increasing the size of the training sample, whenever it is possible. Moreover, the complexity of the model must be adequate to the size of the training sample.

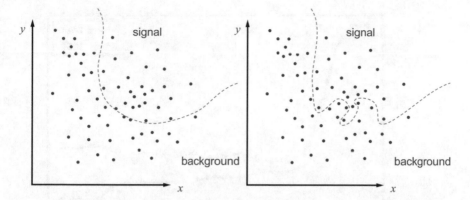

Fig. 11.4 Illustration of overtraining of a machine-learning classification algorithm that exploits two features, represented as the x and y variables corresponding to the horizontal and vertical axes, respectively. Blue points are signal observations, and red points are background in the training sample. The multivariate boundary of the classifier selection is represented as a dashed curve. Left: a smooth selection is based on a regular algorithm output; right: a selection from an overtrained algorithm picks artifacts following statistical fluctuations of the finite-size training sample

- Conversely, the training procedure may try to exploit artifacts due to the limited size of training samples that are not representative of the expected distributions. The effect is called in machine learning, *overtraining*.

An example is illustrated in Fig. 11.4 for an algorithm that classifies points in a two-dimensional xy plane according to two categories represented by points of two different colors. In practice, the algorithm may end up picking or skipping individual observations in the training categories by adopting an irregular selection boundary.

The presence of overtraining may be spotted by comparing the distributions of the algorithm's output and its performances, in terms of efficiency and background rejection, evaluated on the training sample and on another independently generated test sample. For instance, Kolmogorov–Smirnov test (Sect. 10.7) may be used to check the agreement. If the performances evaluated on the training sample are significantly better than the ones evaluated on the test sample, this may indicate the presence of overtraining. In this case, the training of the algorithm and its performances may not be optimal when applied on real data.

11.6 Bias-Variance Trade-Off

What is the optimal degree choice for the fit polynomial? More in general, when should one stop adding new parameters and more complexity to the model to improve the agreement of the fit curve with data? We can address this question by comparing the fit curve obtained from the specific sample used for the fit, or

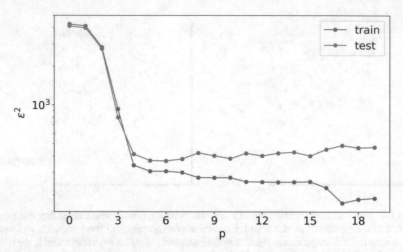

Fig. 11.5 Mean square error evaluated over the train sample (blue) and average of the mean square errors evaluated over 100 independently generated test samples (orange)

training sample, that we call $y_i^\star$, to other test samples, that we call $y_i^{\star\prime}$, independently generated from the same original function.

A Monte Carlo can be implemented to compute the *expected mean squared error*, defined as the average over many randomly generated samples $y_i^{\star\prime}$:

$$\langle \mathcal{C} \rangle = \langle \varepsilon^2 \rangle = \mathbb{E} \left[\frac{1}{N} \sum_{i=1}^{N} \left(y_i^{\star\prime} - \hat{f}(x_i) \right)^2 \right], \tag{11.5}$$

where $\hat{f}(x_i)$ is the value of the function fitted on the training sample, which is compared to $y_i^{\star\prime}$, the test sample value at the point x_i. As a function of p, Fig. 11.5 shows the value of:

- *Training sample*: $\varepsilon_{\text{train}}^2 = \frac{1}{N} \sum_{i=1}^{N} \left(y_i^{\star} - \hat{f}(x_i) \right)^2$ evaluated on the training sample $y_i^\star$, which is the same that is used for the fit.

- *Test samples*: $\langle \varepsilon_{\text{test}}^2 \rangle = \mathbb{E} \left[\frac{1}{N} \sum_{i=1}^{N} \left(y_i^{\star\prime} - \hat{f}(x_i) \right)^2 \right]$, estimated as the mean value

of the averages of the residuals squared computed over 100 test samples $y_i^{\star\prime}$ randomly extracted from the original function. The fit curve is the same for all cases and is fitted on the training sample $y_i^\star$.

The plot shows that the mean square error tends to be systematically lower for the training samples compared to the test sample.

The mean square error for each bin i is defined as the expected value of the difference between the estimate $\hat{f}(x_i)$ and the observed value $y_i^{\star\prime}$ squared:

$$\varepsilon_i^2 = \mathbb{E}\left[\left(\hat{f}(x_i) - y_i^{\star\prime}\right)^2\right]. \tag{11.6}$$

For simplicity of notation, we drop the subscript i in the following, and we define

$\hat{\theta} = \hat{f}(x_i)$, the fit estimate;

$\theta^\star = y_i^{\star\prime}$, the observed value from the test sample;

$\theta = f_{\text{true}}(x_i)$, the true value, computed using the original function f_{true}.

The mean square error (remember that the subscript i has been omitted) is defined as

$$\varepsilon^2 = \mathbb{E}\left[\left(\hat{\theta} - \theta^\star\right)^2\right]. \tag{11.7}$$

The expected value of $\hat{\theta}$, $\mathbb{E}[\hat{\theta}]$, can be inserted in the above expression:

$$\varepsilon^2 = \mathbb{E}\left[\left(\hat{\theta} - \mathbb{E}\left[\hat{\theta}\right] + \mathbb{E}[\hat{\theta}] - \theta^\star\right)^2\right] = \mathbb{E}\left[\left(\hat{\theta} - \mathbb{E}\left[\hat{\theta}\right]\right)^2\right] + \mathbb{E}\left[\left(\mathbb{E}\left[\hat{\theta}\right] - \theta^\star\right)^2\right] +$$
$$+ 2\mathbb{E}\left[\left(\hat{\theta} - \mathbb{E}\left[\hat{\theta}\right]\right)\left(\mathbb{E}\left[\hat{\theta}\right] - \theta^\star\right)\right]. \tag{11.8}$$

The mixed term represents a covariance term. It can be shown that it is null considering that the expected values are calculated independently with respect to variations of $\hat{\theta}$ and of $\theta^\star$, since we are comparing the fit obtained with the train sample to other independently extracted test samples. Therefore, the remaining terms are

$$\varepsilon^2 = \mathbb{E}\left[\left(\hat{\theta} - \mathbb{E}[\hat{\theta}]\right)^2\right] + \mathbb{E}\left[\left(\mathbb{E}[\hat{\theta}] - \theta^\star\right)^2\right]. \tag{11.9}$$

The first term, $\mathbb{E}[(\hat{\theta} - \mathbb{E}[\hat{\theta}])^2]$, is the variance of $\hat{\theta}$. In the second term, we can insert the true value θ:

$$\varepsilon^2 = \text{Var}\left[\hat{\theta}\right] + \mathbb{E}\left[\left(\mathbb{E}\left[\hat{\theta}\right] - \theta + \theta - \theta^\star\right)^2\right] =$$
$$= \text{Var}\left[\hat{\theta}\right] + \mathbb{E}\left[\left(\mathbb{E}\left[\hat{\theta}\right] - \theta\right)^2\right] + \mathbb{E}\left[(\theta - \theta^\star)^2\right] + 2\mathbb{E}\left[\left(\mathbb{E}\left[\hat{\theta}\right] - \theta\right)(\theta - \theta^\star)\right]. \tag{11.10}$$

Again, the mixed covariance term is null. What remains is

$$\varepsilon^2 = \mathbb{V}\text{ar}\left[\hat{\theta}\right] + \mathbb{E}\left[\left(\mathbb{E}\left[\hat{\theta}\right] - \theta\right)^2\right] + \mathbb{E}\left[(\theta - \theta^\star)^2\right]. \tag{11.11}$$

The second term is equal to the *bias* squared:

$$\mathbb{E}\left[\left(\mathbb{E}\left[\hat{\theta}\right] - \theta\right)^2\right] = \left(\mathbb{E}\left[\hat{\theta}\right] - \theta\right)^2 = \mathbb{B}\text{ias}\left[\hat{\theta}\right]^2. \tag{11.12}$$

The last term, finally, is the variance of $\theta^\star$. The decomposition of ε^2 is therefore complete:

$$\boxed{\varepsilon^2 = \mathbb{V}\text{ar}\left[\hat{\theta}\right] + \mathbb{B}\text{ias}\left[\hat{\theta}\right]^2 + \mathbb{V}\text{ar}[\theta^\star].} \tag{11.13}$$

The last term, $\mathbb{V}\text{ar}[\theta^\star]$, is the intrinsic variance of the observed value $\theta^\star$ and can be considered as a *noise* term. It does not depend on the fit model and is therefore not reducible by improving the model. The bias and variance of $\hat{\theta}$, on the other hand, depend on the fit model. It is not possible to reduce both at the same time by varying the degree of the fit polynomial p, and therefore, a value of p must be chosen that gives the optimal compromise between variance and bias.

Figure 11.6 shows a sample of 50 fourth-degree polynomial fits from randomly generated data sets compared to the original curve, and the contributions to the total mean square error due to the bias squared and to the variance, separately.

By varying the polynomial degree p, the average over the N observed points of the bias squared and variance varies as shown in Fig. 11.7. A value of p where bias and variance reach a reasonable trade-off is around 5. For larger values of p, the bias is reduced at the cost of an increasingly large variance, which worsens the total mean squared error and the overall precision of the fit.

Similarly, for machine learning, a model that is too complex compared to the complexity of the data sample may result in an overtraining. Note that we could compute the bias in this example because we knew the original function. This is not always possible in realistic cases.

11.7   Logistic Regression

The best fits encountered so far implement solutions for regression problems. A simple classification problem is addressed by *logistic regression*, presented in this section. We assume a binary classification with only two possible outcomes, encoded as 0 and 1, for a multivariate problem with n input variables $\vec{x} = (x_1, \cdots, x_n)$, usually called *predictors*.

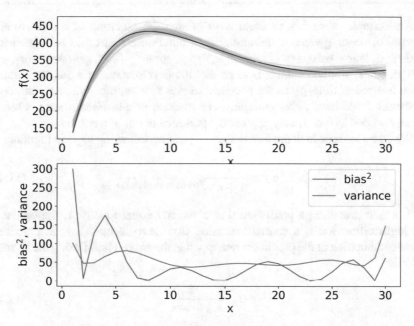

Fig. 11.6 Top: a sample of 50 fourth-degree polynomial fit curves (gray) from randomly generated data sets compared to the originating curve (blue); bottom: bias squared (blue) and variance (orange) contributions to the total square error for all measurement points

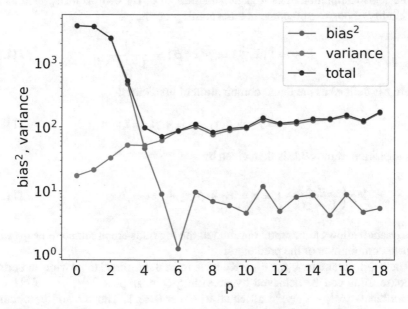

Fig. 11.7 Average over the N observed points of the bias squared (blue) and variance (orange) and total (sum of the two, red) as a function of the polynomial degree p

An example of such a problem is to predict the outcome of a disease, e.g., survival or death, given a certain number of conditions that could be related to a medical situation, outcomes of clinical analyses, comorbidities, genetic factors, age, gender, etc. A similar example is to predict the development of a disease within a given interval of time, given the presence of risk factors, life style, age, etc. Such predictors could have either continuous or discrete values. For instance, a binary value, encoded as 0 or 1, may express the presence or not of a risk factor.

We take as response to the inputs $x_1, \cdots, x_n$ the following logistic function:

$$p(\vec{x}\,;\vec{\beta}) = \frac{1}{1 + e^{-(\beta_0 + \beta_1 x_1 + \cdots + \beta_n x_n)}} \,. \tag{11.14}$$

Rather than providing a prediction that is exactly equal to 0 or 1, the output of the logistic function is a continuous value from zero to one. It is in practice a linear combination of the predictors remapped in the range $]\,0,\ 1\,[$ using the sigmoid function:

$$\sigma(t) = \frac{1}{1 + e^{-t}} \,. \tag{11.15}$$

Equation (11.14) also defines the response of a *perceptron* in an artificial neural network (Sect. 11.10).

The classification can be implemented by applying a selection cut to the output of the logistic function. The logistic function $p(\vec{x}\,;\vec{\beta})$ can be interpreted as the probability of positive outcome in a Bernoulli process:

$$P(y = 1 \,|\, \vec{x}, \vec{\beta}) = p(\vec{x}\,;\vec{\beta}) = \frac{1}{1 + e^{-z}} \,, \tag{11.16}$$

where z is defined as the linear combination of predictors:

$$z = \beta_0 + \beta_1 x_1 + \cdots + \beta_n x_n \,. \tag{11.17}$$

The logarithm of the odds is then given by

$$\log\left(\frac{p(\vec{x})}{1 - p(\vec{x})}\right) = z = \beta_0 + \beta_1 x_1 + \cdots + \beta_n x_n \,. \tag{11.18}$$

This relation allows to interpret our model: the log odds are assumed to be given by a linear combination of the predictors.

The $n + 1$ parameters, $\beta_0, \cdots, \beta_n$, have to be determined to provide an optimal response. This can be achieved by providing N examples $\vec{x}^{\,(1)}, \cdots, \vec{x}^{\,(N)}$ with known labels $y^{(1)}, \cdots, y^{(N)}$ all equal to either 0 or 1. Those could be obtained from a sample of real patients with their clinical history whose disease has a known outcome. The problem can be formulated in terms of maximum likelihood fit.

Assuming that the N examples have labels that are independent and identically distributed Bernoulli variables, the likelihood function for the training sample is

$$L(\vec{x}^{(1)}, \cdots, \vec{x}^{(N)}, y^{(1)}, \cdots, y^{(N)}; \vec{\beta}) = \prod_{i=1}^{N} P(y^{(i)} = 1 \mid \vec{x}^{(i)}; \vec{\beta}) =$$

$$= \prod_{i=1}^{N} p(\vec{x}^{(i)}; \vec{\beta})^{y^{(i)}} (1 - p(\vec{x}^{(i)}; \vec{\beta}))^{1-y^{(i)}}. \tag{11.19}$$

The cost function to be minimized is the negative logarithm of the likelihood function:

$$\mathcal{C}(\vec{\beta}) = -\log L(\vec{\beta}) =$$

$$= -\sum_{i=1}^{N} \left[y^{(i)} \log p(\vec{x}^{(i)}; \vec{\beta}) + (1 - y^{(i)}) \log(1 - p(\vec{x}^{(i)}; \vec{\beta})) \right]. \tag{11.20}$$

This expression is called *cross entropy* and is also to be used for algorithms based on decision trees (see Sect. 11.16). Sometimes it is also referred to as *logistic loss* or *logarithmic loss*. The minimization can be performed numerically using a gradient descent technique, and the parameters $\vec{\beta}$ can be determined with their uncertainties as usual for any maximum likelihood fit.

The interpretation of the fit parameters descends from Eq. (11.18): a unit change in x_i results in a change β_i on the log odds. So if $\beta_i = 0$, the input x_i is irrelevant for the prediction, since the odds are not modified if x_i changes. For instance, when fitting a prediction model for a disease, if the parameter related to gender is compatible with zero, males and females do not have significantly different risks to get the disease.

The performances can be measured on a test sample, independent on the training sample, as probability to have a given predicted value (0 or 1) for each of the true values (0 or 1). This 2×2 matrix is called *confusion matrix*. Diagonal elements are the true positive and true negative probabilities, and the off-diagonal elements are the false positive and false negative probabilities.

11.8 Softmax Regression

Softmax regression generalizes logistic regression for classification problems with multiple categories in number of m. In this case, the possible label values are not 0 and 1, but integer values from 0 to m. The output of the predictor is not a single variable y, with a possible continuous value from 0 or 1, but it is a vector of m variables, $y_1, \cdots, y_m$, each with a possible continuous value from 0 to 1. An observation can be assigned to the category j that has the largest value among the predicted $y_1 \cdots, y_m$. In a good classification, most of the y_i are close to zero, and a single one, y_j, is close to one.

Each predictor has associated a set of parameters $\vec{\beta}^i$, with $i = 1, \cdots, m$. The total number of parameters to be optimized is $m \times (n + 1)$. The sigmoid function is replaced by the *softmax function* defined as follows, assuming that the assigned category index is equal to j:

$$P(y_j = 1 \mid \vec{x}; \vec{\beta}^1, \cdots, \vec{\beta}^m) = p_j(\vec{x}; \vec{\beta}^1, \cdots, \vec{\beta}^m) = \frac{e^{z^{(j)}}}{\sum_{k=1}^{m} e^{z^{(k)}}}, \tag{11.21}$$

where

$$z^{(k)} = \beta_0^k + \beta_1^k x_1 + \cdots + \beta_n^k x_n. \tag{11.22}$$

The cost function to be minimized becomes

$$\mathcal{C}(\vec{\beta}^1, \cdots, \vec{\beta}^m) = -\log L(\vec{\beta}^1, \cdots, \vec{\beta}^m) =$$

$$-\sum_{i=1}^{N} \sum_{k=1}^{m} \left[y_k^{(i)} \log p_k(\vec{x}^{(i)}; \vec{\beta}^1, \cdots, \vec{\beta}^m) + \right.$$

$$\left. + (1 - y_k^{(i)}) \log(1 - p_k(\vec{x}^{(i)}; \vec{\beta}^1, \cdots, \vec{\beta}^m)) \right]. \tag{11.23}$$

Note that the parameter set defined in this way is redundant. If the parameters are transformed according to

$$\vec{\beta}^i \longmapsto \vec{\beta}'^i = \vec{\beta}^i - \vec{\psi}, \tag{11.24}$$

where $\vec{\psi}$ is a vector of constants, it can be easily demonstrated that Eq. (11.21) does not change. For cases like this, the model is said to be *overparameterized*. Therefore, the minimum of the parameter set is not unique. For softmax regression, this issue does not spoil the convergence capabilities of a minimization algorithm using gradient descent. Anyway, one possibility is to set to zero the last parameter set: $\vec{\beta}^m = 0$, and minimize the remaining $(m - 1) \times (n + 1)$ parameters. The special case with $m = 2$, where the second and last sets of parameters are removed, gives the logistic regression.

The confusion matrix for a softmax regression is an $m \times m$ matrix containing the probability to classify as category m' an observation belonging to the category m. Diagonal elements are the probability to correctly classify each category, while off-diagonal elements are the various misclassification probabilities.

11.9 Support Vector Machines

Support vector machines (SVM) [10] are a class of supervised linear classifier algorithms. Given two training samples in an n-dimensional space, the algorithm finds a hyperplane that achieves the best separation between the two samples according to the *maximum margin* criterion, which maximizes the distance between the hyperplane and the nearest point from either sample.

For a binary classification, SVMs have similarities with Fisher's linear discriminant (Sect. 10.4), in the sense that they both separate the parameter space into two halves with a hyperplane. For non-linear problems, the algorithm can map the input features using a possibly non-linear kernel function and then finds the optimal hyperplane in the transformed coordinate space. The intrinsic linearity of support vector machines limits its performance in complex problems, similarly to Fishers' discriminant [11].

11.10 Artificial Neural Networks

Artificial neural networks have been originally inspired by a simplified possible functional model of neuron cells in animal brains, implemented through a computer algorithm. The classifier output is computed by combining the response of multiple *nodes*, each representing a single neuron cell. Nodes are arranged into *layers*. Input features, $(x_1, \cdots, x_n)$, are passed to a first *input layer*, whose output is then passed as input to the next layer, and so on. Finally, the last *output layer* provides the classifier's output, as shown in Fig. 11.8.

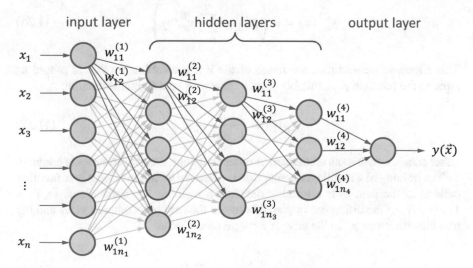

Fig. 11.8 Structure of a multilayer artificial neural network

For a binary classification, the output layer is constituted by a single node providing an output from 0 to 1, like for a logistic regression. For a classification with more possible categories, the output layer can have more nodes, like for the softmax regression, used to encode the possible outcomes of the classification. A classic problem with multiple categories is the identification of handwritten character as one of the letters of the alphabet, or as one of the ten digits from 0 to 9.

Intermediate layers between the input and output layers are called *hidden layers*. Such a structure is also called *feedforward multilayer perceptron*. Before presenting input variables to the input layer, some specific implementations of neural network algorithms apply a suitable linear transformation that adjusts the range to a standard interval, typically [0, 1]. The output of each node is computed as weighted sum of the input variables, with weights that are subject to optimization during the training phase. The weighted sum is then passed to an *activation function* φ, and the output of the kth node of the lth layer is given by

$$y_k^{(l)}(\vec{x}) = \varphi \left(\sum_{j=1}^{n_l} w_{kj}^{(l)} x_j \right) , \tag{11.25}$$

where the weights $w_{kj}^{(l)}$ connect the input node k to the output node j, as in Fig. 11.8. The number of weights in the first layer n_1 is equal to the number of input features n.

The activation function limits the output value within the range [0, 1]. In some cases, a bias, or threshold, w_{k0} is also added as an extra parameter:

$$y_k^{(l)}(\vec{x}) = \varphi \left(w_{k0}^{(l)} + \sum_{j=1}^{n_l} w_{kj}^{(l)} x_j \right) . \tag{11.26}$$

This allows to better adjust the range of the linear combination that is passed as input to the function φ. A suitable choice for φ may be a sigmoid function:

$$\varphi(v) = \frac{1}{1 + e^{-\lambda v}} . \tag{11.27}$$

Other choices, such as an arctangent or other functions, are also commonly adopted.

The training of a multilayer perceptron is achieved by minimizing a loss function defined as the sum over a training data set containing N observations, $\vec{x}_i$, $i = 1, \cdots, N$, of the differences squared between the network's outputs $y(\vec{x}_i)$ and the true classifications y_i. In the case of a single output node:

$$C(w) = \sum_{i=1}^{N} (y_i - y(\vec{x}_i))^2 . \tag{11.28}$$

The loss function depends on the values of the network weights w, which are not explicitly reported in Eq. (11.28), but should appear in the terms $y(\vec{x}_i)$. y_i may represent a category as 0 or 1, in case of a binary classification problem, or may be equal to the expected continuous target value for a regression problem. Regression with neural network is conceptually similar to the fit example discussed in Sect. 11.4, but the functional model is, as seen above, more complex and has many parameters.

Software tools such as Keras and TensorFlow also support the *cross entropy*, already encountered in Sect. 11.7, as loss function for binary classification. For classifications with more than two categories, a *categorical cross entropy* is supported with a *softmax* activation function for the output nodes, similar to what was discussed in Sect. 11.8.

Weights are typically randomly initialized to break the symmetry among different nodes and are then iteratively modified to minimize the loss function after feeding each training observation, or batch of observations, to the network. The minimization usually proceeds via gradient descent [12], where weights are modified at each iteration according to the following formula:

$$w_{ij}^{(l)} \longmapsto w_{ij}^{(l)} - \eta \frac{\partial C(w)}{\partial w_{ij}^{(l)}} \ . \tag{11.29}$$

The parameter η implements the *learning rate* and controls how large is the applied change to the parameters w at each iteration. This method is also called *back propagation* [13], since the value computed as output drives the changes to last node's weights, which, in turn, drive the changes to the weights in the previous layer, and so on.

A more extensive introduction to *artificial neural network* can be found in [14].

11.11 Deep Learning

It has been demonstrated that an artificial neural network with a single hidden layer may approximate any analytical function within any desired precision, provided that the number of neurons is sufficiently large and the activation function satisfies certain regularity conditions [15]. For a binary classification, the ideal function that a neural network should approximate is the likelihood ratio that gives the best possible separation performances according to the Neyman–Pearson lemma.

In practice, the number of nodes required to achieve the desired precision may be very large. Therefore, some performance limitations are unavoidable, considering that the number of nodes may not grow indefinitely in a realistic application. Those limits of neural networks with a single-layer architecture, sometimes defined *shallow neural networks*, made, for several years, boosted decision trees (see Sect. 11.16) a preferred choice over artificial neural networks in many applications in particle physics. At some point, better algorithms, together with more affordable computing power, became available. This allowed to manage neural networks with more hidden layers.

The total number of required nodes to achieve comparable performances with respect to a single-layer architecture is smaller if more hidden layers are added [16]. But adding more hidden layers could make the training process more difficult to converge. In particular, there may be more chances to find an output of one of the activation functions that is close to 0 or 1, which corresponds to a gradient of the loss function that is close to zero. This has the consequence that the gradient descent algorithms may get stuck.

Many of the limitations of more traditional algorithms have been recently over-come thanks to modifications of the network models and the training algorithms. Recent software algorithms are now able to manage several hidden layers and a relatively large numbers of nodes per layer, with up-to-date computing technologies, for cases that were intractable with traditional algorithms. Those techniques are called *deep learning* [17], in the sense that they allow to use deeper network structures in terms of the number of hidden layers. The possibility to manage complex network architectures allows to define more advanced classes of models that can manage a richer variety of input data with many more features compared to traditional neural networks. In this way, deep neural networks can address more complex problems and perform new tasks. For this reason, deep learning has recently become popular for advanced applications such as the capability to identify faces in an image, or human speech in sound recording, and has widely expanded the fields of applicability of machine learning.

One example of improvement of network model is the simplification of activation functions. To improve the learning process with respect to traditional neural net-works, sigmoid activation functions can be replaced by piecewise-linear functions. A commonly adopted function is the very simple *rectified linear unit* (ReLU):

$$\varphi(v) = \max(0, v) . \tag{11.30}$$

The evaluation of a ReLU output requires very limited computing resources and is frequently used in convolutional neural networks (Sect. 11.13) that are very CPU demanding. A smoother version of ReLU was introduced by Softplus [18]:

$$\varphi(v) = \log(1 + e^v) \tag{11.31}$$

and has the property that its derivative is a sigmoid function.

One of the first applications of artificial neural network with several layers and a large number of nodes was implemented to classify handwritten numbers based on samples of training images [19]. The adopted network had $28^2 = 784$ input variables to classify gray scale images made of 28×28 pixels, each representing a single digit. Ten nodes in the output layer allowed to encode the 10 digits from 0 to 9. Five hidden layers were used with 2500, 2000, 1500, 1000, and 500 nodes each. The training method was a standard stochastic gradient descent, and intensive computing power was used. For those interested in trying to implement a similar classifier, the MNIST database of handwritten digits is publicly available online [20].

A few examples of modern deep learning strategies are introduced in the following sections, in particular, convolutional neural networks (Sect. 11.13), recursive neural networks (Sect. 11.14), autoencoders (Sect. 11.20), and generative adversarial network (Sect. 11.22). A more exhaustive reference about deep learning is [21].

11.12 Deep Learning in Particle Physics

In particle physics, deep learning techniques achieve now optimal performances, exceeding shallow neural networks and sometimes boosted decision trees, for the most complex problems. An application in particle physics of deep learning is the classification of substructures in hadronic jets that allow to identify jets produced from $W \rightarrow q\bar{q}'$, $Z \rightarrow q\bar{q}$ or from top-quark decays from jets due to the more common quark and gluon background [22, 23].

Probably, the most interesting and promising improvement achieved with the use of deep learning is the possibility to use directly very basic features as input to the network, such as particles four momenta, instead of working out suitable combinations that achieve an important separation power and are peculiar of a specific analysis channel. For instance, in a search for a resonance, the reconstructed invariant mass of the resonance decay products is a key feature. Its distribution is peaked for the signal, while it is flatter for the background. More complex kinematic variables are optimal for other searches for new particles, typically for those with missing energy due to escaping undetected particles, like in searches of supersymmetric particles.

Shallow neural networks often achieve optimal performances only if they use as input specific kinematic variables, such as the invariant mass for a peaking signal, and have suboptimal performances, instead, when four vectors are provided as input. Conversely, it has been proven that deep learning may achieve optimal performances using either advanced input variables, such as shallow networks, or four momenta directly [24]. For deep networks that use four vectors as input features, a typical number of nodes per layer may be several hundreds, with a number of hidden layers of the order of 5. The optimization of the training procedure may require to adopt particular strategies. For instance, the learning rate parameter λ may be decreased exponentially during the training iterations to have an optimal convergence [24].

The capability of machine-learning algorithms to automatically find the best key features within sets of basic quantities, such as four vectors, may lead to optimal performances with a reduced human effort in implementing data analysis. The first step of a traditional data analysis is the manual human-made identification of key features as inputs for the subsequent signal extraction that typically may require a machine-learning classifier. With deep learning, we may envision, in the forthcoming future, an analysis process where more and more effort is devoted to finding the optimal deep neural network structure, rather than in the preliminary human-made preprocessing of kinematic information.

11.13   Convolutional Neural Networks

Many machine-learning problems, such as image and sound processing, require to manage a very large number of input features. The number of features could be of the order of the number of pixels in an image, and an image from modern photocameras or cell phones may have tens of millions pixels. In order to manage such a large number of inputs, fully connected neural network would become intractable, since the number of parameters in the model would become huge. A special neural network architecture has been developed, called *convolutional neural network* (CNN or ConvNet) [25], which allows to keep the number of parameters in the model at a reasonably manageable level, preserving the capability to manage a large number of input features.

Node arrangement is usually done in more dimensions to adapt to the dimensionality of input data. The individual pixel luminosities may be arranged in a two-dimensional structure, corresponding to the width and height of the bitmap, in the number of pixels. A third dimension may be added to take into account the three RGB color channels. A single time dimension may be suitable for sound processing.

The network architecture can profit of the translational symmetry of the problem to reduce the number of model's parameters. The translational symmetry occurs along two dimensions in the images (width and height) or along the time dimension for sound. For image analysis, this corresponds to inspect local features, e.g., edges or spots that can be identified wherever displaced in an image. Locally identified features may be further arranged in subsequent layer structures to identify more complex features that could be at some depth of the classification, eyes, lips, noses, etc., for a face identification algorithm. Indeed, there is no need to specify a priori which of those features should be identified because the network finds out them automatically during the optimization. It is also well possible that some of the features that the network identifies are not immediately understandable by a human, but this is not an issue.

This strategy provides a more organized approach, as opposite to an enormous fully connected architecture that would consider all possible interconnections among all possible pixels in a way that would also depend on the specific location of the features in an image. The overall CNN architecture is potentially similar to the logical structure of how image processing may proceed in the brain's visual cortex, which is able to perform vision of complex scenes using a limited number of neurons.

The basic ingredients of a CNN are small network units called *filters*, each aiming at identifying a specific feature of the image. Filters are applied to the input at all possible positions, exploiting the translational symmetry. For image recognitions, a filter corresponds to a window having as size a small number of pixels, typically 5×5 or 9×9. The extent of the filter is called *local receptive field*. Pixels in the window end up as inputs into dedicated neurons whose outputs evaluate weighted sums of the pixel luminosity content, plus a possible bias, as in standard neural networks. This operation is applied at all possible positions on the image. The result

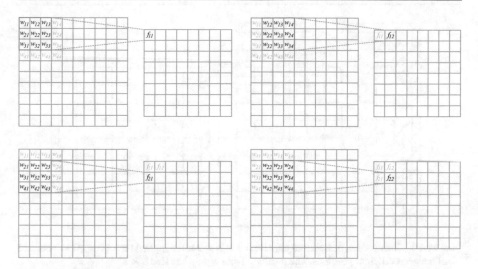

Fig. 11.9 Construction of a feature map for a 3×3 filter applied on a 10×10 pixel image at different positions. The resulting feature map has size 8×8. A grayscale image has been assumed, with one color code instead of the three RGB, reducing the dimensionality of the example

of the operation, called *convolution*, which gives the name to this neural network architecture, is a new matrix determined from the inputs.

The collection of all the outputs of one filter unit, evaluated at all possible image positions, is stored in a new layer called *feature map* and is stored as a two-dimensional array for the considered images processing example. Each filter unit generates in this way a feature map, and more filters can be introduced in the network structure. The construction of a feature map is illustrated in Fig. 11.9 for a two-dimensional case and in Fig. 11.10 for a three-dimensional case. One-dimensional architectures, as for sound processing, proceed similarly.

Usually, several filters are applied, and the set of all feature maps is stored as a three-dimensional array, for two-dimensional feature maps, where the third dimension is a number that identifies the applied filter unit. The feature map array produced at this stage can be used as input to subsequent network layers having a similar structure. A feature map can also be reduced in size by applying a subsampling, called *pooling*. A reduced feature map can then be processed by a subsequent network layer producing an output with a smaller size.

Sequences of layer structures implementing the processing steps described above (convolution, pooling, and the application of rectified linear units) can be repeated several times, giving a deep structure to the network. All the outputs of the last layer of the sequence are finally fed into a smaller-size fully connected network, whose outputs perform the final classification. A simplified structure of a complete convolutional neural network is shown in Fig. 11.11. The training, i.e., the weights optimization, can be performed by stochastic gradient descent or other more performant algorithms.

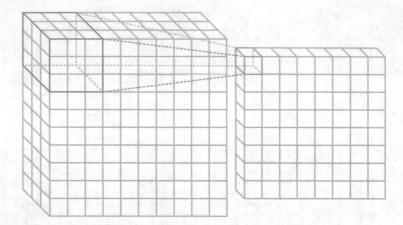

Fig. 11.10 Construction of a feature map for a $3 \times 3 \times 3$ filter applied on a $10 \times 10 \times 3$ colored pixel image at different positions. The resulting feature map has size 8×8

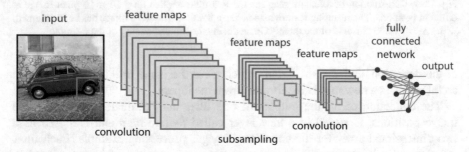

Fig. 11.11 Simplified structure of a convolutional neural network [26]

Convolutional neural networks have a wide range of applications in many fields, as already said, but they also have recent applications in particle physics. For instance, the modular structure of neutrino detectors has been exploited, as for images, to classify different possible neutrino interactions that may occur at any point in the detector [27], automatically identifying the relevant features as patterns in the set of detector signals. The analysis published in the aforementioned reference is implemented using the CAFFE [28] software framework.

11.14   Recursive Neural Networks

Recursive neural networks [29,30] (RvNN) are advanced implementations of neural networks that are able to manage different patterns of information organized, for instance, in the form of trees or graphs, as opposed to fixed-size vectors of features, like in standard neural networks. Examples of structured data are natural language, DNA code, web pages in HTML or XML data structures, etc. Given the intrinsic recursive nature of data, such as a graphs or tree, the same neural network module, with the same values of the weights, can be recursively applied to different substructures within the graph that have the same abstract topology. In this perspective, recursive neural network falls into the domain of deep learning. The data structure is processed in topological order, which reflects the relations within the data structure itself. Relations among elements in a data structure may logically correspond, for instance, to spatial or temporal relations.

If the network structure has a single child per node, like for data sequences, rather than multiple children, as in a tree structure, a recursive neural network is rather called *recurrent neural network* (RNN).

Training methods based on back propagation through the structure may be not optimal for such networks and may exhibit problems of convergence. More sophisticated training methods have been developed, which enhance the speed of convergence but require significant computing power [31].

Recursive neural networks have been used for tracking building problems [32]: the network learns to predict where a hit belonging to a track should be found on the next detector layer, given a set of initial hits. This allows to select the closest hit and build in this way the track.

11.15 Graph Neural Networks

Graph neural networks (GNN) are a category of neural networks that can manage data with a hierarchical structure with nodes and dependencies between pairs of nodes [33]. Those machine-learning architectures have applications in the analysis of social networks, website structures, molecular models, but also in particle physics for applications with non-trivial data structures. GNNs can be thought as an extension of recursive neural networks, with specialized network units that can manage the features of specific categories of nodes.

Data can be structured in a *graph* if it is possible to organize the information into *nodes* (or *vertices*) and *edges*, which connect two different nodes. Each node or edge in a graph can be associated, for training, with a label, typically represented as a real vector. The goal of a GNN is to predict these labels.

In particular, data from a detector at a collider experiment are organized in structures that are different from an image, which would make a convolutional neural network a suitable network structure. Different subdetectors have different shapes and granularities and, unlike the pixel sensor of an ordinary camera, have a different density and geometry of sensitive modules. Moreover, symmetries are not always available. For instance, a calorimeter may have an azimuthal symmetry but nonuniform density of detector units in the pseudorapidity direction. This makes GNN a better architecture for detector reconstruction problems. Given the faster response, compared to ordinary reconstruction algorithms, GNNs have been proposed for real-time event reconstruction and selection in low-latency applications such as high-level trigger processing.

Several applications of GNN in particle physics have been attempted. In track finding [32], a GNN has been trained to classify hits belonging to the same track within a graph containing all possible meaningful hit interconnection. Other applications are, for instance, jet identification [34] and the reconstruction of boosted Higgs boson decays [35]. The significant computational time requires optimization of the code for GPU processing. An overview of applications in particle physics of GNNs can be found in [36].

11.16 Random Forest and Boosted Decision Trees

A *decision tree* is a sequence of selection cuts that are applied in a specified order on a given set of features. In the following, the case of a binary classifier is presented. Each cut splits a data sample into *nodes*, each of which corresponds to a given number of observations classified as one of two categories that are labeled as signal or background. A node of the tree may be further split by applying subsequent cuts. Along the decision tree, the same variable may appear multiple times, depending on the depth of the tree, each time with a different applied cut, possibly even with different inequality direction. Usually, a maximum depth of the tree is set to limit the classifier complexity that, if allowed to grow indefinitely, may easily lead to overtraining.

Nodes in which either signal or background is largely dominant are classified as *leafs*, and no further selection is applied. A node may also be classified as leaf, and the selection path is stopped, in case too few observations per node remain, or in case the total number of identified nodes is too large. Different criteria to stop the growth of a tree have been proposed for various implementations.

Once the selection process is fully defined, each of the branches in a tree represents a sequence of cuts. An example of decision tree is shown in Fig. 11.12. Selection cuts can be tuned to achieve the best split level in each node according to some metric. One possible optimization consists in maximizing for each node the gain of *Gini index* achieved after a splitting. The Gini index is defined as

$$G = P\left(1 - P\right),\tag{11.32}$$

where P is the purity of the node, i.e., the fraction of signal observations. G is equal to zero for nodes containing only signal or background observations. Alternatives to the Gini index are also used as optimization metric. One example is the cross entropy (see Sect. 11.7) equal to

$$E = -\left[P \log P + (1 - P) \log(1 - P)\right].\tag{11.33}$$

The gain due to the splitting of a node A into the nodes B_1 and B_2, which depends on the chosen cut, is given by

$$\Delta I = I(A) - I(B_1) - I(B_2),\tag{11.34}$$

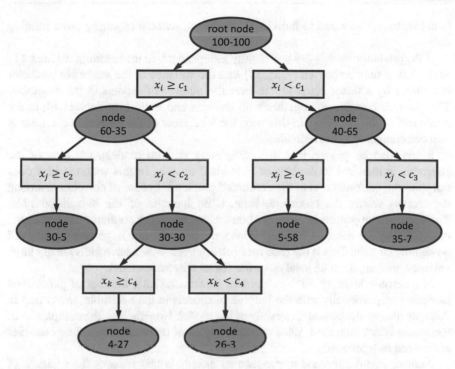

Fig. 11.12 Example of a decision tree. Each node represented as an ellipse contains a different number of signal (left number) and background (right number) observations. The relative amount of signal and background in each node is represented as blue and red areas, respectively. The applied requirements are represented as rectangular boxes

where I denotes the adopted metric: G or E, in case of the Gini index or cross entropy, introduced above. By varying a cut, the optimal gain may be achieved.

A single decision tree can be optimized as described above, but its performances are usually far from the optimal limit set by the Neyman–Pearson lemma. A significant improvement can be achieved by combining the outputs of multiple decision trees into what is usually called a *decision forest*. The *random forest algorithm* [37] consists of growing many decision trees from replicas of the training samples obtained by randomly resampling the input data, where a random subset of the input features is selected. Moreover, the decision tree may only be applied to a random subset of the complete training sample. The final score of the algorithm is given by the average of the predictions (zero or one) provided by each individual tree.

It may be a convenient strategy, in general, to average the output of more predictor models that depend on some random parameter, like for a random forest, where a random resampling is applied. Such approach, where many individual models are combined, is called *ensemble learning*. All models are trained on the same data sample or on a subset of it. In this way, ensemble learning provides a better method, compared to individual predictors, without requiring more training data.

If the predictor models are statistically independent, in the meaning discussed in Sect. 11.6, it may be proven [9, 38, 39] that the variance of the ensemble predictor is reduced by a factor equal to one over the number of models in the ensemble. The variance contributes, together with the bias and noise (see Sect. 11.6), to the total error of the predictor. In this way, the total error of the ensemble classifier is reduced compared to the individual model.

It can also be proven that, for completely random models, the bias of the ensemble is identical to the bias of individual models. In this way, the bias does not spoil the performances of the combined predictor. In case of correlation among the models within the ensemble, instead, an increase of the bias is possible. The ensemble approach is therefore convenient when the predictor uncertainty is dominated by the variance term; otherwise, the presence of a large bias would yield no significant gain. This is the case for decision trees, which have individually large variance, making them an ideal classifier for an ensemble model.

In ensemble learning, such as random forests, the total number of parameters increases proportionally with the number of models in the ensemble, increasing in this way the *representational capacity* of the model. Nonetheless, the complexity of the model is not increased, since multiple copies of the same elementary classifier are trained independently.

Another popular method to produce an ensemble that reduces the variance of a decision tree is the so-called *bagging* (bootstrap aggregation). Several subsets are created from the initial sample applying a limited number of replacements at randomly chosen positions. An individual decision tree is trained from each of such subsets. The prediction of the ensemble is the average of all such predictions from different trees. Since the subsets are not completely independent, the variance reduction happens at the cost of an increase of the bias, that, in most of the cases, can be very small, providing an overall improvement. This technique may be useful when limited training data is available, and training with a completely random sampling is not viable.

Further improvements can be achieved, instead of using the random sampling as for random forests, by iteratively adding to the forest new trees optimized based on observations that have been reweighted according to the score given by the classifier in the previous iteration. The trees in the forest obtained at the end of this iterative procedure are called *boosted decision trees* (BDT) [40], and this procedure, in general, is called *boosting*. The boosting algorithm proceeds according to the following steps:

- Training observations are reweighted using the classifier result from the previous iteration.
- A new tree is built and optimized using the reweighted observations as a training sample.
- A score is given to each tree.

- The output of the final BDT classifier is the weighted average over all the M trees in the forest:

$$y(\vec{x}) = \sum_{k=1}^{M} w_i \, y^{(i)}(\vec{x}) \, . \tag{11.35}$$

Implementations may differ in the way the reweighting is applied.

Among the boosting algorithms, a popular one is the *adaptive boosting* [41, 42] or AdaBoost. With this approach, only observations misclassified in the previous iteration are reweighted by a weight w that depends on the fraction of misclassified observations f in the previous tree, usually equal to

$$w = \frac{1-f}{f} \, , \quad f = \frac{N_{\text{misclassified}}}{N_{\text{tot}}} \, . \tag{11.36}$$

The misclassification fraction f is also used to compute the weights in the combination of individual decision tree outputs:

$$y(\vec{x}) = \sum_{i=1}^{M} \log\left(\frac{1-f^{(i)}}{f^{(i)}}\right) y^{(i)}(\vec{x}) \, . \tag{11.37}$$

With adaptive boosting, observations misclassified in each iteration achieve more importance, and the tree added at the next iteration is more efficient in correctly classifying those observations where the previous iteration failed.

A more recent class of boosted tree classifiers adds gradient descent to the boosting technique and is called *gradient-boosted trees*. At each step of the boosting procedure, the gradient of a cost function is evaluated with respect to the parameters that are introduced to define, for each leaf of the tree, the predictor output for a single tree. Next tree is added to the ensemble to approach the minimum of the cost function. The most popular implementation is XGBoost (extreme gradient boosting) [6], which is being intensively used also in particle physics.

As anticipated, the response of a decision tree is parameterized in a convenient way, such that the loss function and its gradient can be easily expressed and computed as a function of the model's parameters. A feature vector $\vec{x}$ is assigned to a leaf corresponding to an index i given by a function q that implements the decision tree's selection:

$$i = q(\vec{x}) \, . \tag{11.38}$$

The prediction of the decision tree is a function of the leaf index i that we indicate with $g = w_i$. While for a classification problem g should be either zero or one, this more general formalism also allows to use decision trees for regression problems, if the parameters w_i may assume any real value. The prediction g can be written as

$$g = w_{q(\vec{x})} \, . \tag{11.39}$$

If we label with j the index of a decision tree, the prediction of the ensemble is the sum of individual predictions:

$$\hat{y}(\vec{x}) = \sum_{j=1}^{M} g^{(j)}(\vec{x}) = \sum_{j=1}^{M} w_{q(\vec{x})}^{(j)} , \qquad (11.40)$$

where M is the number of decision trees. The cost function is given by the sum of two terms:

- The loss function, which measures the deviation of predictions from true labels, such as a mean square error or a cross-entropy term, as for logistic regression (see Sect. 11.7); it can be written in any case in terms of the parameters $w_i^{(j)}$.
- A regularization term that penalizes the growth in complexity of the model, which otherwise would tend to overtrain the decision tree.

The penalty contribution for each tree, corresponding to one of the functions g, is defined as

$$\Omega(g) = \gamma T + \frac{\lambda}{2} \sum_{i=1}^{T} w_i^2 , \qquad (11.41)$$

where T is the number of leafs, while γ and λ are tunable parameters. The term $\Omega(g)$ regularizes the cost function by introducing a penalty for trees with many of leafs T or with large values of the parameters w_i. More details and a pedagogical tutorial can be found on the XGBoost website, together with the options that allow several choices of loss functions, and various parameters that can be tuned to optimize the network and its training.

Example 11.1 - Comparison of Multivariate Classifiers—I
The performances of different supervised multivariate classifiers can be evaluated and compared using Monte Carlo generated samples. Differences in performances of classifiers can be enhanced if the distributions for the considered categories are not obviously separable with linear cuts, and exhibit overlap. We considered a binary classification problem with two categories that are labeled as signal and background. The feature vector has two dimensions to make the plots visualizable. The distributions generated for signal and background are shown in Fig. 11.13. Only a fraction of the generated data is displayed to avoid overcrowded scatter plots.

The scikit-learn [5] package has been used as implementation of different multivariate classifiers, which have been trained using a subset corresponding to half of the generated Monte Carlo sample. An independent data set, corresponding to the other half of the generated sample, has then been used to test the performances of each classifier. Six methods have been tested with suitable configuration parameters:

(continued)

Example 11.1 (continued)
- Linear classifier, implemented with a support vector machine
- Random forest with a maximum depth of a tree equal to 10
- Boosted decision trees, implemented with XGBoost [6] with a maximum depth of 10
- Neural networks with three configurations:
  - Shallow: a single hidden layer with 50 nodes
  - Middle: two hidden layers with 10 nodes each
  - Deep: three hidden layers with 50, 20, and 10, nodes, respectively

Figure 11.14 shows the ROC curves (see Sect. 10.3) superimposed for the considered classifiers. The distributions of the classifiers' outputs and the dots corresponding to the test samples are also shown in Fig. 11.15 in two dimensions. The distributions of classifiers' outputs for the test samples are shown in Fig. 11.16, compared to the same distributions for the training samples, to verify that no substantial overtraining is present. The linear classifier divides the sample space with a straight line. The orientation of the line is optimized, but clearly the achieved selection performances are suboptimal. The other classifiers all reach comparable performances for this random sample.

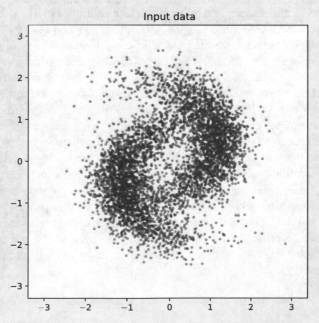

Fig. 11.13 Distribution in two dimensions for signal (blue) and background (red) test samples. Only a fraction of the data is displayed to avoid an overcrowded scatter plot

(continued)

Example 11.1 (continued)

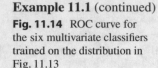

Fig. 11.14 ROC curve for the six multivariate classifiers trained on the distribution in Fig. 11.13

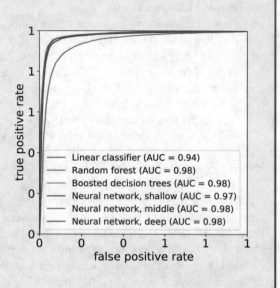

Example 11.2 - Comparison of Multivariate Classifiers—II

A second random sample was generated as shown in Fig. 11.17, where signal and background are distributed in a way similar to a chessboard, with some overlap at the boundary of adjacent square cells. Figure 11.18 shows the ROC curves for the considered classifiers). The distributions of the classifier outputs and the dots corresponding to the test samples are also shown in Fig. 11.19 in two dimensions, and the distribution of the outputs of the classifiers for the test and training samples is shown in Fig. 11.20. One may think that the chessboard distribution favors random forest and boosted decision trees algorithms that are based on cuts on the individual variables, since a selection on the two individual variables is naturally able to separate the points on a chessboard. Neural network has poor performances with a single-layer architecture; it improves with two layers, even if the total number of nodes is reduced compared to the shallow architecture. Optimal performances are fully recovered with a three-layer deep architecture, outperforming all other algorithms, as visible in the ROC curves in Fig. 11.18.

(continued)

Example 11.2 (continued)

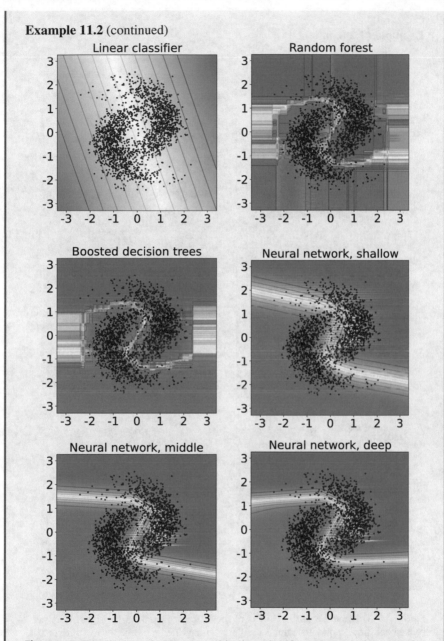

Fig. 11.15 Distribution of different multivariate classifier outputs (color map plus contour levels) compared to the test samples for signal (blue) and background (red)

(continued)

Example 11.2 (continued)

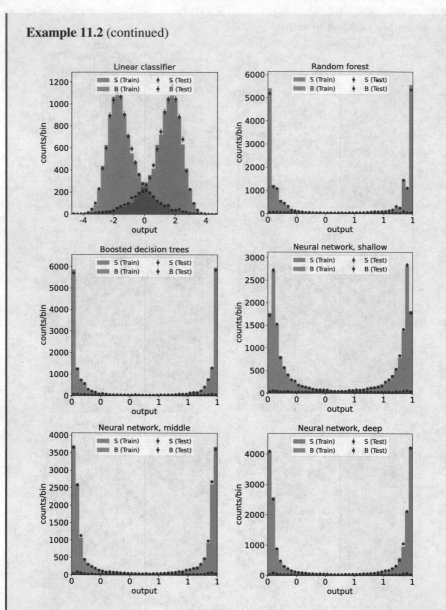

Fig. 11.16 Distribution of the outputs of the different classifiers for signal (blue) and background (red). Solid histograms are the training samples, while dots with error bars are the test samples. The reasonable level of agreement gives indication that no substantial overtraining is present

(continued)

Example 11.2 (continued)

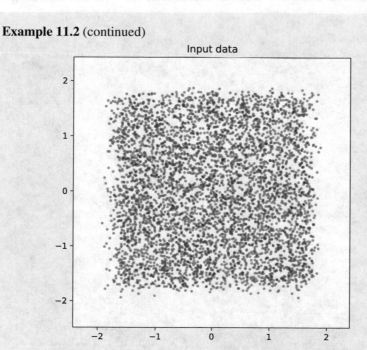

Fig. 11.17 Distribution in two dimensions for signal (blue) and background (red) test samples. Only a fraction of the data is displayed to avoid an overcrowded scatter plot

Fig. 11.18 ROC curve for
the four multivariate
classifiers

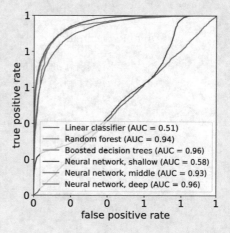

(continued)

Example 11.2 (continued)

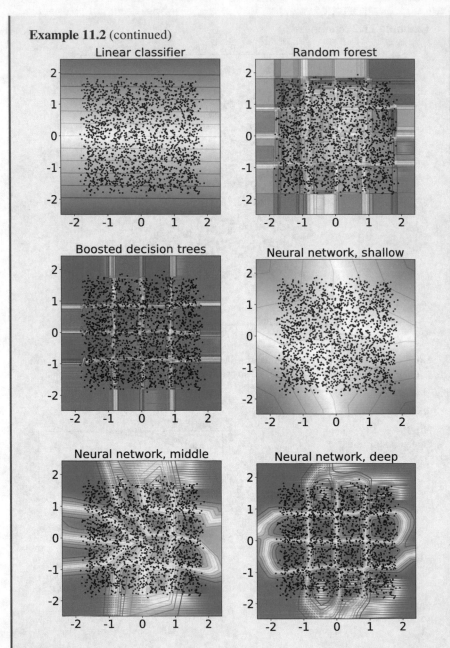

Fig. 11.19 Distribution of different multivariate classifier outputs (color map plus contour levels) compared to the test samples for signal (blue) and background (red)

(continued)

Example 11.2 (continued)

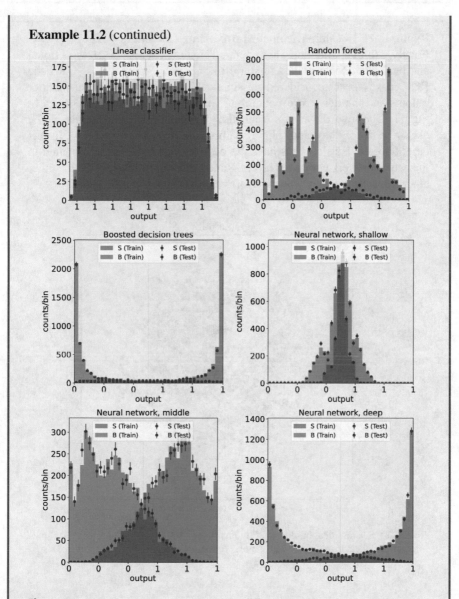

Fig. 11.20 Distribution of the outputs of the different classifiers for signal (blue) and background (red). Solid histograms are the training samples, while dots with error bars are the test samples. The reasonable level of agreement gives indication that no substantial overtraining is present

Example 11.3 - Comparison of Multivariate Classifiers—III

Finally, a third distribution was generated as shown in Fig. 11.21, with signal and background intertwined in spiral-like shapes. Figure 11.22 shows the ROC curves for the considered classifiers. The distributions of the classifier outputs and the dots corresponding to the test samples are also shown in Fig. 11.23 in two dimensions, and the distribution of the outputs of the classifiers for the test and training samples is shown in Fig. 11.24. In this case, boosted decision trees have the best performance, even compared to the neural network with three layers. Random forest follows, while the other algorithms have visibly worse performances.

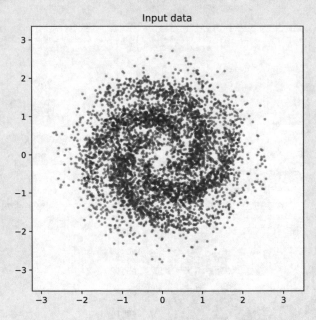

Fig. 11.21 Distribution of different multivariate classifier outputs (color map plus contour levels) compared to the test samples for signal (blue) and background (red)

(continued)

Example 11.3 (continued)

Fig. 11.22 ROC curve for the four multivariate classifiers

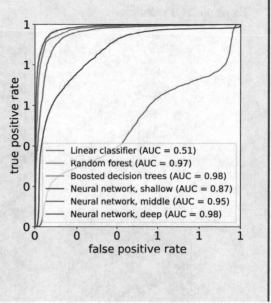

(continued)

Example 11.3 (continued)

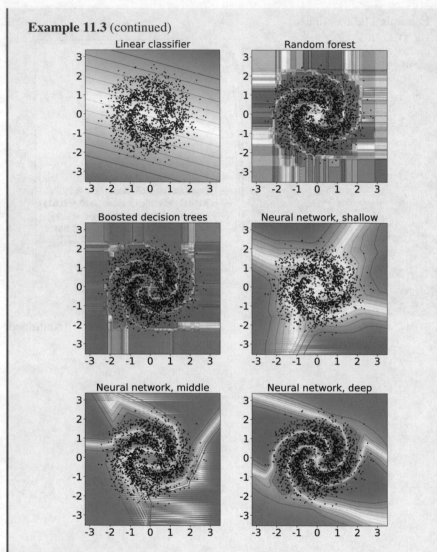

Fig. 11.23 Distribution of different multivariate classifiers compared to the test samples for signal (blue) and background (red)

(continued)

Example 11.3 (continued)

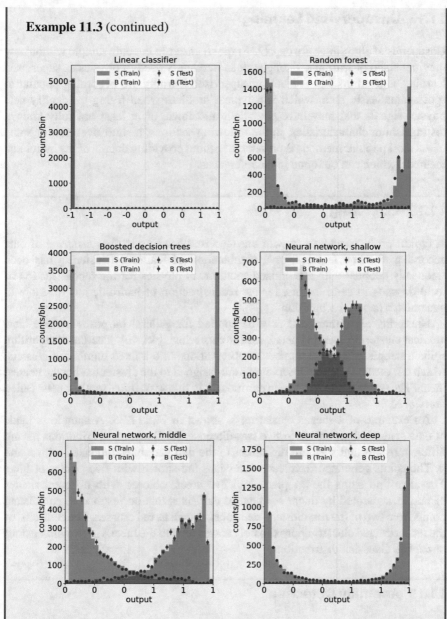

Fig. 11.24 Distribution of the outputs of the different classifiers for signal (blue) and background (red). Solid histograms are the training samples, while dots with error bars are the test samples. The reasonable level of agreement gives indication that no substantial overtraining is present

11.17   Unsupervised Learning

Unsupervised algorithms are capable to spot features in the data sample without any previous stage of learning. Though not as frequently used in physics as supervised learning, algorithms that perform unsupervised learning are becoming popular to spot anomalies in data, which is a typical problem when trying to identify new physics signals that may have potentially unknown, or at least not fully known, distinguishing characteristics. In the following sections, a short overview of some of the most popular methods is presented, without providing details of the individual methods, which can be found in the references.

11.18   Clustering

A typical problem addressed with unsupervised learning is the *clustering* of data according to some features, or a combination of features. Clustering has been frequently adopted with customized solutions in physics for the reconstruction of local deposits in calorimeters, for the reconstruction of hadronic jets or, more in general, for pattern recognition.

Clustering algorithms that may be applied for general purposes include hierarchical clustering [43, 44] and k-means clustering [45, 46]. The latter algorithm splits a sample of a given number of observations into a fixed number of clusters, which is indicated with k. Observations are assigned to the cluster having the nearest mean. The space of the features is partitioned, in this way, into exactly k cells called *Voronoi cells*.

An example of k-means clustering is shown in Fig. 11.25. A sample is made of data generated according to five two-dimensional Gaussian distributions having different means and standard deviations in the two variables, indicated with x and y. The points generated from each Gaussians are shown with five shades of blue. The algorithm splits the xy plane into five areas, colored with different shades of purple, separated by linear segment or half-lines that correspond to equidistant points from two of the five clusters' centroids, shown as red crosses. The centroids of the reconstructed clusters approximately correspond to the means in two dimensions of the five Gaussian distributions.

11.19   Anomaly Detection

Anomaly detection tries to estimate the local density of observations in the multivariate space of the features considered as input and detects an observation, or a small number of observations, that occurs in areas of very low density. Those cases are considered as *outliers*. The *local outlier factor* algorithm, or LOF [47], for instance, measures the local density of observation around a data point and compares its to the local density of its neighbors. An observation may be considered as candidate outlier if it has a significantly lower density compared its neighbors.

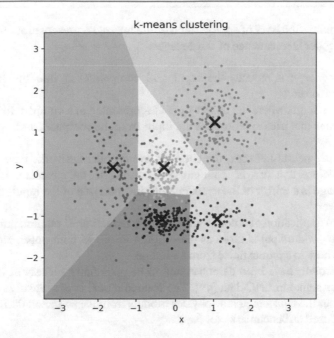

Fig. 11.25 Illustration of the k-means clustering algorithm applied to a data sample generated from five two-dimensional Gaussian distributions

Another possible anomalous signal may be a locally larger density of observations that constitutes a contamination of an otherwise more uniform population.

New algorithms are appearing to identify new physics signals in particle physics data [48], showing the increasing potential interest in the field of anomaly detection.

11.20 Autoencoders

A method that is gaining popularity consists of a feedforward neural network whose output stage is identical to the input, i.e., the network aims at reproducing as output the same data it receives as input. Such network is called *autoencoder*. Anyway, as intermediate stages, the network has layers with a number of nodes smaller than the number of inputs. In this way, it performs a dimensionality reduction or a compression of the information in a way that nonetheless minimizes the loss of capability to reproduce the input, though ideal performances are necessarily lost if a part of the information is not recoverable. The compression is unavoidably lossy, and in this sense, autoencoders cannot be used in place of a lossless compression algorithm.

The layer with the most compact representation of reduced dimensionality is called *latent space*. This layer is hidden within the structure of the model and is not directly observable as the output of the network. The size of the latent space, or *code size*, determines the level of compression of the internal representation. In summary, an autoencoder is a sequence of three stages:

- An *encoder* that reduces the input and compresses it into the latent-space representation
- The *code* stage, where the observation is represented in a compact format
- A *decoder* that tries to reproduce the input from the compact code

As any other neural network, the encoder and decoder networks have layers and nodes structure that can be configured according to the problem. Usually, the decoder stage is a mirror of the encoder, with inverted role of the inputs and outputs of each layer.

A useful application of autoencoders is noise reduction. If random noise is added to the input, and output is taken as the original image without noise, autoencoders may learn how to subtract noise from an image.

Autoencoders have been demonstrated to be potentially successful in detecting dark matter signal in LHC data [49]. The trained model is not sensitive on specific details about the considered new physics models from the so-called dark sector that have been used as benchmarks for the study.

11.21   Reinforcement Learning

Another class of machine-learning algorithm is neither supervised nor unsupervised and falls under the category of *reinforcement learning* [50]. Such algorithms learn from trials and errors but do not receive as input labeled data. They rather receive positive or negative rewards for each action that the algorithm, usually called *agent* in this context, performs during the exploration of the data space.

Reward assignment can be either deterministic, i.e., fully determined by the state where the decision must be taken, or randomly determined with probabilities that depend on the specific state. The evaluation of an action is determined by the sum of all positive and negative rewards. In this way, the algorithm may become smart enough to learn to prefer actions that give initially a negative reward if they are followed by subsequent actions that end up in an overall larger cumulative reward.

Reinforcement learning algorithms are particularly suitable to develop gaming strategies, where a supervised learning algorithm would require a large amount of data labeled with success or failure, corresponding to an entire game, consisting of a large number of possible actions. This is usually unpractical, and a strategy based on trials and errors is much more effective. Given the feedback due to positive or negative reward, reinforcement learning is also different from unsupervised learning, where data are categorized from patterns found internally within the data sample. For each game move, or action, rewards or punishments are given according to a predefined criterion. Initially, random actions are likely to result in negative rewards. The agent learns to explore alternative strategies until those actions are reinforced by positive rewards. The strategy becomes more and more refined the more the agent explores the data space environment.

The transition from one state to the following depends on the action taken by the agent, which in turn is based on the reward received at the previous step. If transition probability depends only on the reward and on the state corresponding to the last step, and not on all the previous states, the decision algorithm is called *Markov decision process*. This hypothesis is similar to what we already encountered for Markov chains, expressed by Eq. (4.53). Most of reinforcement learning algorithms base the decision policy on Markov assumption, which largely simplifies the implementation.

A challenge for reinforcement learning algorithms is that, in order to prefer positive rewards, the algorithm may prefer to take already explored actions. This may limit the capability of the algorithm to explore more territory of the data space. This may lead to a suboptimal learning. A trade-off is therefore necessary between the tendency to explore new states and the tendency to collect positive rewards.

Different algorithms are used to implement reinforcement learning. Algorithms such as SARSA [51] and Q-Learning [52] find optimal policies by maximizing the total estimated rewards. The most advanced reinforcement learning algorithms use deep neural networks in connection to a reinforcement strategy to determine an optimal model to estimate the total expected reward. The Deep Q-learn algorithm [53] has been trained to play arcade games for the ATARI 2600 console outperforming human capabilities using simply raw pixel images as inputs.

Reinforcement learning has been applied in particle physics for jet clustering problems [54, 55].

11.22 Generative Adversarial Network

Neural networks may also be used to generate new data examples once they have been trained on a sample of observations. Newly generated examples are not identical to the ones used for training but plausibly resemble the original ones. The measure of the resemblance is given by the capability of another neural network to discriminate artificially synthesized examples from real ones, i.e., from elements taken from a population of examples that have not been used for training.

A *generative adversarial network*, or GAN [56], is made of two modules that are trained together:

- A generative model, which is trained to generate new examples that resemble a set of original examples used as training inputs
- A discriminator model that is trained to classify examples as either real or artificially generated

In this sense, the discriminator competes against an *adversary*, which is the generator.

Usually, the generative model receives a random-noise input that is used to seed the generation of a new example through a deep neural network stage. The input vectors of variables have the role of the latent space we already encountered when discussing autoencoders (Sect. 11.20). Those provide compressed features that for the input code from which a complete output can be generated.

Batches of real and artificial samples are fed into the discriminator whose parameters are then updated to minimize the probability to misidentify real and artificial inputs. Subsequently, the generator parameters are also updated based on whether the generated samples have been able to be misidentified as real by the discriminator. The two models are, in this way, competing against each other, and both improve during training. The aim is to train the GAN and get as close as possible to an ideal training where the discriminator model misclassifies synthetic examples with 50% probability. This implies that the generator is synthesizing examples that are as credible, to the machine, as real examples.

GAN may suffer from several problems that could limit their performances. In particular, the training may cause an oscillation of the internal model state without convergence. There are situations where gradients of cost functions vanish, stopping the training improvement. Moreover, the model may *collapse*, i.e., it only explores a subset of the population. For instance, when trying to generate plausible handwritten digits, it may always generate a zero. Many improvements have been proposed and implemented, and the technology is still improving [57, 58].

Many applications of GAN use images with a deep CNN model. This allows to generate artificial images that resemble the ones in the training set. For this reason, GANs have also been used to generate photorealistic images that are not recognizable as fakes by humans and to generate artificial art [59] in the form of either images and music [60, 61].

More advanced generation capabilities are developed if labels that specify a category of the desired generated output are also provided together with the training examples. For instance, for a generator of human faces, input labels can specify whether the face belongs to a male or a female, if it has beard, or more. Such networks are referred to as *conditioned GANs* [62]. Application of conditioned GANs may be the generation of images from a text description in human language. Other GAN applications are the transformation of images into different conditions, e.g., day to night, summer to winter, younger to older, or the digital restoration of old and damaged images or videos, an increase of image resolution, noise reduction, blurred image recovery, etc.

The interest in GANs for particle physics applications is to provide generative models as alternatives for computations that would otherwise be very CPU intensive. GANs can potentially provide a fast simulation for an experimental apparatus replacing full Monte Carlo simulations. This could be done either at the level of simulation of individual particle interactions, such as showers in a calorimeter [63], or possibly at the level of full simulation of collision processes [64], replacing entirely, with a machine-learning generator, the more traditional parametric fast simulations that usually require many ad hoc algorithms, studies, and calibrations to tune the numerous response parameters.

11.23 No Free Lunch Theorem

A theorem known as *no free lunch theorem for optimization* [65] states that all optimization algorithms have on average the same performances when the average is evaluated over all possible problems. This theorem implies that there is no single optimization algorithm that outperforms all others and implies that there is no single machine-learning algorithm that outperforms other machine-learning algorithms for either classification or regression problems.

In practice, there is no principle that allows to determine a priori which architecture of machine-learning algorithm performs better compared to others for a given problem. This obliges to test multiple possible methods to select the one that has the best performances for our specific problem. Of course, one may be guided by experience if benchmarks are already available for problems similar to the one we are facing.

References

1. Python programming language. https://www.python.org/
2. TensorFlow, open-source machine-learning library developed by the Google Brain team. https://www.tensorflow.org/
3. Keras, open-source software library for neural networks. https://keras.io/
4. PyTorch, open-source machine-learning library primarily developed by Facebook's AI Research lab. https://pytorch.org/
5. Scikit-learn, open-source machine-learning platform in Python. https://scikit-learn.org
6. XGBoost, gradient boosting library. https://xgboost.readthedocs.io/en/latest/
7. R. Brun, F. Rademakers, ROOT—an object oriented data analysis framework. Proceedings AIHENP96 Workshop, Lausanne (1996). Nucl. Inst. Meth. **A389**, 81–86 (1997). http://root.cern.ch/
8. A. Hoecker, et al., TMVA—toolkit for multivariate data analysis. PoS ACAT 040 (2007). arXiv:physics/0703039
9. P. Mehta, M. Bukov, C.H. Wang, A.G.R. Day, C. Richardson, C.K. Fisher, D.J. Schwab, A high-bias, low-variance introduction to machine learning for physicists. Phys. Rep. **810**, 1–124 (2019). https://doi.org/10.1016/j.physrep.2019.03.001
10. C. Cortes, V. Vapnik, Support-vector networks. Mach. Learn. **20**(3), 273–297 (1995). https://doi.org/10.1007/BF00994018
11. J.C. Obi, A comparative study of the Fisher's discriminant analysis and support vector machines. Eur. J. Eng. Technol. Res. **2**, 8 35–40 (2017). https://doi.org/10.24018/ejeng.2017.2.8.448
12. Y. LeCun, L. Bottou, G.B. Orr, K.R. Müller, *Neural Networks: Tricks of the Trade* (Springer, Berlin, 1998)
13. D.E. Rumelhart, G.E. Hinton, R.J. Williams, Learning representations by back-propagating errors. Nature **323**, 533–536 (1986)
14. C. Peterson, T.S. Rgnvaldsson, *An Introduction to Artificial Neural Networks*. LU-TP-91-23. LUTP-91-23, 14th CERN School of Computing, Ystad, Sweden (1991)
15. H.N. Mhaskar, Neural networks for optimal approximation of smooth and analytic functions. Neural Comput. **8**(1), 164–177 (1996)
16. R. Reed, R. Marks, *Neural Smithing: Supervised Learning in Feedforward Artificial Neural Networks. A Bradford book* (MIT Press, Cambridge, 1999)
17. Y. Le Cun, Y. Bengio, G. Hinton, Deep learning. Nature **521**, 436–444 (2015)

18. C. Dugas, Y. Bengio, F. Bélisle, C. Nadeau, R. Garcia, Incorporating second-order functional knowledge for better option pricing, in *Proceedings of NIPS'2000: Advances in Neural Information Processing Systems* (2001)
19. D.C. Cireşan, et al., Deep, big, simple neural nets for handwritten digit recognition. Neural Comput. **22**, 3207–20 (2010)
20. Y. LeCun, C. Cortes, C.J.C. Burges, The MNIST database of handwritten digits. http://yann.lecun.com/exdb/mnist/
21. I. Goodfellow, Y. Bengio, A. Courville, *Deep Learning* (MIT Press, Cambridge, 2016). http://www.deeplearningbook.org.
22. P. Baldi, K. Bauer, C. Eng, P. Sadowski, D. Whiteson, Jet substructure classification in high-energy physics with deep neural networks. Phys. Rev. **D93**, 094034 (2016)
23. A.J. Larkoski, I. Moult, B. Nachman, Jet substructure at the large Hadron Collider: a review of recent advances in theory and machine learning. Phys. Rep. **841**, 1–63 (2020). https://doi.org/10.1016/j.physrep.2019.11.001
24. P. Baldi, P. Sadowski, D. Whiteson, Searching for exotic particles in high-energy physics with deep learning. Nat. Commun. **5**, 4308 (2014)
25. A. Krizhevsky, I. Sutskever, G.E. Hinton, ImageNet classification with deep convolutional neural networks. Adv. Neural Inform. Proc. Syst. **25**, 1097–1105 (2012)
26. Photo by Angela Sorrentino (2007). http://angelasorrentino.awardspace.com/
27. A. Aurisano, et al., A convolutional neural network neutrino event classifier. JINST **11**, P09001 (2016)
28. Y. Jia, et al., Convolutional architecture for fast feature embedding (2014). arXiv:1408.5093
29. C. Goller, A. Kuchler, Learning task-dependent distributed structure-representations by back-propagation through structure, in *Proceedings of the IEEE International Conference on Neural Networks (ICNN 1996)* (1996), pp. 347–352
30. P. Frasconi, M. Gori, A. Sperduti, A General framework for adaptive processing of data structures. IEEE Trans. Neural. Netw. **9**(5), 768–786 (1998)
31. A. Chinea, Understanding the principles of recursive neural networks: a generative approach to tackle model complexity, in *ICANN 2009*, ed. by C. Alippi, M. Polycarpou, C. Panayiotou, G. Ellinas. Lecture Notes in Computer Science (LNCS), vol. 5768 (Springer, Heidelberg, pp. 952–963, 2009)
32. S. Farrell, et al., Novel deep learning methods for track reconstruction (2018). arXiv:1810.06111. https://arxiv.org/abs/1810.06111
33. F. Scarselli, M. Gori, A.C. Tsoi, M. Hagenbuchner, G. Monfardin, The graph neural network model. IEEE Trans. Neural Netw. **20**(1), 61–80 (2009). https://doi.org/10.1109/TNN.2008.2005605
34. E.A. Moreno, et al., JEDI-net: a jet identification algorithm based on interaction networks. Eur. Phys. J. C **80**, 58 (2020). https://doi.org/10.1140/epjc/s10052-020-7608-4
35. E.A. Moreno, et al., Interaction networks for the identification of boosted $H \rightarrow b\bar{b}$ decays. Phys. Rev. D **102**, 012010 (2020). https://doi.org/10.1103/PhysRevD.102.012010
36. J. Shlomi, P. Battaglia, J.R. Vlimant, Graph neural networks in particle physics. Mach. Learn. Sci. Technol. **2**, 021001 (2020). https://doi.org/10.1088/2632-2153/abbf9a
37. L. Breiman, Random forests. Mach. Learn. **45** 5–32 (2001). http://www.stat.berkeley.edu/~breiman/RandomForests/
38. T.G. Dietterich, et al., Ensemble methods in machine learning. Multiple Classif. Syst. **1857**, 1–15 (2000)
39. G. Louppe, Understanding random forests: From theory to practice (2014). arXiv:1407.7502. https://arxiv.org/abs/1407.7502
40. B.P. Roe, H.J. Yang, J. Zhu, Y. Liu, I. Stancu, G. McGregor, Boosted decision trees as an alternative to artificial neural networks for particle identification. Nucl. Instrum. Meth. **A543**, 577–584 (2005)
41. Y. Freund, R. Schapire, A decision-theoretic generalization of on-line learning and an application to boosting, in *Proceedings of EuroCOLT'94: European Conference on Computational Learning Theory* (1994)

42. Y. Freund, R. Schapire, N. Abe, A short introduction to boosting. J. Jpn. Soc. Artif. Intell. **14**(771–780), 1612 (1999)
43. D. Mullner, Modern hierarchical, agglomerative clustering algorithms (2011). arXiv:1109.2378. https://arxiv.org/abs/1109.2378
44. Z. Bar-Joseph, D.K. Gifford, T.S. Jaakkola, Fast optimal leaf ordering for hierarchical clustering. Bioinformatics **17**(suppl_1), S22–S29 (2001). https://doi.org/10.1093/bioinformatics/17. suppl_1.S22
45. S.P. Lloyd, Least squares quantization in PCM. Technical Report RR-5497, Bell Labs (1957)
46. J.B. MacQueen, Some methods for classification and analysis of multivariate observations, in *Proceedings of the Fifth Berkeley Symposium on Mathematical Statistics and Probability*, vol. 1 (1967), pp. 281–297
47. M.M. Breunig, H.P. Kriegel, R.T. Ng, J. Sander, LOF: identifying density-based local outliers, in *Proceedings of the 2000 ACM SIGMOD International Conference on Management of Data. SIGMOD* (2000), pp. 93–104. https://doi.org/10.1145/335191.335388
48. T. Dorigo, M. Fumanelli, C. Maccani, M. Mojsovska, G.C. Strong, B. Scarpa, RanBox: anomaly detection in the copula space. Comput. Phys. Commun. (2021). https://arxiv.org/abs/2106.05747
49. F. Canelli, A. de Cosa, L. Le Pottier, J. Niedziela, K. Pedro, M. Pierini, Autoencoders for Semivisible Jet Detection. FERMILAB-PUB-21-653-CMS (2021). https://arXiv.org/abs/2112.02864
50. R. Sutton, A.Barto, *Reinforcement Learning: An Introduction* (MIT Press, Cambridge, 1998)
51. G.A. Rummery, M. Niranjan, *On-Line Q-Learning Using Connectionist Systems* (1994)
52. C.J.C.H. Watkins, *Learning from delayed rewards*, Ph.D. Thesis, King's College, London (1989)
53. V. Mnih, K. Kavukcuoglu, D. Silver, A. Graves, I. Antonoglou, D. Wierstra, M. Riedmiller, *Playing Atari with Deep Reinforcement Learning* (DeepMind, 2013)
54. S. Carrazza, F.A. Dreyer, Jet grooming through reinforcement learning. Phys. Rev. D **100**, 014014 (2019)
55. J. Brehmer, S. Macaluso, D. Pappadopulo, K. Cranmer, Hierarchical clustering in particle physics through reinforcement learning, in *Machine Learning and the Physical Sciences Workshop at NeurIPS* (2020). https://arxiv.org/abs/2011.08191
56. I.J. Goodfellow, J. Pouget-Abadie, M. Mirza, B. Xu, D. Warde-Farley, S. Ozair, A. Courville, Y. Bengio, Generative adversarial networks (2014). arXiv:1406.2661 [stat.ML]. https://arxiv.org/abs/1406.2661
57. M. Arjovsky, S. Chintala, L. Bottou, Wasserstein generative adversarial networks (2017). arXiv:1701.07875. https://doi.org/10.48550/arXiv.1701.07875
58. T. Karras, T. Aila, S. Laine, J. Lehtinen, Progressive growing of GANs for improved quality, stability, and variation, in *Proceedings of the 6th International Conference on Learning Representations* (Vancouver, 2018). https://openreview.net/forum?id=Hk99zCeAb
59. I. Santos, L. Castro, N. Rodriguez-Fernandez, Á. Torrente-Patiño, A. Carballal, Artificial Neural networks and deep learning in the visual arts: a review. Neural Comput. Appl. **33**(1), 121–157 (2021). https://doi.org/10.1007/s00521-020-05565-4
60. J. Engel, K.K. Agrawal, S. Chen, I. Gulrajani, C. Donahue, A. Roberts, GANSynth: adversarial neural audio synthesis (2019). https://openreview.net/pdf?id=H1xQVn09FX
61. H.W. Dong, Y.H. Yang, Convolutional generative adversarial networks with binary neurons for polyphonic music generation, in *Proceedings of the 19th International Society for Music Information Retrieval Conference (ISMIR)* (2018). https://salu133445.github.io/musegan/
62. M. Mirza, S. Osindero, Conditional generative adversarial nets (2014). arXiv:1411.1784. https://arxiv.org/abs/1411.1784

63. M. Erdmann, J. Glombitza, T. Quast, Precise simulation of electromagnetic calorimeter showers using a Wasserstein generative adversarial network. Comput. Softw. Big Sci **3**, 4 (2019). https://doi.org/10.1007/s41781-018-0019-7
64. C. Ahdida, et al., SHIP collaboration, Fast simulation of muons produced at the SHiP experiment using generative adversarial networks. JINST **14**, P11028 (2019). https://doi.org/10.1088/1748-0221/14/11/P11028
65. D.H. Wolpert, W.G. Macready, No free lunch theorems for optimization. IEEE Trans. Evol. Comput. **1**(1), 67 (1997)

Discoveries and Limits

12

12.1 Searches for New Phenomena: Discovery and Limits

The goal of many experiments is to search for new physical phenomena. If an experiment provides a convincing measurement of a new signal, the result should be published and claimed as discovery. If the outcome is not sufficiently convincing, in many cases it is nonetheless interesting to quote, as result of the search for a new phenomenon, an upper limit to the yield of the new signal, or limits to some properties of the new signal. For instance, some data analysis could set limits to the mass of a new particle or to its coupling constants. More in general, it is often possible to exclude regions of the parameter space of a new theory that describes the physics of the new signal.

The interpretation of a discovery in the Bayesian and frequentist approaches is very different, as is discussed in the following sections. To provide a quantitative measure of how convincing the result of an experiment is, different possible approaches are possible. Using the Bayesian approach (Chap. 5), the posterior probability, which depends on the experimental data and a subjective prior, can quantify the degree of belief that the new signal hypothesis is true. When comparing two hypotheses, in case of a search for new physics, the presence or absence of signal, Bayes factors (see Sect. 5.11) can be used to quantify how strong the evidence for a new signal is against the background-only hypothesis. In the frequentist approach, the *significance level*, introduced in Sect. 12.3, measures the probability that, in the case of the presence of background-only, a statistical

The original version of the chapter has been revised. Minor changes were made to equations 12.47 and 12.50. A correction to this chapter can be found at https://doi.org/10.1007/978-3-031-19934-9_13

L. Lista, *Statistical Methods for Data Analysis*, Lecture Notes in Physics 1010, https://doi.org/10.1007/978-3-031-19934-9_12

fluctuation in data may produce by chance a measurement of the new signal at least as strong, as the observed in data.

The determination of limits is, in many cases, a complex task whose computation requires, in most of the cases, complex numerical algorithms. Several methods are adopted in particle physics and are documented in the literature to determine limits.

This chapter mainly focuses on frequentist methods. The concepts of p-value and significance level are introduced, and the most popular methods for limit setting are discussed, presenting their main benefits and limitations. The Bayesian approach is only shortly summarized, since all key concepts are already introduced in Chap. 5. The so-called *modified frequentist approach* is also presented, which has been a popular method in particle physics, recently mostly replaced by to use the *profile likelihood* test statistic.

12.2 p-Values and Discoveries

Given an observed data sample, a discovery can be claimed if the sample is sufficiently *inconsistent* with the hypothesis that background-only is present in the data. A test statistic t can be used to quantify the inconsistency of the observation in the hypothesis of the presence of background-only. The background-only hypothesis is typically taken as null hypothesis, H_0.

The probability p that the test statistic t assumes a value greater or equal to the observed one $t^\star$ if H_0 is true, i.e., due to pure background fluctuation, is called p-value. It is assumed here that, by convention, large values of t correspond to a more signal-like sample, but the opposite convention could also be used. The p-value is also defined in Sect. 6.17 in the context of goodness-of-fit tests using the χ^2 as test statistic.

The p-value has, by construction (see Sect. 3.4), a uniform distribution between 0 and 1 for the background-only hypothesis H_0 and tends to have small values in the presence of the signal. To exclude the H_0 hypothesis, it is not necessary to specify the alternative hypothesis H_1, which corresponds to the presence of signal. This hypothesis may be used to define the test statistic t, as discussed below.

If the number of observed events is adopted as test statistic (*counting experiment*), the p-value can be determined as the probability to count a number of events equal to or greater than the observed one assuming the presence of no signal and the expected background level. In this case, the test statistic and the p-value may assume discrete values. In general, the distribution of the p-value is only approximately uniform for a discrete distribution.

Example 12.1 - p-Value for a Poissonian Counting
Figure 12.1 shows the distribution of the number of collision events n due to background-only, which is taken as a Poissonian with an expected number of counts $\nu = b$ equal to 4.5.

(continued)

Example 12.1 (continued)

In case the observed number of events is equal to $n^\star = 8$, the p-value is equal to the probability to observe 8 or more events and is given by

$$p = P(n \geq n^\star) = \sum_{n=n^\star}^{\infty} \text{Pois}(n;\ \nu) = 1 - \sum_{n=0}^{n^\star-1} \text{Pois}(n;\ \nu) =$$

$$= 1 - e^{-4.5} \sum_{n=0}^{7} \frac{4.5^n}{n!}\ .$$

The explicit computation gives a p-value of 0.087.

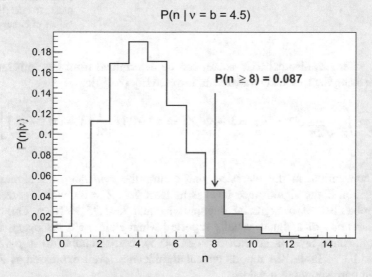

Fig. 12.1 Poisson distribution of the number of collision events n with an expected background yield of $\nu = b = 4.5$. The probability corresponding to $n \geq 8$ (light blue area) is 0.087 and is equal to the p-value, taking the event counting as test statistic

12.3 Significance Level

Instead of quoting a p-value, it is often preferred to report the equivalent number of standard deviations that correspond to an area equal to the p-value under the rightmost tail of a normal distribution. In this way, one quotes the *significance level*

Table 12.1 Significance
levels expressed as $Z\sigma$ and
the corresponding p-values
for typical cases

$Z(\sigma)$	p
1.00	1.59×10^{-1}
1.28	$\mathbf{1.00 \times 10^{-1}}$
1.64	$\mathbf{5.00 \times 10^{-2}}$
2.00	2.28×10^{-2}
2.32	$\mathbf{1.00 \times 10^{-2}}$
3.00	1.35×10^{-3}
3.09	$\mathbf{1.00 \times 10^{-3}}$
3.71	$\mathbf{1.00 \times 10^{-4}}$
4.00	3.17×10^{-5}
5.00	2.87×10^{-7}
6.00	9.87×10^{-10}

Bold values refer to
most frequently used
values of either Z or p

as $Z\sigma$, where Z, also called Z-score, can be determined from the corresponding
p-value using the following transformation (see Eq. (3.48)):

$$p = \int_Z^\infty \frac{1}{\sqrt{2\pi}} e^{-x^2/2}\, dx = 1 - \Phi(Z) = \Phi(-Z) = \frac{1}{2}\left[1 - \mathrm{erf}\left(\frac{Z}{\sqrt{2}}\right)\right].$$

$$(12.1)$$

By convention, in the literature, one claims the *evidence* of a signal under
investigation if the significance level is at least 3σ ($Z = 3$), which corresponds
to a probability of background fluctuation equal to 1.35×10^{-3}. One claims
the *observation* of a signal, usually reported when claiming a *discovery*, in case
the significance level is at least 5σ ($Z = 5$), corresponding to a p-value of
2.87×10^{-7}. Table 12.1 reports typical significance levels expressed as Z-score
and their corresponding p-values.

12.4 Signal Significance and Discovery

Determining the significance level of a signal is only part of the process that leads
to a discovery, in the scientific method. It is worth reporting the following sentence
that was posted by the American Statistical Association (ASA) when publishing
their statement on p-values [1]:

> The p-value was never intended to be a substitute for scientific reasoning. Well-reasoned
> statistical arguments contain much more than the value of a single number and whether that
> number exceeds an arbitrary threshold. The ASA statement is intended to steer research into
> a "post $p < 0.05$ era."

The importance to avoid the confusion between the p-value and the *"probability that the studied hypothesis is true, or the probability that the data were produced by random chance alone"* [1] was also remarked in the physicists community, and the following statement was reported by Cowan et al. [2]:

> It should be emphasized that in an actual scientific context, rejecting the background-only hypothesis in a statistical sense is only part of discovering a new phenomenon. One's *degree of belief* that a new process is present will depend in general on other factors as well, such as the plausibility of the new signal hypothesis and the degree to which it can describe the data.

To evaluate the *"plausibility of a new signal"* and other factors that give confidence in a discovery, the physicist's judgment cannot, of course, be replaced by the statistical evaluation of the p-value alone. In this sense, we can say that a Bayesian interpretation, in terms of credibility, of the very final result is somehow implicitly assumed, even when reporting a frequentist significance level.

12.5 Significance for Poissonian Counting Experiments

A *counting experiment* is a measurement where the number of observed counts $n = n^\star$ is the only considered information. The number of counts n in our data sample is due to a mixture of signal and background processes. The expected total number of counts is $\nu = s + b$, where s and b are the expected numbers of signal and background counts, respectively.

If the expected background b is known, e.g., imagine it could be estimated from theory, or from a control data sample with negligible uncertainty, the main unknown parameter of the problem is the expected signal yield s. The likelihood function for this problem is

$$L(n; \, s, \, b) = \frac{(s + b)^n}{n!} e^{-(s+b)} . \tag{12.2}$$

The number of observed counts $n^\star$ must be compared with the expected number of background events b in the null hypothesis, $s = 0$. If b is sufficiently large, the distribution of n can be approximated with a Gaussian with average b and standard deviation equal to $\sqrt{b}$. An excess in data allows to estimate the signal yield as

$$\hat{s} = n^\star - b, \tag{12.3}$$

which should be compared with the expected standard deviation $\sqrt{b}$. The significance level, in the Gaussian approximation, can be determined with the popular formula:

$$Z = \frac{\hat{s}}{\sqrt{b}} . \tag{12.4}$$

In case the expected background yield b is not known exactly, but the estimate $\hat{b}$ is available with a non-negligible uncertainty $\sigma_{\hat{b}}$, Eq. (12.4) can be modified by adding in quadrature the uncertainty on $\hat{b}$ and the standard deviation of the Poisson distribution, as follows:

$$Z = \frac{\hat{s}}{\sqrt{\hat{b} + \sigma_{\hat{b}}^2}} . \tag{12.5}$$

A better approximation to Eq. (12.4), even for b well below unity, is given by Cowan et al. [2]

$$Z = \sqrt{2 \left[(\hat{s} + b) \log \left(1 + \frac{\hat{s}}{b} \right) - \hat{s} \right]} . \tag{12.6}$$

Equation (12.4) may be obtained from Eq. (12.6) for $\hat{s} \ll b$.

12.6 Significance with Likelihood Ratio

In Sect. 10.9, test statistics based on a likelihood ratios were discussed. A test statistic suitable for a search for a new signal is

$$\lambda(\mu, \vec{\theta}) = \frac{L_{s+b}(\vec{x}_1, \cdots, \vec{x}_N; \mu, \vec{\theta})}{L_b(\vec{x}_1, \cdots, \vec{x}_N; \vec{\theta})} . \tag{12.7}$$

A minimum of $-2 \log \lambda(\mu)$ at $\mu = \hat{\mu}$ indicates the possible presence of a signal having a signal strength equal to $\hat{\mu}$. A significance level must be determined to assess the possibility of an evidence or observation.

The test statistic in Eq. (12.7) is the ratio of two likelihood functions. The denominator is the likelihood function in case the null hypothesis H_0 is true, while the numerator is the likelihood function in case the alternative hypothesis H_1 is true. In particular, the null hypothesis is the special case of the alternative hypothesis with $\mu = 0$. This situation is a case of nested hypotheses, which allows to apply Wilks' theorem, if the likelihood functions are sufficiently regular.

Note that in Eq. (12.7) numerator and denominator are inverted with respect to Eqs. (10.26) and (10.27). For this reason, the minus sign with respect to Eq. (10.27) must be dropped, and we take $2 \log \lambda(\hat{\mu})$ instead of $-2 \log \lambda(\hat{\mu})$ as test statistic. According to Wilks' theorem, the distribution of $2 \log \lambda(\hat{\mu})$ can be approximated with a χ^2 distribution with one degree of freedom, which is the square of a standard normal variable. Therefore, the square root of $2 \log \lambda(\hat{\mu})$ at the minimum gives an approximate estimate of the significance level Z:

$$Z = \sqrt{2 \log \lambda(\hat{\mu})} . \tag{12.8}$$

This significance Z is also called *local significance*, in the sense that it corresponds to a fixed set of values of the parameters $\vec{\theta}$. In case one or more of the parameters $\vec{\theta}$ are estimates from data, the local significance may be affected by the *look elsewhere effect*, which is discussed in Sect. 12.32.

If the background PDF $L_b(\vec{x}; \vec{\theta})$ does not depend on $\vec{\theta}$, for instance, if we have a single parameter θ equal to the mass of an unknown particle, which only affects the signal distribution, we can take the denominator of λ as a constant, and the likelihood ratio $\lambda(\theta)$ is equal, up to a multiplicative factor, to the likelihood function $L_{s+b}(\vec{x}; \mu, \vec{\theta})$. Hence, the maximum likelihood estimate of $\vec{\theta}$, $\vec{\theta} = \hat{\vec{\theta}}$, can also be determined by minimizing $-2\log\lambda(\mu, \vec{\theta})$, which is equal to $-2\log L_{s+b}$ up to a constant. The error matrix, or uncertainty contours on $\vec{\theta}$, can be determined as usual for maximum likelihood estimates (see Sect. 6.10) using the test statistic λ instead of the likelihood function L_{s+b}.

12.7 Significance Evaluation with Toy Monte Carlo

An accurate estimate of the significance corresponding to the test statistic $-2\log\lambda$, or equivalently its negative, can be achieved by generating a large number of Monte Carlo pseudo-experiments assuming the presence of no signal, i.e., setting $\mu = 0$. The distribution of the test statistic for the Monte Carlo sample gives, for a very large number of generated values, its expected distribution. The p-value is equal to the probability that λ is less or equal to the observed value of test statistic $\lambda^\star$:[1]

$$p = P_b(\lambda(\theta) \leq \lambda^\star). \tag{12.9}$$

This can be estimated as the fraction of generated pseudo-experiments for which $\lambda(\theta) \leq \lambda^\star$.

Note that to determine large significance values with sufficient precision, very large Monte Carlo samples are required. The p-value corresponding to a 5σ evidence, for instance, is equal to 2.87×10^{-7} (Table 12.1); hence, samples of the order of $\sim 10^9$ are needed to estimate p with an accuracy better than 10%.

12.8 Excluding a Signal Hypothesis

For signal exclusion, the p-value measures the probability of a signal *underfluctuation*. In practice, compared to the case of a discovery, the null hypothesis and the alternative hypothesis are inverted, when testing the p-values for exclusion.

---

[1] Given the definition of λ in Eq. (12.7), signal-like case tends to have $L_{s+b} > L_b$, hence $\lambda > 1$, which implies $\log\lambda > 0$ and $-2\log\lambda < 0$, and vice versa background-like cases tend to have $-2\log\lambda > 0$. For this reason, small λ values correspond to more signal-like cases. Other choices of test statistics may have the opposite convention.

To exclude a signal hypothesis, usually the applied requirement in terms of p-value is much milder than that for a discovery. Instead of requiring a p-value of 2.87×10^{-7} or less, i.e., the 5σ criterion, upper limits for a signal exclusion are set by requiring $p < 0.05$, which corresponds to a 95% confidence level (CL), or $p < 0.10$, which corresponds to a 90% CL.

12.9   Combined Measurements and Likelihood Ratio

Combining the likelihood ratios of several independent measurements can be performed by multiplying the likelihood functions of individual channels to produce a combined likelihood function (see also Sect. 7.3).

Assume that a first measurement has strong sensitivity to the signal, and it is combined with a second measurement that has low sensitivity. The combined test statistic is given by the product of likelihood ratios from both measurements. Since for the second measurement the $s + b$ and b hypotheses give similar values of the likelihood functions, given the low sensitivity to signal of the second measurement, the likelihood ratio of the second additional measurement is close to one. Hence, the combined test statistic, which is given by the product of the two individual ones, is expected not to differ too much from the one given by the first measurement only. Therefore, the combined sensitivity will not be worsened by the presence of the second measurement, with respect to the case in which only the first measurement is used.

12.10   Definitions of Upper Limit

In the frequentist approach, the procedure to set an upper limit is a special case of confidence interval determination (see Sect. 8.2), typically applied to the unknown signal yield s, or alternatively to the signal strength μ.

In order to determine an upper limit instead of the usual central interval, the choice of the interval corresponding to the desired confidence level $1 - \alpha$, usually 90% or 95%, is fully asymmetric: $[0, s^{\mathrm{up}}[$. This translates into an upper limit that is quoted as

$$s < s^{\mathrm{up}} \text{ at } 95\% \text{ CL (or } 90\% \text{ CL)}.$$

In the Bayesian approach, the interpretation of the upper limit s^{up} is that the credible interval $[0, s^{\mathrm{up}}[$ corresponds to a posterior probability equal to the confidence level $1 - \alpha$.

12.11 Bayesian Approach

The Bayesian posterior PDF for a signal yield s,[2] assuming a prior $\pi(s)$, is given by

$$P(s \mid \vec{x}) = \frac{L(\vec{x} ; s) \pi(s)}{\int_0^\infty L(\vec{x} ; s') \pi(s') \, ds'} . \tag{12.10}$$

The upper limit s^{up} can be computed by requiring that the posterior probability corresponding to the interval $[0, s^{up}[$ is equal to the required confidence level CL, or equivalently that the probability corresponding to $[s^{up}, \infty[$ is equal to $\alpha = 1 - CL$:

$$\boxed{\alpha = \int_{s^{up}}^\infty P(s \mid \vec{x}) \, ds = \frac{\int_{s^{up}}^\infty L(\vec{x} ; s) \pi(s) \, ds}{\int_0^\infty L(\vec{x} ; s) \pi(s) \, ds} .} \tag{12.11}$$

Apart from the technical aspects related to the integral computations and the already mentioned subjectiveness in the choice of the prior $\pi(s)$ (see Sect. 5.12), the above expression poses no fundamental problem, and all the methods to determine s^{up} have already been discussed in Chap. 5.

12.12 Bayesian Upper Limits for Poissonian Counting

In the simplest case of negligible background, $b = 0$, and assuming a uniform prior, $\pi(s)$, the posterior PDF for s has the same expression as the Poissonian probability itself, as demonstrated in Exercise 5.5:

$$P(s \mid n) = \frac{s^n e^{-s}}{n!} . \tag{12.12}$$

In case no count is observed, i.e., $n^\star = 0$, the posterior for s is

$$P(s \mid 0) = e^{-s} , \tag{12.13}$$

and Eq. (12.11) becomes

$$\alpha = \int_{s^{up}}^\infty e^{-s} \, ds = e^{-s^{up}} , \tag{12.14}$$

which can be inverted, and gives the upper limit:

$$s^{up} = -\log \alpha . \tag{12.15}$$

---

[2] The same approach could be equivalently formulated in terms of the signal strength μ.

For the particular cases of $\alpha = 0.05$ (95% CL) and $\alpha = 0.10$ (90% CL), the following upper limits can be set:

$$s < 3.00 \text{ at } 95\% \text{ CL} , \tag{12.16}$$

$$s < 2.30 \text{ at } 90\% \text{ CL} . \tag{12.17}$$

The general case with expected background $b \neq 0$ is addressed by Helene [3]. For non-negligible background, Eq. (12.11) becomes

$$\alpha = e^{-s^{\text{up}}} \frac{\sum_{m=0}^{n^\star} (s^{\text{up}} + b)^m / m!}{\sum_{m=0}^{n^\star} b^m / m!} , \tag{12.18}$$

where $n^\star$ is the observed number of counts. The above expression can be inverted numerically to determine s^{up} for given α, $n^\star$ and b. In case of no background, $b = 0$, Eq. (12.18) becomes

$$\alpha = e^{-s^{\text{up}}} \sum_{m=0}^{n^\star} \frac{(s^{\text{up}})^m}{m!} , \tag{12.19}$$

which, for $n^\star = 0$, gives again Eq. (12.14). The corresponding upper limits in case of negligible background for a different number of observed counts $n^\star$ are reported in Table 12.2. The upper limits from Eq. (12.18) at 90% and 95% CL are shown in Fig. 12.2 for different numbers of observed counts $n^\star$ and a different expected background b.

Table 12.2 Upper limits in the presence of negligible background evaluated under the Bayesian approach for a different number of observed counts $n^\star$

$n^\star$	$1 - \alpha = 90\%$ s^{up}	$1 - \alpha = 95\%$ s^{up}
0	2.30	3.00
1	3.89	4.74
2	5.32	6.30
3	6.68	7.75
4	7.99	9.15
5	9.27	10.51
6	10.53	11.84
7	11.77	13.15
8	12.99	14.43
9	14.21	15.71
10	15.41	19.96

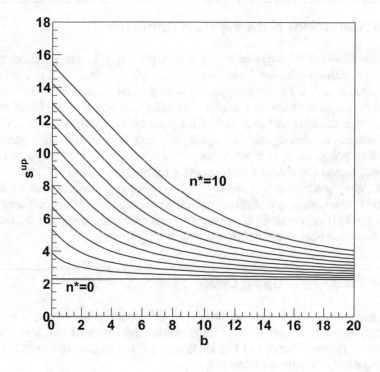

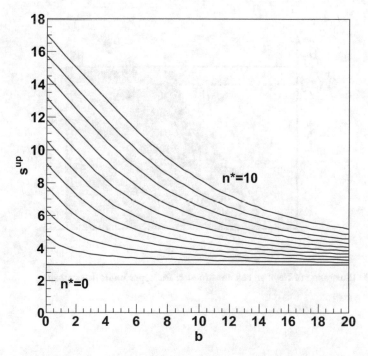

Fig. 12.2 Upper limits at the 90% CL (top) and 95% CL (bottom) to the signal yield s for a Poissonian process using the Bayesian approach as a function of the expected background b and for a number of observed counts $n^\star$, ranging from 0 to 10

12.13 Limitations of the Bayesian Approach

The determination of Bayesian upper limits presented in the previous section assumes a uniform prior for the expected signal yield. Assuming a different prior distribution results in a different upper limit. In general, there is no unique criterion to choose a specific prior that models the complete lack of knowledge about a parameter, in this case the signal yield. This issue is already discussed in Sect. 5.12.

In searches for new signals, the signal yield may be related to other parameters of the theory, e.g., the mass of unknown particles, or specific coupling constants. In that case, should one choose a uniform prior for the signal yield or a uniform prior for the theory parameters? As already said, no unique prescription can be derived from first principles. A possible approach is to choose more priors that reasonably model one's ignorance about the unknown parameters and verify that the obtained upper limits are not too sensitive to the choice of the prior.

12.14 Frequentist Upper Limits

Frequentist upper limits can be computed by inverting the Neyman belt (see Sect. 8.2) for a parameter θ using fully asymmetric intervals for the observed variable x. This is illustrated in Fig. 12.3, which is the equivalent of Fig. 8.1 when adopting a fully asymmetric interval.

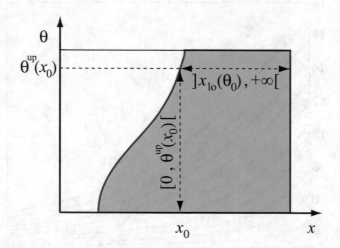

Fig. 12.3 Illustration of Neyman belt construction for upper limits determination

Assuming that the Neyman belt is monotonically increasing, the choice of the intervals $]x^{\mathrm{lo}}(\theta_0), +\infty[$ for x as a function of θ_0 leads to a confidence interval $[0, \theta^{\mathrm{up}}(x_0)[$ for θ, given a measurement $x_0 = x^\star$, which corresponds to the upper limit:

$$\theta < \theta^{\mathrm{up}}(x^\star). \qquad (12.20)$$

In the most frequent cases, the parameter θ is a signal yield s or a signal strength μ.

12.15 Frequentist Upper Limits for Counting Experiments

Similarly to Sect. 12.12, the case of a counting experiment with negligible background is analyzed first, under the frequentist approach. The probability to observe n events with an expectation s is given by a Poisson distribution:

$$P(n;\ s) = \frac{e^{-s} s^n}{n!}. \qquad (12.21)$$

An upper limit to the expected signal yield s can be set by using the observed number of counts, $n = n^\star$, as test statistic. The values of s for which the probability to observe $n^\star$ events or less, i.e., the p-value, is below $\alpha = 1 - \mathrm{CL}$ are excluded. For $n^\star = 0$, we have

$$p = P(0;\ s) = e^{-s}, \qquad (12.22)$$

and the condition $p > \alpha$ gives

$$p = e^{-s} > \alpha \qquad (12.23)$$

or, equivalently,

$$s < -\log \alpha = s^{\mathrm{up}}. \qquad (12.24)$$

For $\alpha = 0.05$ or $\alpha = 0.1$, upper limits are

$$s < 3.00 \text{ at } 95\% \text{ CL}, \qquad (12.25)$$

$$s < 2.30 \text{ at } 90\% \text{ CL}. \qquad (12.26)$$

Those results accidentally coincide with the ones obtained under the Bayesian approach in Eqs. (12.16) and (12.17). The numerical coincidence of upper limits computed under the Bayesian and frequentist approaches for the simple but common case of a counting experiment may lead to confusion. There is no intrinsic reason for which limits evaluated under the two approaches should coincide, and in general, with very few exceptions, like in this case, Bayesian and frequentist limits do not coincide numerically. Moreover, regardless of their numerical value, the interpretation of Bayesian and frequentist limits is very different, as already discussed several times.

Note that if the true value for the signal yield is $s = 0$, then the interval $[0, s^{up}[$ covers the true value with a 100% probability, instead of the required 90% or 95%, similarly to what was observed in Exercise 8.2. The extreme overcoverage is due to the discrete nature of the counting problem and may appear as a counterintuitive feature.

12.16 Frequentist Limits in Case of Discrete Variables

When constructing the Neyman belt for a discrete variable n, like in a Poissonian case, it is not always possible to find an interval $\{n^{lo}, \ldots, n^{up}\}$ that has exactly the desired coverage because of the intrinsic discreteness of the problem. This issue was already introduced in Sect. 8.5, when discussing binomial intervals. For discrete cases, it is possible to take the smallest interval that has a probability greater or equal to the desired confidence level. Upper limits determined in those cases are *conservative*, i.e., the procedure ensures that the probability content of the confidence belt is greater than or equal to $1 - \alpha$, i.e., it overcovers.

Figure 12.4 shows an example of Poisson distribution corresponding to the case with $s = 4$ and $b = 0$. Using a fully asymmetric interval as ordering rule, the

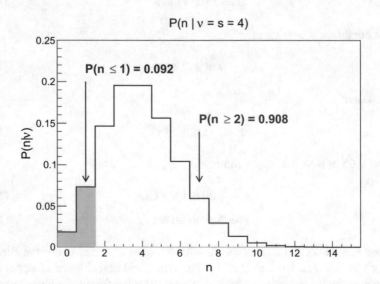

Fig. 12.4 Poisson distribution of the total number of counts $n = s + b$ for $s = 4$ and $b = 0$. The white bins show the smallest possible fully asymmetric confidence interval, $\{2, 3, 4, \cdots\}$ in this case, that gives at least the required coverage of 90%

interval $\{2, 3, \cdots\}$ of the discrete variable n corresponds to a probability $P(n \geq 2) = 1 - P(0) - P(1) = 0.9084$ and is the smallest interval that has a probability greater or equal to a desired confidence level of 0.90. In fact, the interval $\{3, 4, \cdots\}$ would have a probability $P(n \geq 3)$ less than 90%, while enlarging the interval to $\{1, 2, 3, \cdots\}$ would give a probability $P(n \geq 1)$ even larger than $P(n \geq 2)$.

If $n = n^{\star}$ events are observed, the upper limit s^{up} is given by the inversion of the Neyman belt, which corresponds to

$$s^{\mathrm{up}} = \inf \left\{ s : \sum_{m=0}^{n^{\star}} P(m; s) < \alpha \right\} . \tag{12.27}$$

The simplest case with $n^{\star} = 0$ gives the result shown in Sect. 12.15.

Consider a case with non-negligible background, $b \neq 0$, and take again $n^{\star}$ as the test statistic. Even the assumption $s = 0$ or possibly unphysical values $s < 0$ could be excluded if data have large underfluctuations, which are improbable put possible, according to a Poisson distribution with an expected number of events b, or $b - |s|$, for a negative s, such that the p-value is less than the required 0.1 or 0.05. The possibility to exclude parameter regions where the experiment should be insensitive, $s = 0$ in this case, or even unphysical regions, e.g.: $s < 0$, is rather unpleasant to a physicist.

From the pure frequentist point of view, moreover, this result potentially suffers from the flip-flopping problem (see Sect. 8.6): if one decides a priori to quote an upper limit, Neyman's construction with a fully asymmetric interval leads to the correct coverage, but if the choice is to switch from fully asymmetric to central intervals in case a significant signal is observed, this would produce an incorrect coverage.

12.17   Feldman–Cousins Unified Approach

The Feldman–Cousins approach, introduced in Sect. 8.7, provides a continuous transition from central intervals to upper limits, avoiding the flip-flopping problem. Moreover, it ensures that no unphysical parameter value, $s < 0$ in our case, is excluded. In the considered Poissonian counting problem, the 90% confidence belt obtained with the Feldman–Cousins approach is shown in Fig. 12.5 taking $b = 3$ as example. The intervals computed in case of no background, $b = 0$, are reported in Table 12.3 for different values of the number of observed events $n^{\star}$. Figure 12.6 shows the value of the 90% CL upper limit computed using the Feldman–Cousins approach as a function of the expected background b for different values of $n^{\star}$.

Comparing Table 12.3 with 12.2, which reports the Bayesian results, Feldman–Cousins upper limits are in general numerically larger than Bayesian limits, unlike the case considered in Sect. 12.15 of upper limits from a fully asymmetric Neyman belt, where frequentist and Bayesian upper limits coincide. In particular, for $n^{\star} = 0$, the 90% CL upper limits are 2.30 and 2.44, respectively, and the 95% CL upper

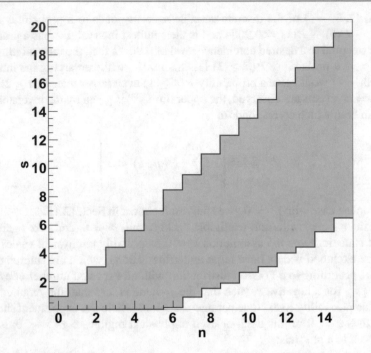

Fig. 12.5 Confidence belt at the 90% CL for a Poissonian process using the Feldman–Cousins approach for $b = 3$

Table 12.3 Upper and lower limits in the presence of negligible background ($b = 0$) with the Feldman–Cousins approach

$n^\star$	$1 - \alpha = 90\%$		$1 - \alpha = 95\%$	
	s^{lo}	s^{up}	s^{up}	s^{lo}
0	0.00	2.44	0.00	3.09
1	0.11	4.36	0.05	5.14
2	0.53	5.91	0.36	6.72
3	1.10	7.42	0.82	8.25
4	1.47	8.60	1.37	9.76
5	1.84	9.99	1.84	11.26
6	2.21	11.47	2.21	12.75
7	3.56	12.53	2.58	13.81
8	3.96	13.99	2.94	15.29
9	4.36	15.30	4.36	16.77
10	5.50	16.50	4.75	17.82

limits are 3.00 and 3.09. Anyway, as remarked before, the numerical comparison of those upper limits should not lead to any fundamental conclusion, since the interpretation is very different for frequentist and Bayesian limits.

A peculiar feature of Feldman–Cousins upper limits is that, for $n^\star = 0$, a larger expected background b corresponds to a more stringent, i.e., lower, upper limit, as can be seen in Fig. 12.6 (lowest curve). This feature is absent in Bayesian limits

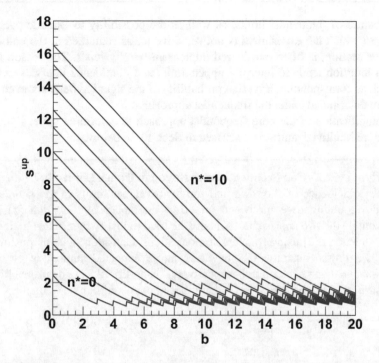

Fig. 12.6 Upper limits at 90% CL to the signal s for a Poissonian process using the Feldman–Cousins method as a function of the expected background b and for a number of observed events $n^\star$ ranging from 0 to 10

that do not depend on the expected background b for $n^\star = 0$ (see Fig. 12.2). This dependence of upper limits on the expected amount of background is somewhat counterintuitive: imagine two experiments, say A and B, performing a search for a rare signal. Both experiments are designed to achieve a very low background level, but A can reduce the background level more than B, say $b = 0.01$ and $b = 0.1$ expected counts for A and B, respectively. If both experiments observe zero counts, which is for both the most likely outcomes, the experiment that achieves the most stringent limit is the one with the largest expected background, B in this case, i.e., the one that has the worse expected performances.

The Particle Data Group published in their review the following sentence about the interpretation of frequentist upper limits for what concerns the difficulty to interpret a more stringent limit for an experiment with worse expected background, in case no count is observed:

> The intervals constructed according to the unified [Feldman Cousins] procedure for a Poisson variable n consisting of signal and background have the property that for $n = 0$ observed events, the upper limit decreases for increasing expected background. This is counter-intuitive since it is known that if $n = 0$ for the experiment in question, then no background was observed, and therefore one may argue that the expected background should not be relevant. The extent to which one should regard this feature as a drawback is a subject of some controversy [4].

This feature of frequentist limits, as well as the possibility to exclude parameter values to which the experiment is not sensitive to, as remarked at the end of the previous section, is often considered unpleasant by physicists. The reason is that human intuition tends to interpret upper limits and confidence intervals, more in general, as corresponding Bayesian probability of the signal hypothesis, even when they are determined under the frequentist approach.

A modification of the pure frequentist approach to determine upper limits that have more intuitive features is discussed in Sect. 12.18 below.

Example 12.2 - Can Frequentist and Bayesian Upper Limits Be Unified?
The coincidence of Bayesian and frequentist upper limits in the simplest event counting case motivated an effort attempted by G. Zech [5] to reconcile the two approaches. Namely, he tried to justify the limits obtained by Helene in [3] using frequentist motivations. Consider the superposition of two Poissonian processes having s and b expected number of events from signal and background, respectively. Using Eq. (2.71), the probability distribution for the total observed number of counts $n^\star$ can be written as

$$P(n;\, s,\, b) = \sum_{\substack{n_s = 0 \\ n_b = n^\star - n_s}}^{n^\star} P(n_s;\, s)\, P(n_b;\, b)\,, \qquad (12.28)$$

where P represents a Poissonian distribution.

Zech proposed to modify the background term of the sum in Eq. (12.28), $P(n_b;\, b)$, to take into account that the observation of $n^\star$ events should put a constraint on the possible values of n_b, which can only range from 0 to $n^\star$. In this way, $P(n_b;\, b)$ was replaced with

$$P'(n_b;\, b) = P(n_b;\, b) \left/ \sum_{n_b'=0}^{n^\star} P(n_b';\, b)\right.\,. \qquad (12.29)$$

This modification leads to the same results obtained by Helene in Eq. (12.18), which apparently indicates a possible convergence of Bayesian and frequentist approaches.

This derivation was later criticized by Highland and Cousins [6] who demonstrated that the modification introduced by Eq. (12.29) produces an incorrect coverage. Zech himself then admitted the non-rigorous application of the frequentist approach [7]. This attempt could not provide a way to conciliate the Bayesian and frequentist approaches, which, as said, have completely different interpretations. Anyway, Zech's intuition anticipated the formulation of the *modified frequentist approach* that is discussed in Sect. 12.18, which is nowadays widely used in particle physics.

12.18   Modified Frequentist Approach: The CL$_s$ Method

The concerns about frequentist limits discussed at the end of Secs. 12.16 and 12.17 have been addressed with the definition of a procedure that was adopted for the first time for the combination of the results obtained by the four LEP experiments, Aleph, Delphi, Opal, and L3, in the search for the Higgs boson [8]. The approach consists in a modification of the pure frequentist approach with the introduction of a conservative corrective factor to the p-value that cures the counterintuitive peculiarities mentioned above. In particular, it avoids the possibility to exclude, due to pure statistical fluctuations, parameter regions where the experiment is not sensitive to, and, if zero counts are observed, a higher expected background does not correspond to a more stringent limit, as with the Feldman–Cousins approach.

The so-called *modified frequentist approach* is illustrated in the following using the test statistic adopted in the original proposal, introduced in Sect. 10.9, which is the ratio of the likelihood functions evaluated under two different hypotheses: the presence of signal plus background, H_1, corresponding to the likelihood function L_{s+b}, and the presence of background only, H_0, corresponding to the likelihood function L_b:

$$\lambda(\vec{\theta}) = \frac{L_{s+b}(\vec{x}; \vec{\theta})}{L_b(\vec{x}; \vec{\theta})} . \tag{12.30}$$

Different test statistics have been applied after the original formulation. The profile likelihood (Eq. (10.29)) is to date the most used. The method described in the following is valid for any test statistic.

The likelihood ratio in Eq. (12.30) can also be written introducing the signal strength μ separately from the other parameters of interest $\vec{\theta}$, as in Eq. (10.36):

$$\lambda(\mu, \vec{\theta}) = e^{-\mu s(\vec{\theta})} \prod_{i=1}^{N} \left(\frac{\mu s(\vec{\theta}) f_s(\vec{x}_i; \vec{\theta})}{b(\vec{\theta}) f_b(\vec{x}_i; \vec{\theta})} + 1 \right) , \tag{12.31}$$

where the functions f_s and f_b are the PDFs for signal and background of the variables $\vec{x}$ that constitute our data sample. The negative logarithm of the test statistic is given by Eq. (10.37), reported below:

$$-\log \lambda(\mu, \vec{\theta}) = \mu s(\vec{\theta}) - \sum_{i=1}^{N} \log \left(\frac{\mu s(\vec{\theta}) f_s(\vec{x}_i; \vec{\theta})}{b(\vec{\theta}) f_b(\vec{x}_i; \vec{\theta})} + 1 \right) . \tag{12.32}$$

In order to quote an upper limit using the frequentist approach, the distribution of the test statistic λ, or equivalently $-2 \log \lambda$, in the hypothesis of signal plus background, has to be known, and the p-value corresponding to the observed value $\lambda = \lambda^\star$, denoted below as p_{s+b}, has to be determined as a function of the parameters of interest μ and $\vec{\theta}$. The proposed modification to the purely frequentist approach consists of finding two p-values corresponding to both the H_1 and H_0 hypotheses. Below, for simplicity of notation, the set of parameters $\vec{\theta}$ also includes μ, which is omitted:

$$p_{s+b}(\vec{\theta}) = P_{s+b}(\lambda(\vec{\theta}) \geq \lambda^\star), \tag{12.33}$$

$$p_b(\vec{\theta}) = P_b(\lambda(\vec{\theta}) \leq \lambda^\star). \tag{12.34}$$

From those two probabilities, the following quantity can be derived:

$$\mathrm{CL}_s(\vec{\theta}) = \frac{p_{s+b}(\vec{\theta})}{1 - p_b(\vec{\theta})}. \tag{12.35}$$

Upper limits are determined excluding the range of the parameters of interest for which $\mathrm{CL}_s(\vec{\theta})$ is lower than the conventional confidence level, typically 95% or 90%. For this reason, the modified frequentist approach is often referred to as the *CL$_s$ method* [26]. In most of the cases, the probabilities $p_{s+b}(\vec{\theta})$ and $p_b(\vec{\theta})$ in Eqs. (12.33) and (12.34) are not trivial to obtain analytically and are determined most often using pseudo-experiments generated by Monte Carlo. An example of the outcome of this numerical approach is shown in Fig. 12.7.

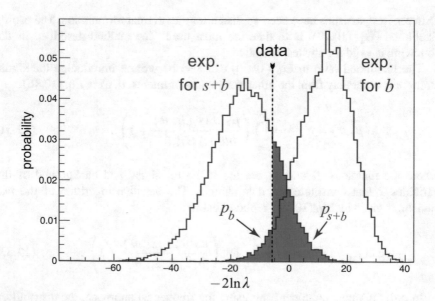

Fig. 12.7 Example of evaluation of CL$_s$ from pseudo-experiments. The distribution of the test statistic $-2 \log \lambda$ is shown in blue assuming the signal-plus-background hypothesis and in red assuming the background-only hypothesis. The black line shows the value of the test statistic measured in data, $\lambda^\star$, and the shaded areas represent p_{s+b} (blue) and p_b (red). CL$_s$ is determined as $p_{s+b}/(1 - p_b)$

The modified frequentist approach does not provide the desired coverage, which would be required in a purely frequentist approach, but does not suffer from the counterintuitive features of frequentist upper limits. CL$_s$ limits have convenient statistical properties:

- Limits are conservative from the frequentist point of view. In fact, since $p_b \leq 1$, we have that $\mathrm{CL}_s(\vec{\theta}) \geq p_{s+b}(\vec{\theta})$. Hence, the intervals obtained with the CL_s method *overcovers*, and the corresponding limits are less stringent than purely frequentist ones.
- Unlike upper limits obtained using the Feldman–Cousins approach, if no count is observed, CL$_s$ upper limits do not depend on the expected amount of background. This feature is also common with Bayesian upper limits.

If the distributions of the test statistic λ, or equivalently $-2\log\lambda$, for the two hypotheses H_0 (b) and H_1 ($s + b$) are well separated, as in Fig. 12.8, top, in case

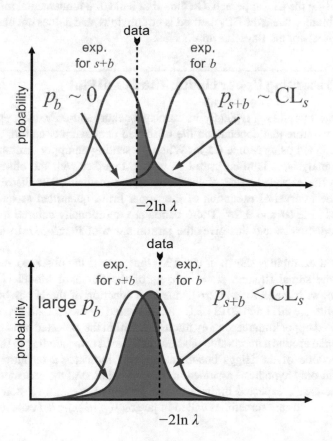

Fig. 12.8 Illustration of the application of the CL$_s$ method in cases of well separated distributions of the test statistic $-2\log\lambda$ for the $s + b$ and b hypotheses (top) and in case of largely overlapping distributions (bottom) where the experiment has poor sensitivity to the signal

H_1 is true, then p_b has large chance to be very small, and consequently, $1 - p_b \simeq 1$ and $CL_s \simeq p_{s+b}$. If this is the case, the CL_s limit is almost identical to the purely frequentist upper limit based on the p-value p_{s+b}.

If the two distributions have large overlap, as Fig. 12.8, bottom, there is an indication that the experiment has low sensitivity on the signal. In this case, if, due to a statistical fluctuation, p_b is large, then $1 - p_b$ at the denominator in Eq. (12.35) is small, preventing CL_s to become too small, and therefore to reject cases in which the experiment has poor sensitivity.

For a simple Poissonian counting experiment with expected signal s and a background b, using the likelihood ratio from Eq. (10.40), it is possible to demonstrate analytically that the CL_s approach leads to results identical to the Bayesian ones (Eq. (12.18)) that are also identical to the results of the method proposed by G. Zech [5], discussed in Exercise 12.2. In general, in many realistic applications, CL_s upper limits are numerically close to Bayesian upper limits assuming a uniform prior, but of course the Bayesian interpretation of upper limits cannot be applied to limits obtained using the CL_s approach. On the other hand, the fundamental interpretation of limits obtained using the CL_s method is not obvious, and it does not match neither the frequentist nor the Bayesian approaches.

12.19 Presenting Upper Limits: The Brazil Plot

Under some hypothesis, typically in case of background-only, the upper limit is a random variable that depends on the observed data sample, and its distribution can be predicted using Monte Carlo. When presenting an upper limit as the result of a data analysis, it is often useful to report, together with the observed upper limit, also the expected value of the limit or the median of its distribution, and possibly the interval of excursion of the upper limit, quantified as quantiles that correspond to $\pm 1\sigma$ and $\pm 2\sigma$. These bands are traditionally colored in green and yellow, respectively, and this gives the jargon name of *Brazil plot* to this kind of presentation.

A typical example is shown in Fig. 12.9 that reports the observed and expected limits to the signal strength $\mu = \sigma/\sigma_{SM}$ of the Standard Model Higgs boson production, with the $\pm 1\sigma$ and 2σ bands as a function of the Higgs boson mass, obtained with the 2011 and 2012 LHC data, reported by the ATLAS experiment [9]. The observed upper limit reasonably fluctuates within the expected band but exceeds the $+2\sigma$ band around a mass value of about 125 GeV, corresponding to the presently measured value of the Higgs boson mass. This indicates a deviation from the background-only hypothesis assumed in the computation of the expected limit.

In some cases, expected limits are also presented assuming a nominal signal yield. Those cases are sometimes called in jargon *signal-injected* expected limits.

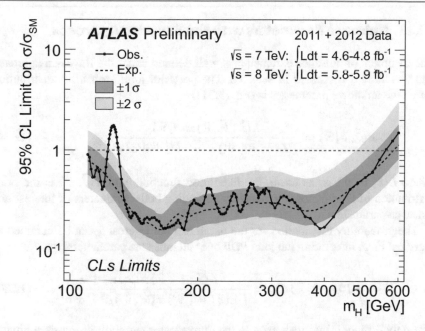

Fig. 12.9 Upper limit reported as Brazil plot in the context of the search for the Higgs boson at the LHC by the ATLAS experiment. The expected limit to the signal strength $\mu = \sigma/\sigma_{SM}$ is shown as a dashed line, surrounded by the $\pm 1\sigma$ (green) and $\pm 2\sigma$ (yellow) bands. The observed limit is shown as a solid line. All mass values corresponding to a limit below $\mu = 1$ (dashed horizontal line) are excluded at the 95% confidence level. Reprinted from [9], © ATLAS Collaboration. Reproduced under CC-BY-4.0 license

12.20 Nuisance Parameters and Systematic Uncertainties

In most of realistic cases, a test statistic contains some parameters that are not of direct interest of our measurement but are *nuisance parameters* needed to precisely describe the PDF model of our data sample, as introduced in Sect. 1.11. In the following, the parameter set is split into two subsets: parameters of interest, $\vec{\theta} = (\theta_1, \cdots, \theta_h)$, and nuisance parameters, $\vec{\nu} = (\nu_1, \cdots, \nu_k)$. For instance, if we are only interested in the measurement of signal strength μ, we have $\vec{\theta} = (\mu)$; if we want to measure, instead, both the signal strength and a new particle's mass m, we have $\vec{\theta} = (\mu, m)$.

12.21 Nuisance Parameters with the Bayesian Approach

The treatment of nuisance parameters is well defined under the Bayesian approach
and was already discussed in Sect. 5.6. The posterior joint probability distribution
for all the unknown parameters is (Eq. (5.31))

$$P(\vec{\theta}, \vec{v} \mid \vec{x}) = \frac{L(\vec{x}; \vec{\theta}, \vec{v}) \, \pi(\vec{\theta}, \vec{v})}{\int L(\vec{x}; \vec{\theta}', \vec{v}') \, \pi(\vec{\theta}', \vec{v}') \, \mathrm{d}^h\theta' \mathrm{d}^k v'}, \tag{12.36}$$

where $L(\vec{x}; \vec{\theta}, \vec{v})$ is, as usual, the likelihood function, and $\pi(\vec{\theta}, \vec{v})$ is the prior
distribution of the unknown parameters, including both parameters of interest and
nuisance parameters.

The probability distribution for the parameters of interest $\vec{\theta}$ can be obtained as
marginal PDF, integrating the joint PDF over all nuisance parameters:

$$P(\vec{\theta} \mid \vec{x}) = \int P(\vec{\theta}, \vec{v} \mid \vec{x}) \, \mathrm{d}^k v = \frac{\int L(\vec{x}; \vec{\theta}, \vec{v}) \, \pi(\vec{\theta}, \vec{v}) \, \mathrm{d}^k v}{\int L(\vec{x}; \vec{\theta}', \vec{v}') \, \pi(\vec{\theta}', \vec{v}') \, \mathrm{d}^h\theta' \mathrm{d}^k v'}. \tag{12.37}$$

The only difficulty that may arise is the numerical integration in multiple dimensions, but the Bayesian approach poses no conceptual problems. A particularly
performant class of algorithms in those cases is based on Markov chain Monte
Carlo [10], introduced in Sect. 4.11.

12.22 Hybrid Treatment of Nuisance Parameters

The treatment of nuisance parameters under the frequentist approach is more
difficult to perform rigorously with the test statistic L_{s+b}/L_b (Eq. (12.7)). Cousins
and Highland [11] proposed to adopt the same approach used for the Bayesian
treatment and to determine an approximate likelihood function by integrating Eqs.
(10.33) and (10.35) over the nuisance parameters:

$$L_{s+b}(\vec{x}_1, \cdots, \vec{x}_N \mid \mu, \vec{\theta}) =$$

$$= \frac{1}{N!} \int e^{-(\mu s(\vec{\theta}, \vec{v}) + b(\vec{\theta}, \vec{v}))} \prod_{i=1}^{N} \left(\mu s(\vec{\theta}, \vec{v}) \, f_s(\vec{x}_i; \vec{\theta}, \vec{v}) + b(\vec{\theta}, \vec{v}) \, f_b(\vec{x}_i; \vec{\theta}, \vec{v}) \right) \mathrm{d}^k v,$$
$$\tag{12.38}$$

$$L_b(\vec{x}_1, \cdots, \vec{x}_N \mid \vec{\theta}) = \frac{1}{N!} \int e^{-b(\vec{\theta}, \vec{v})} \prod_{i=1}^{N} b(\vec{\theta}, \vec{v}) \, f_b(\vec{x}_i; \vec{\theta}, \vec{v}) \, \mathrm{d}^k v. \tag{12.39}$$

This so-called *hybrid* Bayesian–frequentist approach does not ensure an exact
frequentist coverage [12]. It has been proven on simple models that the results are
numerically close to Bayesian limits assuming a uniform prior [13].

Likelihood functions determined with the hybrid approach have been used in the combined search for the Higgs boson at LEP [8] in conjunction with the modified frequentist approach (see Sect. 12.18).

12.23 Event Counting Uncertainties

For an event counting problem, if the number of background counts is known with some uncertainty, the PDF of the background estimate $\hat{b}$ can be modeled as a function of the true unknown expected background b, $P(\hat{b}; b)$. The likelihood functions, which depend on the parameter of interest s and the unknown nuisance parameter b, can be written as

$$L_{s+b}(n, \hat{b}; s, b) = \frac{e^{-(s+b)}(s+b)^n}{n!} P(\hat{b}; b), \tag{12.40}$$

$$L_b(n, \hat{b}; b) = \frac{e^{-b}b^n}{n!} P(\hat{b}; b). \tag{12.41}$$

In order to eliminate the dependence on the nuisance parameter b, the hybrid likelihoods, using the Cousins–Highland approach, can be written as

$$L_{s+b}(n, \hat{b}; s) = \int_0^\infty \frac{e^{-(s+b)}(s+b)^n}{n!} P(\hat{b}; b) \, db, \tag{12.42}$$

$$L_b(n, \hat{b}) = \int_0^\infty \frac{e^{-b}b^n}{n!} P(\hat{b}; b) \, db. \tag{12.43}$$

In the simplified assumption that $P(\hat{b}; b)$ is a Gaussian function, the integration can be performed analytically [14]. This approximation may be valid only if the uncertainty on the background prediction is small. Otherwise, the PDF $P(\hat{b}; b)$ may extend with non-negligible probability to unphysical negative values of b, which are included in the integration range.

In order to avoid such cases, the use of distributions whose range is limited to positive values is often preferred. For instance, a log normal distribution (Sect. 3.10) or a gamma distribution (Sect. 3.12) is usually preferred to a Gaussian. For such more complex cases, the integration should proceed numerically, with potential computing performance penalties.

12.24 Upper Limits Using the Profile Likelihood

A test statistic that accounts for nuisance parameters and allows to avoid the hybrid Bayesian–frequentist approach is the *profile likelihood*, defined in Eq. (10.29):

$$\lambda(\mu) = \frac{L(\vec{x} \mid \mu, \hat{\hat{\vec{\theta}}}(\mu))}{L(\vec{x} \mid \hat{\mu}, \hat{\vec{\theta}})}, \tag{12.44}$$

where $\hat{\mu}$ and $\hat{\vec{\theta}}$ are the best fit values of μ and $\vec{\theta}$ corresponding to the observed data sample, and $\hat{\hat{\vec{\theta}}}(\mu)$ is the best fit value of $\vec{\theta}$ obtained for a fixed value of μ. All parameters are treated as nuisance parameters, except for μ that is the only parameter of interest, in this case. A convenient test statistic is

$$t_\mu = -2 \log \lambda(\mu) \,. \tag{12.45}$$

See Exercise 12.5 for a concrete application.

A scan of t_μ as a function of μ reveals a minimum at the value $\mu = \hat{\mu}$. The minimum value $t_\mu(\hat{\mu})$ is equal to zero by construction. An uncertainty interval for μ can be determined as discussed in Sect. 6.12 from the excursion of t_μ around the minimum $\hat{\mu}$. The intersections of the curve with a straight line corresponding to $t_\mu = 1$ give the interval extremes.

The profile likelihood is introduced to satisfy the conditions required by Wilks' theorem (see Sect. 10.8), according to which, if μ corresponds to the true value, then t_μ follows a χ^2 distribution with one degree of freedom. If the true hypothesis is $\mu = 0$, then $t_0 = -2 \log \lambda(0)$ follows approximately a χ^2 distribution with one degree of freedom; therefore, the significance level at which the null hypothesis $\mu = 0$ can be excluded is, approximately,

$$Z \simeq \sqrt{t_0} = \sqrt{-2 \log \lambda(0)} \,. \tag{12.46}$$

Usually, the addition of nuisance parameters broadens the shape of the profile likelihood as a function of the parameter of interest μ compared with the case where nuisance parameters are not added. As a consequence, the uncertainty on μ increases with the addition of nuisance parameters in the test statistic that are required to model sources of systematic uncertainties. Compared with the Cousins–Highland hybrid method, the profile likelihood is more statistically sound from the frequentist point of view. In addition, no numerical integration is needed, which is usually a more CPU intensive task compared with the minimizations required for the profile likelihood evaluation.

Since the profile likelihood is based on a likelihood ratio, according to the Neyman–Pearson lemma (see Sect. 10.5), it has optimal performances for what concerns the separation of the two hypotheses assumed in the numerator and in the denominator of Eq. (12.44).

The test statistic t_μ can be used to determine p-values corresponding to the different hypotheses on μ to determine upper limits or significance. Those p-values can be computed in general by generating sufficiently large Monte Carlo pseudo-samples, but in many cases, asymptotic approximations allow a much faster evaluation, as discussed in Sect. 12.30.

12.25 Variations of the Profile Likelihood Test Statistic

Different variations of the profile likelihood definition have been adopted for various data analysis cases. A review of the most popular test statistics is presented in [2] where approximate formulae, valid in the asymptotic limit of a large number of measurements, are provided to simplify the computation. The main examples are reported in the following.

12.26 Test Statistic for Positive Signal Strength

In order to enforce the condition $\mu \geq 0$, since a signal yield cannot have negative values, the test statistic $t_\mu = -2 \log \lambda(\mu)$ in Eq. (12.45) can be modified as follows:

$$
\tilde{t}_\mu = -2 \log \tilde{\lambda}(\mu) =
\begin{cases}
-2 \log \dfrac{L(\vec{x} \mid \mu, \hat{\hat{\theta}}(\mu))}{L(\vec{x} \mid \hat{\mu}, \hat{\hat{\theta}})} & \hat{\mu} \geq 0 \,, \\[3mm]
-2 \log \dfrac{L(\vec{x} \mid \mu, \hat{\hat{\theta}}(\mu))}{L(\vec{x} \mid 0, \hat{\hat{\theta}}(0))} & \hat{\mu} < 0 \,.
\end{cases}
\tag{12.47}
$$

In practice, the estimate of μ is replaced with zero if the best fit value $\hat{\mu}$ is negative, which may occur in case of a downward fluctuation in data.

12.27 Test Statistic for Discovery

In order to assess the presence of a new signal, the hypothesis of a positive signal strength μ is tested against the hypothesis $\mu = 0$. This is done using the test statistic $t_\mu = -2 \log \lambda(\mu)$ evaluated for $\mu = 0$. The test statistic $t_0 = -2 \log \lambda(0)$, anyway, may reject the hypothesis $\mu = 0$ in case a downward fluctuation in data would result in a negative best fit value $\hat{\mu}$. A modification of t_0 has been proposed that is only sensitive to an excess in data, by setting the test statistic to zero in case of negative $\hat{\mu}$ [2]:

$$
q_0 =
\begin{cases}
-2 \log \lambda(0) & \hat{\mu} \geq 0 \,, \\
0 & \hat{\mu} < 0 \,.
\end{cases}
\tag{12.48}
$$

The p-value corresponding to the test statistic q_0 can be evaluated using Monte Carlo pseudo-samples that only simulate background processes. The distribution of q_0 has a component, proportional to the Dirac's delta $\delta(q_0)$, i.e., it has a spike at $q_0 = 0$, corresponding to all the cases that give a negative $\hat{\mu}$.

12.28 Test Statistic for Upper Limits

Similarly to the definition of q_0, one may not want to consider upward fluctuations in data in order to exclude a given value of μ in case the best fit value $\hat{\mu}$ is greater than the assumed value of μ. In order to avoid those cases, the following modification of t_μ has been proposed:

$$q_\mu = \begin{cases} -2 \log \lambda(\mu) & \hat{\mu} \le \mu \,, \\ 0 & \hat{\mu} > \mu \,. \end{cases} \tag{12.49}$$

The distribution of q_μ presents a spike at $q_\mu = 0$ corresponding to the cases that give $\hat{\mu} > \mu$.

12.29 Higgs Test Statistic

Both cases considered for Eqs. (12.47) and (12.49) are taken into account in the test statistic adopted for Higgs search at the LHC:

$$\tilde{q}_\mu = \begin{cases} -2 \log \dfrac{L(\vec{x} \mid \mu, \hat{\hat{\theta}}(\mu))}{L(\vec{x} \mid 0, \hat{\hat{\theta}}(0))} & \hat{\mu} < 0 \,, \\[2ex] -2 \log \dfrac{L(\vec{x} \mid \mu, \hat{\hat{\theta}}(\mu))}{L(\vec{x} \mid \hat{\mu}, \hat{\theta})} & 0 \le \hat{\mu} \le \mu \,, \\[2ex] 0 & \hat{\mu} > \mu \,. \end{cases} \tag{12.50}$$

A null value replaces $\hat{\mu}$ in the denominator of the profile likelihood for the cases where $\hat{\mu} < 0$, as for $\tilde{t}_\mu$, in order to protect against unphysical values of the signal strength. In order to avoid spoiling upper limit performances, as for q_μ, the cases of upward fluctuations in data are not considered as evidence against the assumed signal hypothesis. Therefore, if $\hat{\mu} > \mu$, the test statistic is set to zero.

12.30 Asymptotic Approximations

Asymptotic approximations for the test statistics presented in the previous sections can be obtained using Wilks' theorem and approximate formulae due to Wald [15], as discussed in a comprehensive way in [2]. The test statistic defined in Eq. (12.45) can be approximated as:

$$-2 \ln \lambda(\mu) = \frac{(\mu - \hat{\mu})^2}{\sigma^2} + \mathcal{O}\left(1/\sqrt{N}\right) \,. \tag{12.51}$$

The last term can be neglected for large sample size N, and a parabolic approximation of the test statistic as a function of μ is obtained. The estimate $\hat{\mu}$ is approximately distributed according to a Gaussian with mean μ' and standard deviation σ, therefore the test statistic is distributed according to a *noncentral χ^2 distribution*, which is known, and depends on the noncentrality parameter Λ:

$$\Lambda = \frac{(\mu - \mu')^2}{\sigma^2}. \tag{12.52}$$

For the special case where $\mu = \mu'$ the noncentrality parameter is $\Lambda = 0$, and the distribution becomes a χ^2 distribution with one degree of freedom, consistently with Wilks' theorem. Assuming $\mu' = 0$, one recovers Eq. (12.46), which gives the asymptotic approximation for the significance level when using the test statistic for discovery q_0 (Eq. (12.48)):

$$Z_0 \simeq \sqrt{q_0}. \tag{12.53}$$

12.31 Asimov Data Sets

The application of Eq. (12.51) in order to obtain an approximation of the test statistic and its distribution requires the estimate of the variance of $\hat{\mu}$, σ^2. One way to estimate σ^2 is to find a data sample A such that the estimated value of μ is equal to the true value: $\hat{\mu} = \mu'$. For such a sample, one gets:

$$q_{\mu, A} = -2 \ln \lambda_A(\mu) \simeq \frac{(\mu - \mu')^2}{\sigma^2} = \Lambda, \tag{12.54}$$

which allows to determine the estimate of σ^2:

$$\sigma_A^2 = \frac{(\mu - \mu')^2}{q_{\mu, A}}. \tag{12.55}$$

Such sample is called *Asimov data set* defined originally as follows [2]:

> We define the Asimov data set such that when one uses it to evaluate the estimators for all parameters, one obtains the true parameter values.

In practice, the values of the random variable that characterize the data sample are set to their respective expected values. In particular, all variables that represent yields in the data sample, e.g., all entries in a binned histogram, are replaced with their expected values, which may also be non-integer values. Nuisance parameters are set at their nominal values. The use of a single *representative data set* in asymptotic formulae avoids the generation of typically very large sets of Monte Carlo pseudo-experiments, reducing significantly the computation time. While in the past Asimov data sets have been used as a pragmatic and CPU-efficient solution, a mathematical motivation is provided by Wald's approximation adopted for the evaluation of asymptotic formulae. The asymptotic approximation for the distribution of $\tilde{q}_\mu$ (Eq. (12.50)), for instance, can be computed in terms of the Asimov data set as follows:

$$
f(\tilde{q}_\mu \mid \mu) = \frac{1}{2}\delta(\tilde{q}_\mu) +
\begin{cases}
\dfrac{1}{2\sqrt{2\pi}}\dfrac{1}{\sqrt{\tilde{q}_\mu}}e^{-\tilde{q}_\mu/2} & 0 < \tilde{q}_\mu \leq \mu^2/\sigma^2 \,, \\[2ex]
\dfrac{1}{\sqrt{2\pi}\,(2\mu/\sigma)}\exp\left[-\dfrac{1}{2}\dfrac{(\tilde{q}_\mu + \mu^2/\sigma^2)^2}{(2\mu/\sigma)^2}\right] & \tilde{q}_\mu > \mu^2/\sigma^2 \,,
\end{cases}
$$

$$(12.56)$$

where $\frac{1}{2}\delta(\tilde{q}_\mu)$ is the Dirac's delta function component that models the cases in which the test statistic is set to zero. From Eq. (12.56), implies that the median significance, for the hypothesis of background only, can be written as the square root of the test statistic evaluated at the Asimov data set:

$$
\text{med}[Z_\mu \mid 0] = \sqrt{\tilde{q}_{\mu A}} \,.
$$

$$(12.57)$$

For a comprehensive treatment of asymptotic approximations, again, refer to [2]. For practical applications, asymptotic formulae are implemented in the ROOSTATS library [17], released within the ROOT [18] framework. A common software interface to most of the commonly used statistical methods allows to easily switch from one test statistic to another and to perform computations using either asymptotic formulae or Monte Carlo pseudo-experiments. Bayesian and frequentist methods are both implemented and can be easily compared.

Example 12.3 - Bump Hunting with the L_{s+b}/L_b Test Statistic

A classic bump hunting case is presented in this example. The analysis aims at identifying an excess modeled as a broad peak on top of a continuous background. The signal significance is determined using the ratio of likelihood functions in the two hypotheses $s+b$ and b as test statistic. This test statistic was traditionally used by experiments at LEP and Tevatron, and later it has become less popular in favor of profile likelihood, which is more frequently used at the LHC.

The systematic uncertainty on the expected background yield is added in the following Exercise 12.4. Finally, an application of the profile likelihood test statistic is presented in Exercise 12.5 for a similar case.

The Data Model

A data sample, generated with a simplified parametric Monte Carlo, reproduces a typical spectrum of a reconstructed mass m. The spectrum is compared with two hypotheses:

1. Background-only, assuming an exponential distribution of the mass m

---

(continued)

Example 12.3 (continued)

2. Background plus signal, with a Gaussian signal on top of the exponential background

The expected distributions in the two hypotheses are shown in Fig. 12.10 superimposed to the data histogram. The spectrum is subdivided into $B = 40$ bins.

Likelihood Function

The likelihood function for a binned distribution is the product of Poisson distributions corresponding to the number of entries observed in each bin, $\vec{n} = (n_1, \cdots, n_B)$:

$$L(\vec{n} \mid \vec{s}, \vec{b}, \mu, \beta) = \prod_{i=1}^{B} \text{Pois}\,(n_i \mid \beta b_i + \mu s_i)\,. \tag{12.58}$$

The expected distributions for signal and background are modeled as $\vec{s} = (s_i, \cdots, s_B)$ and $\vec{b} = (b_i, \cdots, b_B)$, respectively, and are determined from the expected binned signal and background distributions.

The normalizations of $\vec{s}$ and $\vec{b}$ are given by theory expectations, and variations of the normalization scales are modeled with the extra parameters μ and β for signal and background, respectively. The parameter of interest is the signal strength μ. β has the same role as μ for the background yield and in this case is a nuisance parameter.

In this example, $\beta = 1$ is assumed, which corresponds to a negligible uncertainty on the expected background yield.

Test Statistic

The test statistic based on the likelihood ratio L_{s+b}/L_b can be written as

$$q = -2\log\left(\frac{L(\vec{n} \mid \vec{s}, \vec{b}, \mu, \beta = 1)}{L(\vec{n} \mid \vec{s}, \vec{b}, \mu = 0, \beta = 1)}\right)\,. \tag{12.59}$$

In the assumption $\beta = 1$, the test statistic q in Eq. (12.44) depends on no nuisance parameter. Compared to the profile likelihood in (Eq. 10.29):

$$\lambda(\mu) = \frac{L_{s+b}(\vec{n} \mid \mu, \hat{\hat{\vec{\theta}}}(\mu))}{L_{s+b}(\vec{n} \mid \hat{\mu}, \hat{\vec{\theta}})}\,, \tag{12.60}$$

considering that no nuisance parameters $\vec{\theta}$ are present, the test statistic q is equal to the profile likelihood $-2\log\lambda(\mu)$ up to a constant term (Eq. (10.42)):

$$q = -2\log\lambda(\mu) + 2\log\lambda(0)\,. \tag{12.61}$$

(continued)

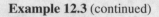

Example 12.3 (continued)

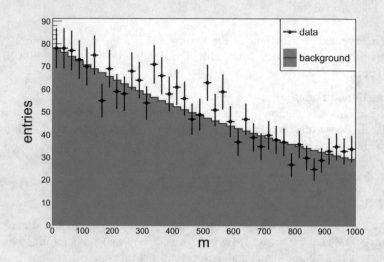

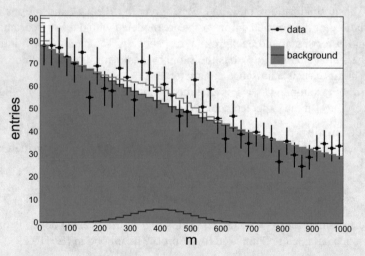

Fig. 12.10 Monte Carlo generated data sample superimposed to an exponential background model (top) and to an exponential background model plus a Gaussian signal (bottom)

Significance Evaluation

The distributions of the test statistic q for the background-only and for the signal-plus-background hypotheses are shown in Fig. 12.11, where 100,000 pseudo-experiments have been generated in each hypothesis. The observed value of q is closer to the bulk of the $s + b$ distribution than to the b

(continued)

Example 12.3 (continued)

distribution. The p-value corresponding to the background-only hypothesis can be determined as the fraction of pseudo-experiments having a value of q lower than the one observed in data.

In the distributions shown in Fig. 12.11, 375 out of 100,000 toy samples have a value of q below the one corresponding to the considered data sample. Therefore, the p-value is equal to 3.75‰. Considering that the evaluation of the p-value with a Monte Carlo is a binomial process, the p-value is determined with an uncertainty of $\pm 0.19‰$. A significance $Z = 2.7$ is determined from the p-value using Eq. (12.1).

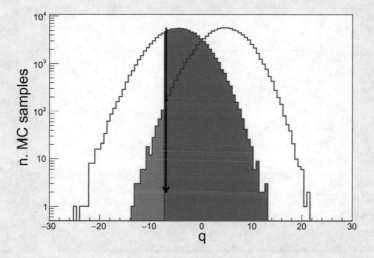

Fig. 12.11 Distribution of the test statistic q for the background-only hypothesis (blue) and the signal-plus-background hypothesis (red). The value determined with the considered data sample (black arrow) is superimposed. p-values can be determined from the shaded areas of the two PDFs

The significance can also be approximately determined from the scan of the test statistic as a function of the parameter of interest μ in Fig. 12.12. The minimum value of q is reached for $\mu = \hat{\mu} = 1.24$ and can be used to determine the significance in the asymptotic approximation. If the null hypothesis ($\mu = 0$ assumed in the denominator of Eq. (12.59)) is true, then Wilks' theorem holds, giving the approximate expression:

$$Z \simeq \sqrt{-q_{\min}} = 2.7 \,,$$

in agreement with the estimate obtained with the toy generation.

(continued)

Example 12.3 (continued)

Considering the range of μ where q exceeds the minimum value by not more than one unit, the uncertainty interval for μ can be determined as $\hat{\mu} = 1.24^{+0.49}_{-0.48}$, reflecting the very small asymmetry of test statistic curve.

Note that in Fig. 12.12 the test statistic is zero for $\mu = 0$ and reaches a negative minimum for $\mu = \hat{\mu}$, while the profile likelihood (Eq. (12.44)) has a minimum value of zero.

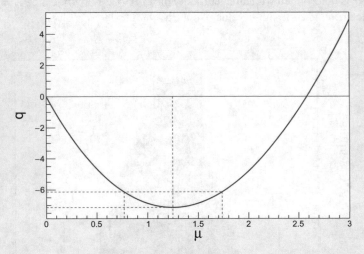

Fig. 12.12 Scan of the test statistic q as a function of the parameter of interest μ

Example 12.4 - Adding Systematic Uncertainty with the L_{s+b}/L_b Approach

Let us modify Exercise 12.3 assuming the background normalization is known with a 10% uncertainty, corresponding to an estimate of the nuisance parameter $\beta = \hat{\beta} \pm \delta\hat{\beta} = 1.0 \pm 0.1$. The extreme cases where $\beta = 0.9$ or $\beta = 1.1$ are shown in Fig. 12.13, superimposed to the data histogram.

Assuming a generic distribution $P(\hat{\beta} \,|\, \beta)$ for the estimated background yield $\hat{\beta}$, given the unknown true value β, the test statistic in Eq. (12.59) can be modified as follows to incorporate the effect of the uncertainty on β:

$$q = -2\log\left(\frac{\displaystyle\sup_{-\infty < \beta < +\infty} L(\vec{n}\,|\,\vec{s},\vec{b},\mu,\beta)}{\displaystyle\sup_{-\infty < \beta < +\infty} L(\vec{n}\,|\,\vec{s},\vec{b},\mu = 0,\beta)} \right), \tag{12.62}$$

(continued)

Example 12.4 (continued)

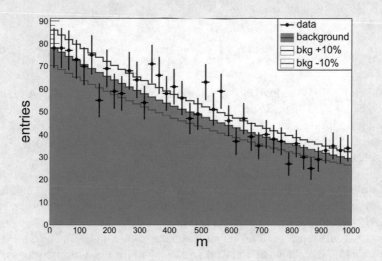

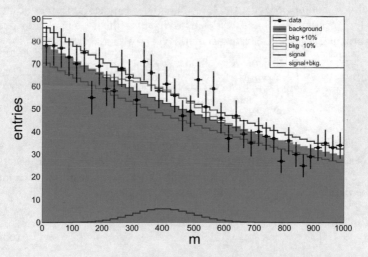

Fig. 12.13 The considered data sample superimposed to an exponential background model (top) and to an exponential background model plus a Gaussian signal (bottom) adding a ±10% uncertainty to the background normalization. Background excursions of ±10% correspond to the histograms shown in purple and green

where the likelihood function now also takes into account the nuisance parameter β:

$$L(\vec{n} \mid \vec{s}, \vec{b}, \mu, \beta) = L_0(\vec{n} \mid \vec{s}, \vec{b}, \mu, \beta) \, L_\beta(\hat{\beta} \mid \beta), \tag{12.63}$$

(continued)

Example 12.4 (continued)

where

$$L_0(\vec{n} \mid \vec{s}, \vec{b}, \mu, \beta) = \prod_{i=1}^{B} \text{Pois } (n_i \mid \beta b_i + \mu s_i) \qquad (12.64)$$

and $P(\hat{\beta} \mid \beta)$ is the PDF of $\hat{\beta}$ given the unknown true value of β. A typical choice for $P(\hat{\beta} \mid \beta)$ may be a Gaussian distribution, with the inconvenience that it could also lead to negative unphysical values of $\hat{\beta}$. Alternatively, a log normal distribution (see Sect. 3.10) or gamma distribution (see Sect. 3.12) could be used, which constrain $\hat{\beta}$ to be positive.

The numerical evaluation can also be simplified assuming a uniform distribution of $\hat{\beta}$ within the given uncertainty interval. Remember that half the range of a uniform distribution is larger than the corresponding standard deviation by a factor $\sqrt{3}$, so the uncertainty we take here, $\hat{\beta} \pm \delta\hat{\beta} = 1.0 \pm 0.1$, does not represent a $\pm 1\sigma$ interval but a $\pm \sqrt{3}\sigma$ interval. See Sect. 3.7, Eq. (3.45). The choice of a uniform distribution for the nuisance parameter is also sometimes adopted for some specific theoretical uncertainties, such as for uncertainties due to the sensitivity of a theory calculation due to renormalization scale. The test statistic can be evaluated as

$$q = -2 \log \left(\frac{\displaystyle\sup_{0.9 \le \beta \le 1.1} L(\vec{n} \mid \vec{s}, \vec{b}, \mu, \beta)}{\displaystyle\sup_{0.9 \le \beta \le 1.1} L(\vec{n} \mid \vec{s}, \vec{b}, \mu = 0, \beta)} \right). \qquad (12.65)$$

The scan of this test statistic is shown in Fig. 12.14. Compared with the case where no uncertainty was included, the shape of the test statistic curve is now broader, and the minimum is less deep, resulting in a larger uncertainty: $\mu = 1.40^{+0.61}_{-0.60}$, and a smaller significance: $Z = 2.3$.

(continued)

Example 12.4 (continued)

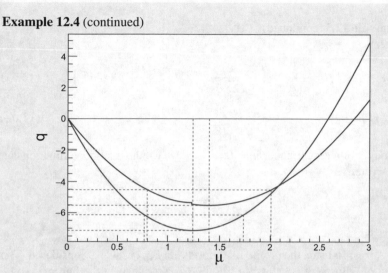

Fig. 12.14 Scan of the test statistic q as a function of the parameter of interest μ including systematic uncertainty on β (red) compared with the case with no uncertainty (blue)

Example 12.5 - Bump Hunting with Profile Likelihood
Similarly to Exercise 12.3, a pseudo-sample is randomly extracted accord-ing to a Gaussian signal with yield $s = 40$ centered at a value $m = 125\,\mathrm{GeV}$ on top of an exponential background with yield $b = 100$, as shown in Fig. 12.15. While in the previous Exercises 12.3 and 12.4 a binned likelihood was used, in this case, an unbinned likelihood function is used, and the signal yields s are determined from a fit to data. This exercise is implemented using the ROOSTATS library [17] within the ROOT [18] framework.

For simplicity, all parameters in the model are fixed, i.e., are considered as constants known with negligible uncertainty, except the background yield b, which is assumed to be known with an estimate $\hat{b}$, whose PDF is a log normal described by a nuisance parameter β, as detailed in the following, with a standard deviation parameter σ_β. The likelihood function for a single measurement m, according to the considered model, only depends on two parameters, s and β, and has the following expression:

$$L(m;\ s,\ \beta) = L_0(m;\ s,\ b = \hat{b}\,e^\beta)\,L_\beta(\beta;\ \sigma_\beta)\,, \qquad (12.66)$$

(continued)

Example 12.5 (continued)
where

$$L_0(m;\ s,\ b) = \frac{e^{-(s+b)}}{n!}\left(s\frac{1}{\sqrt{2\pi}\sigma}e^{-(m-\mu)^2/2\sigma^2} + b\lambda\,e^{-\lambda m}\right)\,, \qquad (12.67)$$

$$L_\beta(\beta;\ \sigma_\beta) = \frac{1}{\sqrt{2\pi}\sigma_\beta}e^{-\beta^2/2\sigma_\beta^2}\,. \qquad (12.68)$$

For a set of measurements $\vec{m} = (m_1, \cdots, m_N)$, the likelihood function can be written as

$$L(\vec{m}\,;\ s,\ \beta) = \prod_{i=1}^{N} L(m_i;\ s,\ \beta)\,. \qquad (12.69)$$

As test statistic, the profile likelihood λ in Eq. (12.44) is considered. The scan of $-\log\lambda(s)$ is shown in Fig. 12.16. Note that, in the figure, $-\log\lambda$ is the default visualization choice provided by the ROOSTATS library in ROOT, which differs by a factor of 2 with respect to the choice of $-2\log\lambda$ adopted in the previous examples.

The profile likelihood was first evaluated assuming $\sigma_\beta = 0$ (no uncertainty on b, blue curve) and then assuming $\sigma_\beta = 0.3$ (red curve), which corresponds to a $\pm 30\%$ uncertainty on $\hat{b}$. The minimum value of $-\log\lambda(s)$, unlike Fig. 12.14, is equal to zero for both cases. The red curve, where the uncertainty on β was added, is broader than the blue curve. The signal yield s can be estimated as the minimum of the curve, and its uncertainty is given by the intersection of the curve with a horizontal line corresponding to $-\log\lambda(s) = 0.5$ (green line in the figure). The broadening of the curve due to the addition of systematic uncertainty on b reflects into a larger total uncertainty on the estimate on s.

The significance of the observed signal can be determined using Wilks' theorem. Assuming $\mu = 0$ (null hypothesis), the quantity $q_0 = -2\log\lambda(0)$

(continued)

Example 12.5 (continued)

is distributed, approximately, as a χ^2 with one degree of freedom. The significance can be evaluated within the asymptotic approximation as

$$Z \simeq \sqrt{q_0}. \tag{12.70}$$

q_0 is twice the intercept of the curve in Fig. 12.16 with the vertical axis, which gives $Z \simeq \sqrt{2 \times 6.66} = 3.66$ in case of no uncertainty on b, and $Z \simeq \sqrt{2 \times 3.93} = 2.81$, if the uncertainty on b is added. The effect of the uncertainty on the background yield, in this example, reduces the significance below the 3σ evidence level.

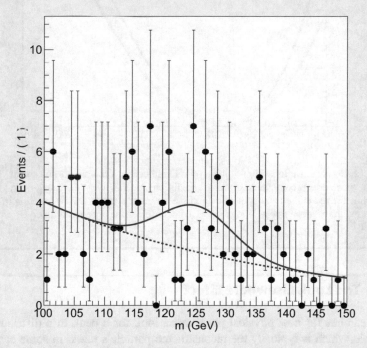

Fig. 12.15 Example of pseudo-experiment generated with a Gaussian signal on top of an exponential background. The assumed distribution for the background is shown as a red dashed line, while the distribution for signal plus background is shown as a blue solid line

(continued)

Example 12.5 (continued)

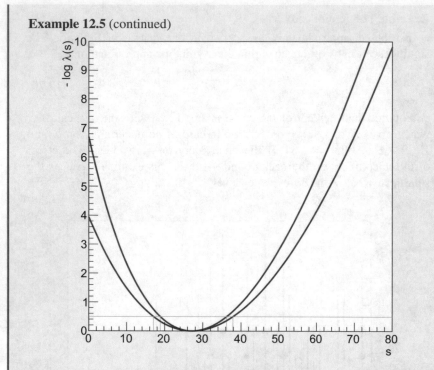

Fig. 12.16 Negative logarithm of the profile likelihood as a function of the signal yield s. The blue curve is computed assuming a negligible uncertainty on the background yield b, while the red curve is computed assuming a 30% uncertainty. The intersection of the curves with the green line at $-\log \lambda(s) = 0.5$ determines the uncertainty intervals on s

12.32 The Look Elsewhere Effect

Many searches for new physical phenomena look for a peak in a distribution of a variable, which is typically the reconstructed particle's mass. In some cases, the location of the peak is known, like in searches for rare decays of a known particle, such as $B_s \to \mu^+\mu^-$. But this is not the case in the search for new particles, such as the Higgs boson discovered at the LHC, or yet-to-be-discovered particles.

If an excess in data, compared with the background expectation, is found at *any* mass value, the excess could be interpreted as a possible signal of a new resonance at the observed mass. Anyway, the peak could be produced either by the presence of a real new signal or by a background fluctuation. One way to compute the significance of the new signal is to use the p-value corresponding to the measured test statistic q assuming a fixed value m_0 of the resonance mass m. In this case, the significance is called *local significance*. Given the PDF $f(q \mid m, \mu)$ of the adopted test statistic

q, the local p-value is the probability to have a value of q greater or equal to the observed value q^*:

$$p(m_0) = P(q \geq q^*) = \int_{q^*(m_0)}^{\infty} f(q \mid m_0, \, \mu = 0) \, dq \, . \tag{12.71}$$

$p(m_0)$ is the probability that a background fluctuation, assuming a fixed value of the mass $m = m_0$, results in a value of q greater or equal to the observed value $q^*(m_0)$.

The probability of a background fluctuation at *any* possible mass value in the range of interest is called *global p-value*. It is, in general, larger than the local p-value. Therefore, if the local p-value is mistakenly interpreted as global p-value, the global p-value is underestimated, and consequently, if the local significance is mistakenly interpreted as *global significance*, the global significance is overestimated. In general, the effect of the reduction of significance, when moving from local to global evaluation, in case one or more parameters of interest are determined from data, is called *look elsewhere effect*.

More in general, when an experiment is looking for a signal where one or more parameters of interest $\vec{\theta}$ are unknown,[3] e.g., could be both the mass and the width, or other properties of a new particle, in the presence of an excess in data with respect to the background expectation, the unknown parameters can be determined from the data sample itself. The local p-value of the excess is

$$p_{\text{loc}}(\vec{\theta}_0) = \int_{q^*(\theta_0)}^{\infty} f(q \mid \vec{\theta}_0, \, \mu = 0) \, dq \, . \tag{12.72}$$

The global p-value can be computed using, as test statistic, the largest value of the estimator over the entire parameter range:

$$q_{\text{glob}} = \sup_{\substack{\theta_i^{\min} < \theta_i < \theta_i^{\max}, \\ i=1,\cdots,m}} q(\vec{\theta}, \, \mu = 0) = q(\hat{\vec{\theta}}, \, \mu = 0) \, , \tag{12.73}$$

where $\hat{\vec{\theta}}$ denotes the set of parameters of interest that maximize $q(\vec{\theta}, \, \mu = 0)$. The global p-value can be determined from the distribution of the test statistic q_{glob} assuming background-only, given the observed value q_{glob}^*:

$$p_{\text{glob}} = P(p_{\text{glob}} \geq p_{\text{glob}}^*) = \int_{q_{\text{glob}}^*}^{\infty} f(q_{\text{glob}} \mid \mu = 0) \, dq_{\text{glob}} \, . \tag{12.74}$$

Even if the test statistic q is derived, as usual, from a likelihood ratio, in this case, Wilks' theorem cannot be applied because the values of the parameters $\vec{\theta}$ determined

---

[3] Here, as in Sect. 5.9, we denote $\vec{\theta}$ as parameters of interest. Other possible nuisance parameters $\vec{\nu}$ are dropped for simplicity of notation. The signal strength parameter μ will be explicitated, and it is not included in the set of parameters $\vec{\theta}$.

from data are undefined for $\mu = 0$. Consider, for instance, a search for a resonance: in case of background-only ($\mu = 0$), the test statistic would no longer depend on the resonance mass m. In this case, the two hypotheses assumed at the numerator and the denominator of the likelihood ratio in the test statistic considered in Wilks' theorem hypotheses are not nested [19].

The distribution of q_{glob} from Eq. (12.73) can be computed with Monte Carlo samples. Large significance values, corresponding to very low p-values, require considerably large random pseudo-samples, which demand large CPU time. Therefore, an approximate method has been developed, as presented in the next section.

12.33 Trial Factors

An approximate way to determine the global significance taking into account the look elsewhere effect is reported in [20], relying on the asymptotic behavior of likelihood ratio estimators. As we have seen in the previous section, the global significance is in general lower than the local significance. A correction factor $F \geq 1$ needs to be applied to the local p-value in order to obtain the global p-value and is called *trial factor*:

$$F = \frac{p_{glob}}{p_{loc}} . \tag{12.75}$$

The trial factor is related to the width of the peak, which depends on the experimental resolution and on the intrinsic resonance width. Empirical evaluations of the trial factor, when the mass is determined from data, give a factor F typically proportional to the ratio of the search range R_m and the peak width σ_m [21]:

$$F \simeq \frac{R_m}{\sigma_m} . \tag{12.76}$$

For a single parameter m, the global test statistic is (Eq. (12.73))

$$q_{glob} = q(\hat{m}, \mu = 0) . \tag{12.77}$$

It is possible to demonstrate [22] that the probability that the test statistic q_{glob} is greater than a given value u, used to determine the global p-value, is bound by the following inequality:

$$\boxed{p_{glob} = P(q(\hat{m}, \mu = 0) > u) \leq P(\chi^2 > u) + \langle N_u \rangle .} \tag{12.78}$$

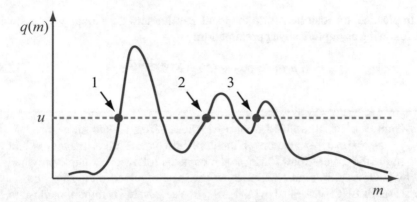

Fig. 12.17 Illustration of upcrossing of a test statistic $q(m)$ counted in order to determine $\langle N_{u_0} \rangle$. In this example, we have $N_u = 3$

The two terms on the right-hand side of Eq. 12.78 are determined as follows:

- The term $P(\chi^2 > u)$ is related to the local p-value. It derives from an asymptotic approximation of the local test statistic:

$$q_{\text{loc}}(m) = q(m, \mu = 0) \tag{12.79}$$

  as a χ^2 with one degree of freedom. A test statistic based on the profile likelihood $q = t_\mu$ (Eq. (12.45)) with $\mu = 0$ has been assumed in Eq. (12.78); in case of a test statistic for discovery q_0 (see Sect. 12.27), the term $P(\chi^2 > u)$ achieves an extra factor $1/2$. The inequality in Eq. (12.78) may be considered as an equality asymptotically.
- The term $\langle N_u \rangle$ in Eq. (12.78) is the average number of *upcrossings*, i.e., the expected number of times the local test statistic curve $q_{\text{loc}}(m)$ crosses a horizontal line at a given level $q = u$ with a positive derivative. An example of the evaluation of the number of upcrossings for a specific curve is illustrated in Fig. 12.17.

$\langle N_u \rangle$ can be evaluated using Monte Carlo as average value over many samples, but the value of $\langle N_u \rangle$ could be very small, depending on the level u. In those case very, very large Monte Carlo samples would be required for a numerical evaluation. Fortunately, a scaling law allows to extrapolate a value $\langle N_{u_0} \rangle$ evaluated for a different level u_0 to the desired level u:

$$\boxed{\langle N_u \rangle = \langle N_{u_0} \rangle \, e^{-(u-u_0)/2} .} \tag{12.80}$$

One can take a convenient value of u_0 such that a not too large number of pseudo-experiments are sufficient to provide an estimate of $\langle N_{u_0} \rangle$ with good precision; then $\langle N_u \rangle$ can be determined using Eq. (12.80) preserving a good numerical precision.

In practice, one can transform the local p-value into the global p-values using the following asymptotically approximation:

$$p_{\mathrm{glob}} \simeq p_{\mathrm{loc}} + \langle N_{u_0} \rangle \, e^{-(u-u_0)/2} \,. \qquad (12.81)$$

Example 12.6 - Simplified Look Elsewhere Effect Calculation

An approximate evaluation of the look elsewhere effect may even not require the use of Monte Carlo, as shown in the following example provided by E. Gross [23]. Figure 12.18 shows the local p-value for the Higgs boson search at LHC performed by ATLAS [9]. The p-value is minimum close to the mass $m_{\mathrm{H}} = 125\,\mathrm{GeV}$ and corresponds to a local significance of about 5σ, according to the red scale at the right of the plot, which reports the equivalent significance Z for a given p-value, reported on the left scale.

Instead of generating Monte Carlo samples, an estimate of $\langle N_{u_0} \rangle$ can be obtained using the single observed test statistic curve as a function of m_{H} and counting the number of upcrossings. As test statistic curve, we can take the p-value curve in Fig. 12.18 and express it as equivalent significance level squared Z^2.

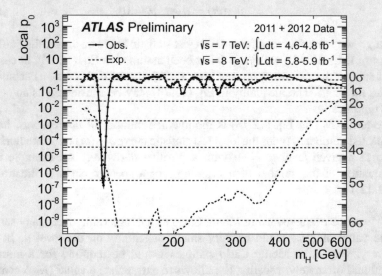

Fig. 12.18 Local p-value as a function of the Higgs boson mass in the search for the Higgs boson at the LHC performed by ATLAS. The solid line shows the observed p-value, while the dashed line shows the median expected p-value according to the prediction of a Standard Model Higgs boson corresponding to a given mass value m_{H}. Reprinted from [9], © ATLAS Collaboration. Reproduced under CC-BY-4.0 license

---

(continued)

Example 12.6 (continued)

As convenient level, $u_0 = 0$ can be taken, corresponding to the red 0σ curve, or equivalently to a p-value $p_0 = 0.5$. The number of times the black solid curve crosses the red dashed 0σ curve with positive derivative can be counted from the figure, and it is equal to $N_0 = 9$. So we can give the approximate estimate:

$$\langle N_0 \rangle = 9 \pm 3 \,.$$

$\langle N_u \rangle$ for $u \simeq 5^2$, corresponding to the minimum p-value, can be determined from $\langle N_0 \rangle$ using the scaling law in Eq. (12.80):

$$\langle N_{5^2} \rangle = \langle N_0 \rangle \, e^{-(5^2-0)/2} = (9 \pm 3)\, e^{-25/2} \simeq (3 \pm 1) \times 10^{-5} \,.$$

The local p-value, corresponding to 5σ, is about 3×10^{-7}. From Eq. (12.81), the global p-value is, approximately,

$$p_{\text{glob}} \simeq 3 \times 10^{-7} + 3 \times 10^{-5} \simeq 3 \times 10^{-5} \,,$$

which corresponds to a global significance of about 4σ, to be compared with the local 5σ.

The trial factor is, within a $\sim 30\%$ accuracy,

$$F = \frac{p_{\text{glob}}}{p_{\text{loc}}} \simeq \frac{3 \times 10^{-5}}{3 \times 10^{-7}} = 100 \,.$$

12.34   Look Elsewhere Effect in More Dimensions

In some cases, more than one parameter is determined from data. For instance, an experiment may measure both the mass and the width of a new resonance, which are both not predicted by theory. In more dimensions, the look elsewhere correction proceeds in a way similar to one dimension, but in this case, the test statistic depends on more than one parameter. Equation (12.78) and the scaling law in Eq. (12.80) are written in terms of the average number of upcrossing, which is only meaningful in one dimension. The generalization to more dimensions can be obtained by replacing the number of upcrossings with the *Euler characteristic*, which is equal to the number of disconnected components minus the number of holes in the multidimensional sets of the parameter space defined by $q_{\text{loc}}(\vec{\theta}) > u$ [24]. Examples of sets with different values of the Euler characteristic are shown in Fig. 12.19.

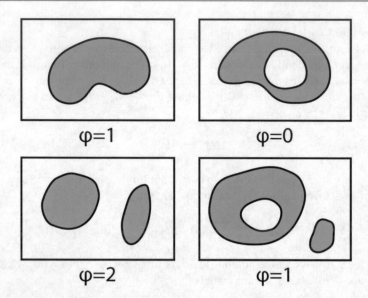

Fig. 12.19 Examples of sets with different values of the Euler characteristic φ, defined as the number of disconnected components minus the number of holes. Top left: $\varphi = 1$ (one component, no holes), top right: $\varphi = 0$ (one component, one hole), bottom left: $\varphi = 2$ (two components, no holes), bottom right: $\varphi = 1$ (two components, one hole)

The expected value of the Euler characteristic $\langle\varphi(u)\rangle$ for a multidimensional random field has a dependency on the level u that depends in turn on the dimensionality D as follows:

$$\langle\varphi(u)\rangle = \sum_{d=0}^{D} N_d \, \rho_d(u) \,, \tag{12.82}$$

where the functions $\rho_0(u), \cdots, \rho_D(u)$ are characteristic of the specific random field, and N_d are constants. For a χ^2 field with $D = 1$, Eq. (12.82) gives the scaling law in Eq. (12.80). For a two-dimensional χ^2 field, which is, for instance, the case when measuring from data both the mass and the width of a new resonance, Eq. (12.82) becomes

$$\langle\varphi(u)\rangle = \left(N_1 + N_2\sqrt{u}\right) e^{-u/2} \,. \tag{12.83}$$

The expected value $\langle\varphi(u)\rangle$ can be determined, typically with Monte Carlo, at two values of u, $u = u_1$ and $u = u_2$. Once $\langle\varphi(u_1)\rangle$ and $\langle\varphi(u_2)\rangle$ are determined, N_1 and N_2 can be found by inverting the system of two equations given by Eq. (12.83) for the two values $\langle\varphi(u_1)\rangle$ and $\langle\varphi(u_2)\rangle$.

The magnitude of the look elsewhere effect in more dimensions may be important. A search for a new resonance decaying in two photons at the Large Hadron Collider by the ATLAS and CMS collaborations raised quite high attention at the end of 2015 because of an excess corresponding to a resonance mass of about 750 GeV. The ATLAS collaboration quoted a local significance of 3.9σ, but the look elsewhere effect, due to the measurement of both the mass and width of the resonance from data, reduced the global significance to 2.1σ [25].

References

1. R.L. Wasserstein, N.A. Lazar, The ASA's statement on p-values: context, process, and purpose. Am. Stat. **70**, 129–133 (2016)
2. G. Cowan, K. Cranmer, E. Gross, O. Vitells, Asymptotic formulae for likelihood-based tests of new physics. Eur. Phys. J. **C71**, 1554 (2011)
3. O. Helene, Upper limit of peak area. Nucl. Instrum. Meth. **A212**, 319 (1983)
4. C. Amsler et al., The review of particle physics. Phys. Lett. **B667**, 1 (2008)
5. G. Zech, Upper limits in experiments with background or measurement errors. Nucl. Instrum. Meth. **A277**, 608 (1989)
6. V. Highland, R. Cousins, Comment on "upper limits in experiments with background or measurement errors" [Nucl. Instrum. Meth. **A277**, 608–610 (1989)]. Nucl. Instrum. Meth. **A398**, 429 (1989)
7. G. Zech, Reply to "comment on "upper limits in experiments with background or measurement errors" [Nucl. Instrum. Meth. **A277**, 608–610 (1989)]". Nucl. Instrum. Meth. **A398**, 431 (1989)
8. G. Abbiendi et al., Search for the standard model Higgs boson at LEP. Phys. Lett. **B565**, 61–75 (2003)
9. ATLAS collaboration, Observation of an Excess of Events in the Search for the Standard Model Higgs boson with the ATLAS detector at the LHC. ATLAS-CONF-2012-093 (2012)
10. B. Berg, *Markov Chain Monte Carlo Simulations and Their Statistical Analysis* (World Scientific, Singapore, 2004)
11. R. Cousins, V. Highland, Incorporating systematic uncertainties into an upper limit. Nucl. Instrum. Meth. **A320**, 331–335 (1992)
12. V. Zhukov, M. Bonsch, Multichannel number counting experiments, in *Proceedings of PHYSTAT2011* (2011)
13. C. Blocker, Interval estimation in the presence of nuisance parameters: 2. Cousins and Highland method. CDF/MEMO/STATISTICS/PUBLIC/7539 (2006)
14. L. Lista, Including Gaussian uncertainty on the background estimate for upper limit calculations using Poissonian sampling. Nucl. Instrum. Meth. **A517**, 360 (2004)
15. A. Wald, Tests of statistical hypotheses concerning several parameters when the number of observations is large. Trans. Am. Math. Soc. **54**, 426–482 (1943)
16. I. Asimov, Franchise, in *The Complete Stories*, vol. 1, ed. by I. (Broadway Books, New York, 1990)
17. Grégory Schott for the RooStats Team, RooStats for searches, in *Proceedings of PHYSTAT2011* (2011). https://twiki.cern.ch/twiki/bin/view/RooStats
18. R. Brun, F. Rademakers, Root—an object oriented data analysis framework, in *Proceedings AIHENP96 Workshop, Lausanne (1996)*. Nuclear Instruments and Methods, vol. A389 (1997), pp. 81–86. http://root.cern.ch/
19. G. Ranucci, The profile likelihood ratio and the look elsewhere effect in high energy physics. Nucl. Instrum. Meth. **A661**, 77–85 (2012)
20. E. Gross, O. Vitells, Trial factors for the look elsewhere effect in high energy physics. Eur. Phys. J. **C70**, 525 (2010)
21. E. Gross, O. Vitells, Talk at Statistical issues relevant to significance of discovery claims (10w5068), Banff, AB (2010)
22. R. Davies, Hypothesis testing when a nuisance parameter is present only under the alternative. Biometrika **74**, 33 (1987)

23. E. Gross, *Proceedings of the European School of High Energy Physics 2015* (2015)
24. O. Vitells, E. Gross, Estimating the significance of a signal in a multi-dimensional search. Astropart. Phys. **35** 230–234 (2011)
25. ATLAS Collaboration, Search for resonances in diphoton events at $\sqrt{s} = 13$ TeV with the ATLAS detector. JHEP **09** 001 (2016)
26. A. Read, Modified frequentist analysis of scarch results (the CL$_s$ method), in *Proceedings of the 1st Workshop on Confidence Limits*, CERN (2000)

Correction to: Statistical Methods for Data Analysis: With Applications in Particle Physics

Correction to:
Chapter 6 and Chapter 12 in:
L. Lista, *Statistical Methods for Data Analysis*,
Lecture Notes in Physics 1010,
https://doi.org/10.1007/978-3-031-19934-9

This book was inadvertently published with the below listed errors, which have been corrected now.

Abbreviations:

p. = page
Eq. = Equation

Corrections:

In Section 6.16, the following correction was done.

p. 150, Eq. 6.94: The term $\frac{1}{N-D}$ was changed to $\frac{N}{N-D}$.

In Section 6.18, the following corrections were done.

p. 152, Eq. 6.99: The denominator $\mu_i(\theta_1, \cdots, \theta_m)$ was changed to n_i.
p. 152, p. 153: Eq. 102 and Eq. 103 were inverted.
p. 152, at the bottom and p. 153 at the beginning: the two italicized words "*Pearson's*" and "*Neyman's*" were inverted.

The updated version of these chapters can be found at
https://doi.org/10.1007/978-3-031-19934-9_6
https://doi.org/10.1007/978-3-031-19934-9_12

© The Author(s), under exclusive license to Springer Nature Switzerland AG 2023   C1
L. Lista, *Statistical Methods for Data Analysis*, Lecture Notes in Physics 1010,
https://doi.org/10.1007/978-3-031-19934-9_13

In Section 12.26, the following correction was done.

p. 303, Eq. 12.47: In the denominator, $\hat{\vec{\theta}}$ was changed to $\hat{\hat{\vec{\theta}}}$.

In Section 12.29, the following correction was done.

p. 304, Eq. 12.50: In the first row denominator, $\hat{\vec{\theta}}$ was changed to $\hat{\hat{\vec{\theta}}}$. In the second row, the argument of $\hat{\vec{\theta}}(\mu)$ at denominator was changed to $\hat{\vec{\theta}}$, since it does not depend on μ.

Index

© The Author(s), under exclusive license to Springer Nature Switzerland AG 2023   325
L. Lista, *Statistical Methods for Data Analysis*, Lecture Notes in Physics 1010,
https://doi.org/10.1007/978-3-031-19934-9

Printed in the United States
by Baker & Taylor Publisher Services